P9-DXK-111
3 2052 00606 1025

DISCARD

The
Young Oxford Companion
to

MAPS AND MAPMAKING

The Young Oxford Companion to MAPS AND MAPMAKING

Rebecca Stefoff

Oxford University Press
New York • Oxford

To Nancy Toff, with appreciation

Oxford University Press

Oxford New York
Athens Auckland Bangkok Bombay
Calcutta Cape Town Dar es Salaam Delhi
Florence Hong Kong Istanbul Karachi
Kuala Lumpur Madras Madrid Melbourne
Mexico City Nairobi Paris Singapore
Taipei Tokyo Toronto

and associated companies in

Berlin Ibadan

Published by Oxford University Press, Inc.
200 Madison Avenue, New York, New York 10016

Design: Sandy Kaufman
Layout: Valerie Sauers
Picture Research: Martin Baldessari, Wendy P. Wills
Consultant: Joseph Stoltman, Professor of Geography, Western Michigan University

Library of Congress Cataloging-in-Publication Data
Stefoff, Rebecca
The young Oxford companion to maps and mapmaking / Rebecca Stefoff
p. cm.
Includes bibliographical references and index.
ISBN 0-19-508042-4
1. Cartography—Encyclopedias, Juvenile. 2. Discoveries in geography—Encyclopedias, Juvenile. [1. Cartography—Encyclopedias. 2. Maps—Encyclopedias. 3. Discoveries in geography—Encyclopedias. 4. Explorers—Encyclopedias.] I. Title.
GA102.S74 1995
912—dc20 94-7900
CIP

9 8 7 6 5 4 3 2 1

Printed in Hong Kong
on acid-free paper

On the cover: *(top left) A compass rose from* Seaman's Secrets, *a 1596 sailor's manual; (top right) image of the world's oceans produced from a NASA satellite in 1992; (bottom) a world map from Karel Allard's* Atlas Minor, *1696.*

Frontispiece: *The Mapparium, located in the Christian Science Publishing Society building in Boston, Massachusetts.*

The compass rose that appears at the beginning of each entry is taken from John Ogilby's 1675 map of the route from London to Chicester.

CONTENTS

HOW TO USE THIS BOOK

Journey over all the universe
in a map, without the expense
and fatigue of traveling,
without suffering the inconveniences
of heat, cold, hunger, and thirst.

—Miguel de Cervantes, *Don Quixote de la Mancha*

The subject of mapmaking is constantly expanding as we learn more about maps from the past, develop new tools to make maps today, and envision terrains that we hope to map in the future.

No single book can cover every aspect of cartography, as the study of maps and mapmaking is called, but this book is a general guide. It provides information about maps and their makers and is also a "road map" that outlines areas you may want to explore more fully.

The term *mapping* has taken on a host of new meanings in the late 20th century. Biologists speak of mapping chromosomes, astronomers of mapping stellar radiation, and physicists of mapping chaos. *The Young Oxford Companion to Maps and Mapmaking* is primarily about what most people think of when they hear the word *maps:* the geographic mapping of the world, closely related to exploration. It is about how humankind has pictured the world over the centuries and how that ever-changing picture—shaped by belief, imagination, painstaking exploration, and science—has appeared on maps and globes.

I have tried, however, to indicate the broadening scope of mapping today and some of the directions it is likely to take in the future. I have also tried to suggest some of the major ideas in current cartography. One very important aspect of cartographic study involves new ways of looking at old maps. Scholars who examine the social and political meanings of maps have shown us that even a map whose geography has been out of date for 400 years can still teach us something new about our cultural attitudes, our ways of looking at the world, and our history.

The articles in this *Companion* are arranged alphabetically, so you can look up words, concepts, or names as you come across them in other reading. Many of the entries contain cross-references directing you to related entries. For example, an entry on a general topic may refer you to more specific and detailed entries; after reading the Navigation article, you will find cross-references to Astrolabe, Dead reckoning, Global Positioning System (GPS), and Orienteering, among others. An entry on something specific, such as an individual mapmaker, may guide you to entries that offer a broader context. Sometimes, you may find that the *Companion* deals with information under a different article name than what you looked up. The book will then refer you to the proper article. For example, if you look up Conic projection, you will find the notation "SEE Projections." If you cannot find an article on a particular subject, look in the index, which will guide you to the relevant articles. All people are listed alphabetically by last name; for example, the entry for John Cabot is listed as Cabot, John, under C.

This book contains entries in the broad categories listed below.

Biographies: Mapmakers and geographers appear in individual entries by name or in entries such as Ancient and medieval mapmakers, Chinese mapmakers, and Dutch mapmakers. There are also biographies of many of the explorers whose journeys enlarged our knowledge of the world's geography, with emphasis upon their contributions to mapmaking.

Important maps: Some entries deal with maps that have historical or geographic significance, such as the Gough map, the La Cosa map, and the Vinland map.

Exploration and expeditions: To read about important exploring expeditions, you might look up the Lewis and Clark Expedition or the Challenger Expedition. Exploration is also discussed in the entries that deal with specific areas of the world and in the biographies of the explorers who journeyed there.

Geographic and cartographic organizations: These agencies are described in entries such as U.S. Geological Survey; Admiralty, British; and Académie des Sciences.

Cartographic and geographic terms: On the more technical side of mapmaking, you can look up cartographic concepts and terms such as Cadastral map, Compass rose, Hachure, Isogram, Projections, and Scale and geographic terms such as Antipodes, Latitude, and Hemisphere.

Mapmaking techniques: Activities and technologies related to mapmaking are discussed in entries on Computers in mapmaking, Engraving, and Surveying, for example.

Types and uses of maps: To read about the different kinds of maps and how they are used, you might look up Geologic maps, Space mapping, Undersea mapping, or similar entries.

Regions of the world: An entry for each major region of the world summarizes the exploration and mapping of that area. Examples are Asia, mapping of; Australia, mapping of; and Pacific Ocean, mapping of.

The back of this book contains some additional resources. For an overview of the main events in the history of mapmaking, see Appendix 1: Important Dates in the History of Mapmaking. Appendix 2: Organizations and Publications Related to Maps lists organizations connected with cartography and the magazines and newsletters that they publish. Museum collections are listed in Appendix 3: Collections and Exhibits of Maps. If you want to learn more about maps and mapmaking, there are lists of books and articles in the Further Reading guide as well as in the Further Reading suggestions that appear at the end of some articles.

Try to have an atlas or globe handy when you use the *Companion.* You may want to compare the old maps pictured in this book with a current map, or you may find a large map useful as a hands-on example of cartographic concepts.

Maps are everywhere around us. Today we see them not just in schoolrooms and newspapers but on umbrellas, shower curtains, and baseball caps. And every one of these maps has something to tell us. Whether you are using this book for research or simply browsing in it because you are fascinated by maps, I hope it adds to your appreciation of maps and their makers.

R.S.

THE YOUNG OXFORD COMPANION TO MAPS AND MAPMAKING

Académie des Sciences

THE FRENCH ACADEMY of Sciences was founded in the mid-17th century when the leading scholars and scientists of France began meeting informally to share the results of their studies and discuss the state of knowledge. In 1699 they were invited to hold their meetings in the library at the royal palace in Paris; this gave the organization a more formal structure and an official status. Many cartographers worked for or were sponsored by the academy. Its single biggest contribution to cartography—the topographic survey of France by the Cassini family—took most of the 18th century and was the first large-scale land survey carried out using scientific principles.

SEE ALSO

Cassini family

FURTHER READING

Konvitz, Josef W. *Cartography in France, 1660 to 1848: Science, Engineering, and Statecraft.* Chicago: University of Chicago Press, 1987.

Admiralty, British

THE BOARD of Admiralty—usually called simply the Admiralty—is the government department that handles naval matters in Great Britain. At the beginning of the 19th century, the Admiralty was put in charge of coastal and hydrographic surveys—that is, maps of coastlines and of the bottoms of rivers, bays, and harbors. These surveys had previously been carried out by a combination of individual mariners, private commercial companies such as the British East India Company, and naval officers. Since the early 19th century, the Admiralty has produced sea charts not just for British waters and coasts but for the whole world; these charts are used by pilots of many nations.

The Admiralty has played a role in other aspects of mapmaking as well by organizing expeditions such as William Dampier's mission to explore the seas north of Australia (1699–1702) and Captain James Cook's three voyages in the Pacific Ocean (1768–79). The Admiralty also awarded cash prizes to scientists and craftsmen who developed new navigational tools such as the marine chronometer, invented by John Harrison in the 18th century.

Aerial mapping

AERIAL MAPPING is mapmaking with the help of flight, usually using pictures taken from an airplane. The term is limited to flight within the earth's atmosphere using balloons or airplanes; the gathering of images from space by satellite is called remote sensing.

Aerial mapping began soon after photography was invented in the 1830s. Army officers took photographs from towers and mountains for use in mapping the terrain below, and in 1858 a French photographer went aloft in a balloon and brought back recognizable photographs of the village over which he had flown. But flight, photography, and mapping did not really merge until World War I (1914–18). By then, special cameras had been developed for aerial photography, and photos taken from planes were used for military reconnaissance. After the war, the technique was adapted to mapmaking. The key to aerial mapmaking was the photogrammetric camera, which was developed in the 1920s and 1930s. It was designed to take

During World War II, the British Royal Air Force used maps made from aerial photographs. Pilots on both sides took aerial pictures of combat zones, enemy cities, and strategic areas.

pictures from which precise location measurements could be obtained. Photogrammetric mapmaking used a series of pictures at fixed intervals. When such a series of photographs was taken from a moving plane and the speed of the plane was known, the distance between points on the ground could be calculated quite accurately. Using photogrammetric equipment, flight crews took pictures that were painstakingly converted into maps. Because scale was not uniform in the photographs, complex calculations as well as observations on the ground were needed to eliminate distortions caused by the tilt of the camera or by variations in distance between the camera and the ground.

During World War II (1939–45) aerial mapping was used to produce new maps of strategically important areas such as the Caribbean Sea and the Panama Canal. With the skill gained during the war, dozens of aerial surveying and mapping companies sprang into existence in the United States and Europe during the 1940s and 1950s. From the air, they mapped parts of South America, Africa, and the Middle East that had never before been surveyed in detail. Aerial surveyors endured unusual on-the-job hazards, including spears thrown at their planes by startled Africans and rifle shots from suspicious operators of illegal liquor stills in the Appalachian Mountains of the United States. But their maps grew ever more accurate, thanks to a camera invented in 1956 by Russell Bean of the U.S. Geological Survey (USGS). Called the orthophotoscope, this camera took pictures that could be converted directly into maps in which scale would be correct, not distorted as in earlier photomaps. Maps produced this way are called orthophotographic maps, or orthophoto maps (*ortho* is a Greek word meaning "correct."). They combine the visual qualities of a photograph with the uniform scale, contour lines, and other information required of a map. When a number of individual photos are joined together to form an image of a larger area, the result is called a photomosaic.

By the 1980s, aerial photogrammetry had become the standard way of

Plattsburgh, New York, in 1939. Such aerial maps are now being used to document the spread of urban development.

producing nearly all new maps. In recent years, remote sensing from satellites has taken on a greater role, but aerial mapping is still widely used.

SEE ALSO

Photogrammetry; Remote sensing; Satellites

Aeronautical maps

AERONAUTICAL MAPS are charts designed for the use of airplane pilots and navigators. They show land features that can be recognized from the air, and they aid aerial navigators by highlighting potentially dangerous changes in elevation, high structures such as towers, and emergency landing strips.

Afonso, Diogo

PORTUGUESE NAVIGATOR

• *Active: 15th century*

DIOGO AFONSO was a member of the Portuguese royal household in the 15th century, when Portugal was busy exploring sea routes south along the west coast of Africa. He was one of many ship captains sent out by his uncle, Prince Henry the Navigator, to probe and map the West African coastline. In 1446 he visited the coast of what is now Mauritania in Africa. In another voyage that same year he explored the Cape Verde Islands, off the West African coast, and added the central and western islands of that group to the map.

SEE ALSO

Africa, mapping of; Henry, Prince of Portugal

Africa, mapping of

THE EARLIEST known maps made in Africa date from the 13th century B.C. They were made by the ancient Egyptians and covered their territories in the northeastern part of the continent. Centuries later, mapmakers made maps of northern Africa that showed not just Egypt but also Phoenician, Greek, and Roman colonies on the North African coast.

Ptolemy, a Greek geographer who worked in Egypt in the 2nd century A.D., represented North Africa fairly well in his maps but guessed wrong about the southern part of the continent. He pictured it widening to a broad landmass that stretched around the world instead

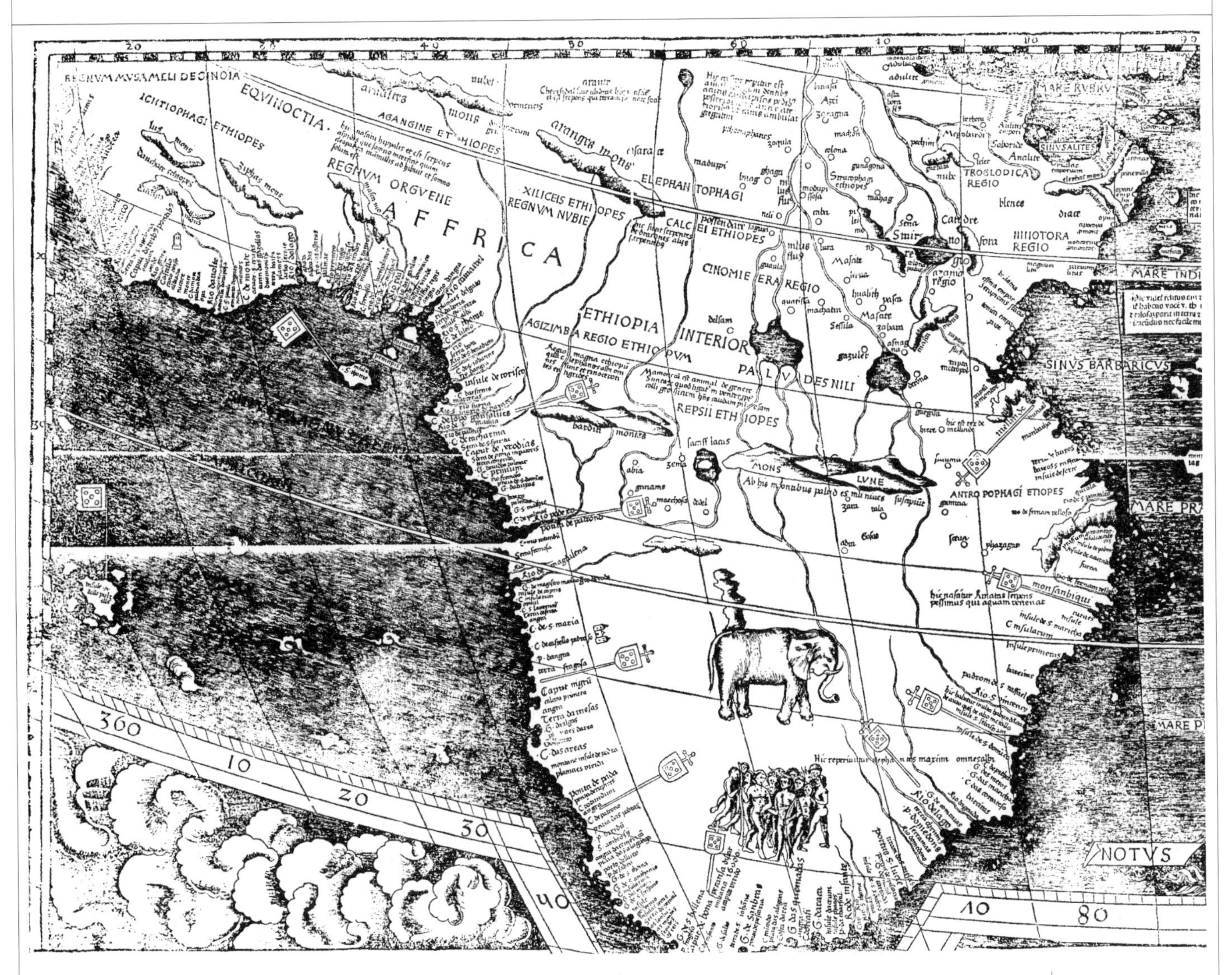

On Martin Waldseemüller's 1507 map of Africa, flags and posts bearing the emblem of Portugal—five circles within a square—indicate the landfalls made by the Portuguese as they worked their way around the Cape of Good Hope.

of narrowing to a rounded point. Later, European geographers were to adopt Ptolemy's notion that the southern part of the African continent reaches eastward to enclose the Indian Ocean, turning it into a large lake or landlocked sea. Ptolemy left the interior of Africa almost entirely blank, except for a few lakes and mountains around the source of the Nile River that were to haunt the imaginations of later explorers.

The Islamic medieval mapmakers of Egypt and North Africa continued to draw Africa on the Ptolemaic model, with one exception—most Islamic geographers, especially those in Arabia, were familiar with the southern Indian Ocean. They knew that that ocean is not a landlocked sea. When they reproduced Ptolemy's map of the world, therefore, they drew southern Africa extending off the map and out of the known world, but not stretching east to surround the Indian Ocean. Furthermore, because a network of caravan routes and Indian Ocean sea trade routes linked all parts of the Islamic empire, Islamic scholars in Africa, Arabia, and Asia knew more than Europeans did about the Sahara Desert, the Muslim kingdoms just south of the desert, and the coast of East Africa, where Arab trading posts were established. But the central, western, and southern parts of the continent were as much a mystery to the Arabs as to the Europeans.

In the 14th century, Portuguese navigators began to piece together the map of the West African coastline. By the end of the 15th century they had rounded the tip of Africa and sailed north along its eastern coast. The general outline of Africa was thus known by the dawn of the 16th century. Printed maps showing the continent in something approaching its correct shape began to appear in 1505. Africa appears on Martin Waldseemüller's *Carta Marina (Marine Chart)* of 1516, in which the outline of the continent is well drawn. It is studded with a string of Portuguese place names, marking the trail of Portuguese exploration (Waldseemüller seems to have had access to secret Portuguese charts). Waldseemüller was not deterred by the fact that neither he nor anyone else knew anything about Africa beyond a narrow coastal strip: He filled the interior with rivers, mountains, elephants, kings on thrones, and other imaginative elements—all based on sheer guesswork.

Sebastian Münster's 1540 woodblock map of Africa is less accurate in its outline but even more imaginative in its treatment of the interior than Waldseemüller's map. In addition to the elephants and kings, Münster populated Africa with "Monoculi" (people with a single eye in the middle of their foreheads). He correctly showed the Arab colony of Malindi on Africa's east coast, but just inland from it he placed the utterly mythical kingdom of Prester John, a Christian priest-king. This mix of fact and fiction persisted on maps of Africa throughout the 17th century. Cartographers hated the idea of blank spaces on their maps, and most of Africa was still a blank to Europeans. So they created a geography of the interior based on hints from travelers and ancient legends and on guesswork, and they adorned it with eye-catching, entertaining images. Such excesses of cartographic creativity were satirized by Anglo-Irish writer Jonathan Swift (1667–1740), who wrote

> So geographers, in Afric-maps,
> With savage-pictures fill their gaps;
> And o'er uninhabitable downs,
> Place elephants for want of towns.

One of the best-known and most decorative 17th-century maps of Africa was made by the Dutch cartographer Willem Janszoon Blaeu in 1617. It appeared in many editions of Blaeu's atlas. The map featured fairly accurate pictures of elephants, ostriches, and monkeys roaming the continent—along with highly imaginative pictures of sea beasts frolicking in the offshore waters. Square panels at the sides illustrated the native inhabitants of various regions. A row of pictures along the top of the map illustrated Alexandria, Tunis, Mozambique, and other cities and trading posts on the African coast. In 1626 the British mapmaker John Speed produced a very similar map that also was reprinted many times. Although these 17th-century maps lacked the monsters and monarchs of earlier maps, they contained their share of myths. Somehow mapmakers seemed reluctant to admit that they did not know everything about geography; they felt compelled to fill every inch of their maps with rivers, lakes, and place names, even though most of these features and names were purely imaginary.

In the 18th century, a new cartographic style changed the map of Africa, wiping away the guesswork and myth. Led by Guillaume De L'Isle and Jean Baptiste d'Anville, French mapmakers called "scientific cartographers" made maps that showed only verifiable information. If these mapmakers were not persuaded by reliable sources that a place existed, they did not put it on their maps. For the first time, large blank

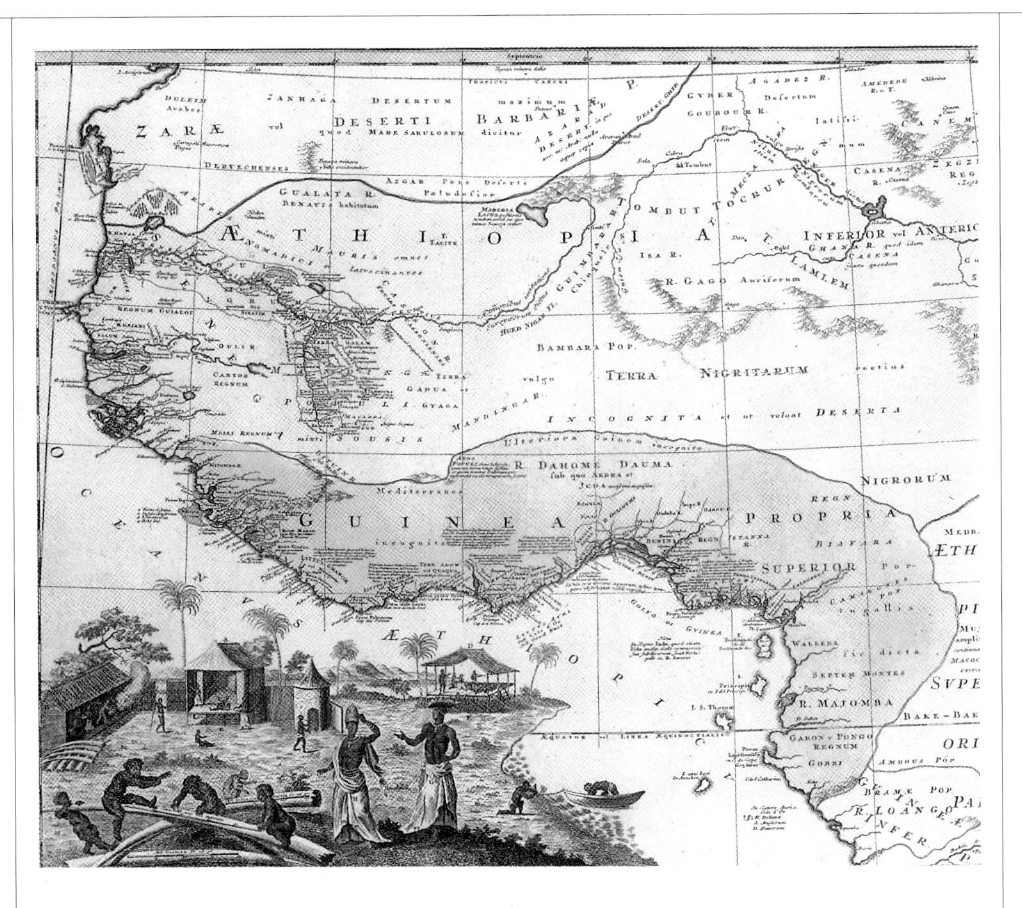

A 1743 map of West Africa, showing much detail along the coast. The blank spaces in the interior are summed up by the word incognita *(unknown).*

spaces began appearing on the map of Africa—a symbol of how much of the continent remained to be explored by Europeans. This trend in mapmaking continued into the 19th century, when many spare, almost empty maps were made by explorers. These maps inspired Joseph Conrad to put the following words into the mouth of Marlow, the narrator of his 1902 tale *Heart of Darkness:*

> Now when I was a little chap I had a passion for maps. I would look for hours at South America, or Africa, or Australia, and lose myself in the glories of exploration. At that time there were many blank spaces on the earth, and when I saw one that looked particularly inviting on a map (but they all look that) I would put my finger on it and say, "When I grow up I will go there."

Marlow's words are a powerful tribute to the lure of maps, but as *Heart of Darkness* showed, the "glories of exploration" sometimes had disastrous results for the native peoples whose homelands were "explored." Mapmakers in the 18th and 19th centuries achieved a new level of geographic honesty but, as modern historians of cartography have pointed out, those blank spaces on the map seemed to Europeans to be up for grabs. Exploration was often followed by territorial claims and economic exploitation.

Nineteenth-century maps of Africa demonstrate two trends. First, the blank spaces are slowly and painstakingly filled in with geographic detail as the journeys of explorers such as Mungo Park, René-Auguste Caillié, Heinrich Barth, David Livingstone, Henry Morton Stanley, Richard Burton, and John Hanning Speke make their way onto new maps.

In the accompanying trend, Africa is carved up like a pie by the European

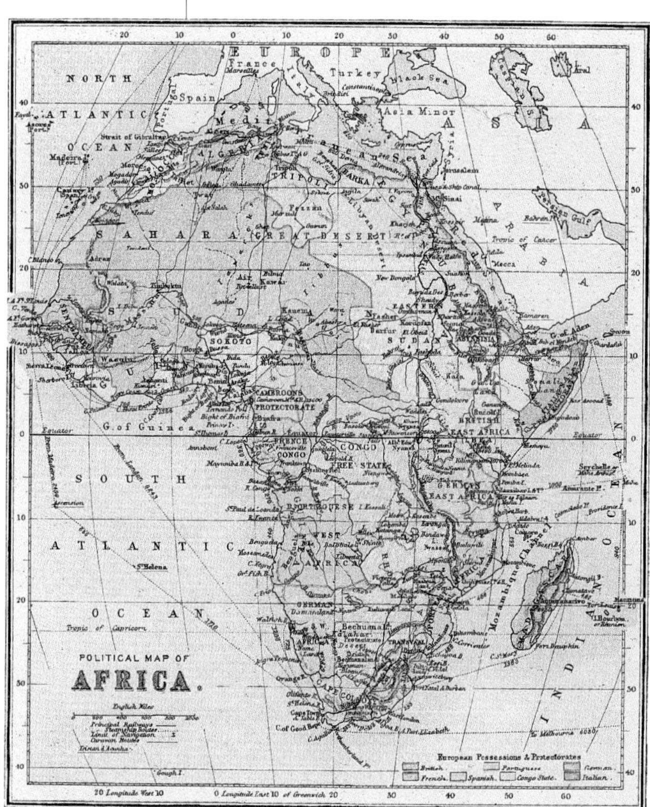

In the late 19th century, Africa was carved up like a pie by the nations of Europe. This map from the early 20th century is color-coded to show the territorial claims of seven colonial powers.

powers, who claim large tracts of the continent as overseas territories. The 1880s and 1890s saw the "scramble for Africa," in which France, Portugal, Great Britain, Belgium, Germany, and Italy laid formal claim to regions that had been opened by their explorers. By 1914, the heyday of European imperialism in Africa, there were only two independent African states: Liberia and Abyssinia (now Ethiopia). European flags flew everywhere else. Maps of the period showed not just geography but also ownership. On some maps, each nation's territories were shown in a different color, so that Britons or Germans could see at a glance the extent of their African holdings.

In the 20th century, colonialism has all but vanished from Africa, and the chief trend in the mapping of the continent has been political rather than geographic. As colonies gained independence and new nations were created, maps had to be updated frequently to record changes in borders, place names, and sovereignty.

SEE ALSO

Baker, Samuel and Florence; Barth, Heinrich; Bruce, James; Burton, Sir Richard Francis; Caillié, René-Auguste; Dias, Bartolomeu; Gama, Vasco da; Henry, Prince of Portugal; Ibn Battuta; Livingstone, David; Prester John; Speke, John Hanning; Stanley, Sir Henry Morton

FURTHER READING

Harley, J. B., and David Woodward, eds. *The History of Cartography*. Vol. 2, *Cartography in Traditional Islamic and South Asian Societies.* Chicago: University of Chicago Press, 1992.

Tooley, Ronald V. "Africa." In *Landmarks of Mapmaking,* 147–90. 1976. Reprint. New York: Dorset, 1989.

African Association

THE AFRICAN ASSOCIATION, more formally known as the Association for Promoting the Discovery of the Interior Parts of Africa, was founded in England in 1788 by Sir Joseph Banks, a well-known botanist. Banks had made several journeys of scientific exploration, including a trip to the Pacific Ocean with Captain James Cook. He was interested in Africa and founded the Association to fill in the "wide extended blank" of the African map. The association promoted exploration by giving money and encouragement to explorers.

The main goal of the Association was to trace the course of the Niger River, which flows through West Africa. During the 1790s, the Association sent several explorers to the Niger basin; the best-known of these is Mungo Park. In 1831 the Association was merged into the Royal Geographical Society.

SEE ALSO

Park, Mungo

Agnese, Giovanni Battista

16TH-CENTURY ITALIAN MAPMAKER

- *Born: 1514*
- *Died: 1564*

PRACTICALLY NOTHING is known about the life of Giovanni Battista Agnese, one of 16th-century Italy's most productive makers of portolan charts, which are maps of seas and coastlines made to help sailors navigate. He probably spent most of his life in Genoa, although he may have lived and worked for a time in Venice.

Agnese's portolan charts are small in scale and highly decorated, made of costly parchment with lettering in silver and gold paint. They were probably made for the libraries of wealthy collectors rather than for use at sea. About 70 of Agnese's portolan atlases, or collections of charts, remain in existence. Scholars have tried in vain to identify Agnese's source of geographic information, for his charts reveal a wealth of knowledge about new discoveries. For example, charts he made between 1538 and 1548 show that Baja California is a peninsula. At that time most geographers and mapmakers believed Baja to be an island. Only a few Spanish geographers had learned that it was a peninsula, but Agnese knew of this discovery. Agnese's earlier collections of charts are devoted to the coasts of the Mediterranean and Black seas, but his later work covers the new territories that were being explored by the Spanish and Portuguese in South America, India, and Indonesia.

SEE ALSO

Alarçon, Hernando de; Portolan

Agricultural maps

AGRICULTURAL MAPS are special-purpose maps that communicate information about the cultivation of crops. They may be designed for direct use by farmers and planters, or they may have a broader educational purpose. Either way, they are generally concerned with soil and the things that grow in it (although maps of climate and weather also have agricultural uses).

Agricultural mapping began in the United States in 1899, when the U.S. Department of Agriculture (USDA) launched a soil survey that eventually covered the whole country. Information about the chemical composition and depth of soil, plotted on local, state, regional, or national maps, helps agriculturalists determine what they can grow where and how much of it they can expect to harvest. Similar surveys have been made in other countries, and the USDA now publishes a map called "Soils of the World" that shows at a glance broad areas of clay, sand, topsoil, loess, and other kinds of earth.

Another type of agricultural map concerns the crops that are grown in the soil. Geography and social studies textbooks often include world maps that

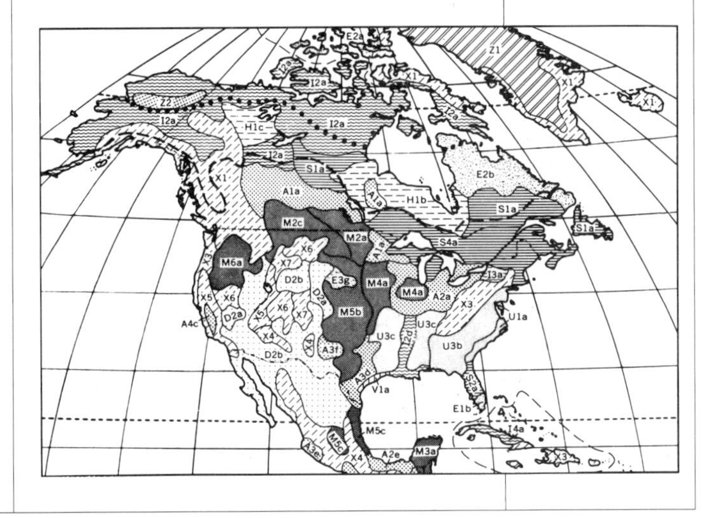

Soil types in North America. More detailed soil maps have been made of smaller areas, even of single farms.

show, for example, the areas where rice, corn, and wheat are the principal grain crops. An agricultural map might show the distribution of a particular crop, such as cotton, across the world, or the distribution of different crops, such as grapes, apples, wheat, and soybeans, within a single country. Agricultural maps of great detail and sophistication, plotting scores of different crops right down to the level of individual fields, are now available. These are mostly based on information obtained from aerial photography or from satellites and may take the form of photomaps, in which the outlines of countries, states, or smaller regions are printed over a photographic mosaic.

SEE ALSO

Aerial mapping; Satellites

Alarçon, Hernando de

SPANISH CONQUISTADOR

- *Born: 1500, Trujillo, Spain*
- *Died: date and place unknown*

HERNANDO DE ALARÇON was one of many Spanish conquistadors, adventurers who sought fortune and glory in the Americas. In 1540 he was given command of a fleet that ventured from Acapulco, on the west coast of Mexico, north along the California coast. He made a map that accurately portrayed the coastline and showed that Baja, or Lower, California is a peninsula, not an island, as had been believed. For more than a century, though, many mapmakers ignored his discovery, continuing to draw California as an island.

Alarçon made one of the first maps to show the true shape of Lower California.

SEE ALSO

Americas, mapping of

FURTHER READING

Schwarz, Seymour I., and Ralph E. Ehrenberg. *The Mapping of America.* New York: Abrams, 1980.

Albi world map

THE ALBI MAP is the oldest known map of western Europe. It is preserved in the library of Albi, a town in France. The map was drawn about A.D. 750; nothing is known of its maker. The geography of the map is crude and filled with errors. The world appears to be a rounded oblong surrounded by water, and both the Red Sea and the Ganges River flow into the Mediterranean. Yet the Albi map is a valuable picture of how Europeans in the time of Charlemagne—the early Middle Ages—may have viewed their world.

SEE ALSO

Ancient and medieval mapmakers

Allard family

DUTCH MAPMAKERS

THE ALLARD FAMILY was involved in map publishing in Amsterdam, in what is now the Netherlands, during the 17th and early 18th centuries. Hugo Allard the elder, an engraver and publisher who started the family business, died in 1691. His son Karel Allard (1648–1709), also an engraver and publisher, inherited the business and then, in

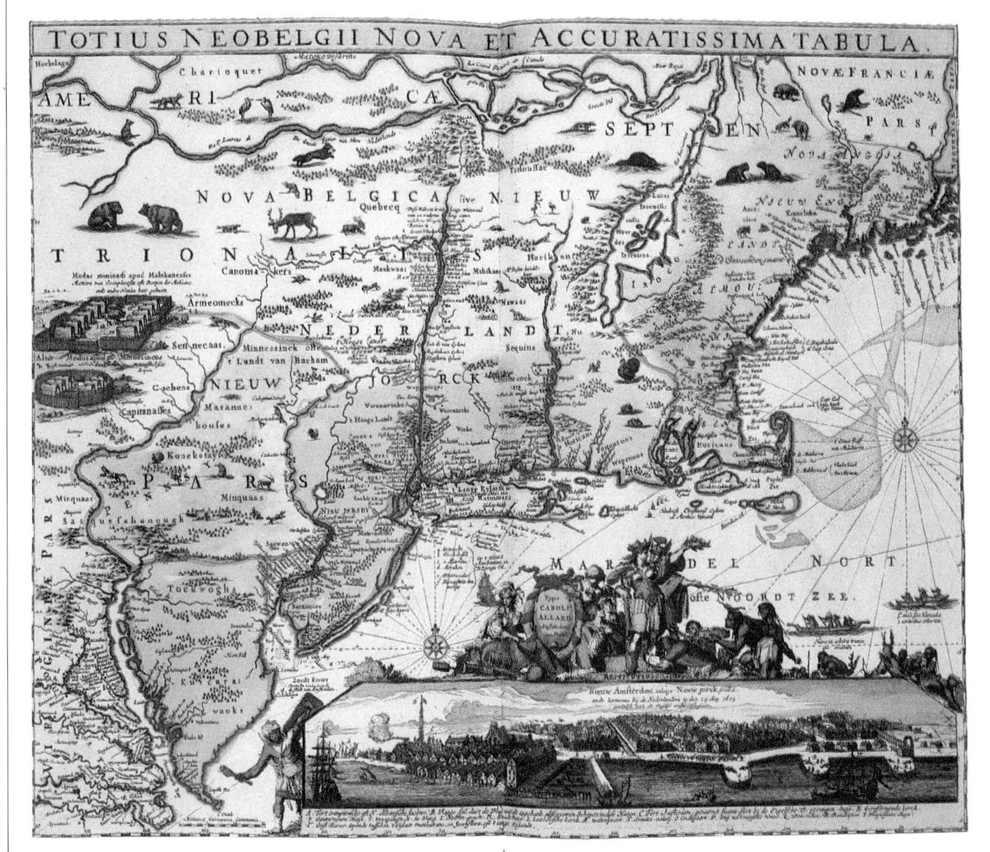

Hugo Allard's 1673 map shows the North American coastline from Maine to Virginia, but it was mainly intended to celebrate the Dutch claim to New Netherland (now known as New York).

1706, turned it over to his son Abraham Allard (1676–1725). Another of Karel's sons, Hugo Allard the younger (1673–91), also published some of his father's maps.

Some of the Allards' notable maps show how maps were copied and reused by mapmakers. In 1650 Hugo Allard the elder published a world map in two hemispheres that was based on a 1594 map by Peter Plancius. Allard improved the map in some ways. He removed the large, mythical continent that Plancius had placed in the southern seas, and he added some Dutch discoveries in Asia. However, he mistakenly showed California as an island, an error shared by many cartographers.

In 1673 Allard published a map called "A New and Exact Map of All New Netherland." It was a map of New York, printed in honor of the Dutch capture of that colony from the British. Allard's map was not original; it was a modified version of a 1650 map by another Dutch mapmaker named Jan Jansson. It even included Jansson's original illustrations of Indian settlements and animals. (Such borrowing, which often was little more than outright copying, was standard practice among both mapmakers and book publishers in the days before copyright laws.) One new feature that Allard added to the map was a drawing of New Amsterdam (present-day New York City) that showed the city bustling and prosperous under Dutch control—and ready to defend itself with smoking guns and rows of marching soldiers.

Karel Allard's 1680 map of Africa is another example of the thrifty use of previously published material. The map was decorated with inset pictures of West African scenes—including scantily clad villagers dancing and, rather improbably, playing Greek-style lyres—that were copied from 16th-century Dutch maps.

American mapmakers

NATIVE AMERICANS were making and using maps long before Europeans arrived in the Americas. Many Native American cultures possessed a high degree of geographic knowledge. Whether as nomadic hunters in the forests of North America or as city-dwelling traders in Mexico's Aztec Empire, they traveled and collected information from other travelers. The Europeans who began arriving in the Americas at the beginning of the 16th century often relied on Native American guides or maps to help them get around. Many early maps by explorers or settlers reflect information obtained from Indians. For example, in 1694 Lawrence Van den Bosh, an English priest, made a map of the lower Mississippi River that he claimed was based on information he received from a "French Indian."

Colonial American mapmaking began with a map of the Virginia colony sketched in 1608 by Robert Tindall, one of the colonists. The map shows Jamestown, which had been founded the year before, and also shows Indian villages along the James and York rivers. John Smith, founder of the colony, produced the first engraved map of Virginia in 1612; it was printed in Europe.

The first map to be drawn, carved, and printed in America was made in 1677 by Boston printer John Foster to illustrate a book about settlers' wars with the Native Americans. The map was a woodcut, printed from a wood block, and it covered New England; north is at the right of the map. Foster proudly claimed that his map was "the first that was ever here cut." The first printed plan of an American city appeared in 1683, when Thomas Holme engraved a map of Philadelphia. This map was widely circulated in England, for it appeared in a pamphlet published by William Penn, the founder of the Pennsylvania colony, aimed at encouraging colonists to move to Pennsylvania. As the American colonies grew, issues of land ownership and boundary marking became important, and local surveys were carried out up and down the Atlantic coast. George Washington, later to become the first President of the United States, worked as a surveyor and produced a map of Virginia based on his survey; it was copied many times and exists in many versions. Regional charts and local surveys were important, but in the 18th century colonists such as Cyprian Southack (sometimes spelled Southwick) and Lewis Evans also began producing maps of wider areas. Southack's "New Chart of North America," published in Boston in 1717, was the first map published in America to show most of the known continent. It is also the oldest known map from a copper engraving to be published in America. It features the entire Atlantic seaboard.

Evans's 1749 map of the British colonies in North America contains a vast amount of information. Although some of his geographic details are wrong—for example, he greatly extends the Allegheny mountain range and labels it "Endless Mountains"—he includes scientific facts about the climate, weather, and wildlife of various regions. Another type of scientific map was made in 1769 or 1770 by patriot and printer Benjamin Franklin, working with a Nantucket sea captain named Timothy Folger. Franklin, founder of a postal service, wondered why it took longer for mail from Europe to reach America than it did for mail from America to reach Europe. He investigated and discovered what sailors had known for some time: that a current of warm water flowed swiftly along the coast of North America

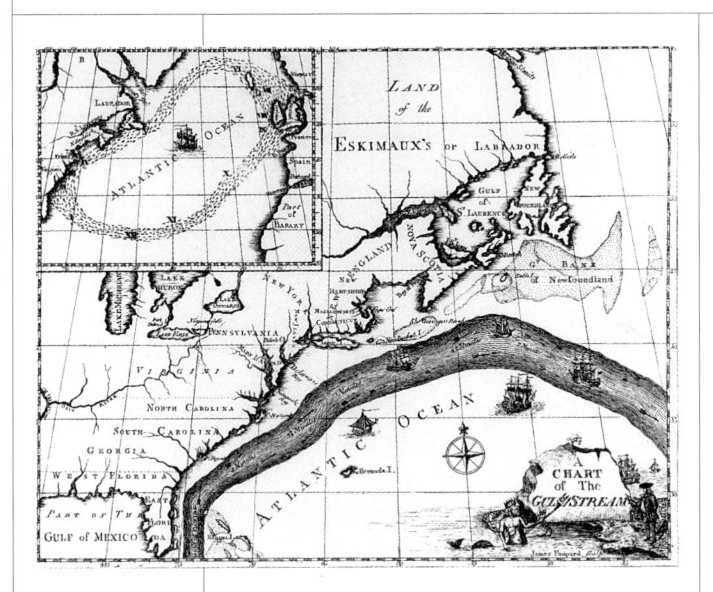

The first known map of the Gulf Stream, made by Benjamin Franklin in 1786. The inset map in the upper left corner shows where herring, a prized food fish, could be found in the North Atlantic.

and then across the North Atlantic Ocean, hastening the passage of ships bound from American to European ports. Franklin and Folger made the oldest known map to show this current, now called the Gulf Stream. The map was lost until 1978, when two copies of it were found in Paris.

The first map of the United States made entirely in America was issued in 1784 by silversmith-turned-printer Abel Buell of New Haven, Connecticut. Irish immigrant Matthew Carey established himself as a publisher in Philadelphia in 1790, and five years later he issued the first American atlas. The maps of the United States were engraved by American artists, but Carey bought the plates of foreign maps from European cartographers. The *New American Atlas* was produced in 1832 by Philadelphia publisher Henry Schenk Tanner. In 1839 an atlas of the United States was published by David H. Burr, the official geographer to the House of Representatives in Washington, D.C. Burr's was the most complete atlas of the United States that had yet been compiled; it covered American states, territories, and settlements from the Atlantic to the Pacific.

In the 1840s, scientific and thematic maps began to appear. American cartographers produced geologic, topographic, and military maps of the country as the nation's borders reached ever westward. The most prolific American mapmakers of the 19th century were associated with government surveys of the West. One who deserves special note is Karl Preuss, a German-born cartographer who accompanied John Charles Frémont on several expeditions in the 1840s. Preuss's maps of Frémont's explorations in the Oregon Territory were the most accurate maps of the American West in the 1840s and 1850s. In the second half of the century, the mapping of the West was completed by surveys under Clarence King, George Wheeler, John Wesley Powell, and Ferdinand Vandiveer Hayden. These surveys merged to form the United States Geological Survey, which has been the chief agent of topographic mapping in the United States throughout the 20th century. Today, maps are also produced by other arms of the government, such as the Defense Department and the Department of Agriculture, and by private organizations, chief among which is the National Geographic Society.

SEE ALSO

Americas, mapping of; Buell, Abel; Carey, Matthew; Frémont, John Charles; Hayden, Ferdinand Vandiveer; King, Clarence; National Geographic Society; Native American mapmakers; Powell, John Wesley

FURTHER READING

Goss, John. *The Mapping of North America: Three Centuries of Map-making, 1500–1860.* Secaucus, N.J.: Wellfleet, 1990.

Luebke, Frederick C., Frances W. Kaye, and Gary E. Moulton, eds. *Mapping the North American Plains.* Norman, Okla., and London: University of Oklahoma Press, 1987.

Schwarz, Seymour I., and Ralph E. Ehrenberg. *The Mapping of America.* New York: Abrams, 1980.

Wheat, James C., and Christian Brun. *Maps Printed in America before 1800: A Bibliography.* New Haven, Conn.: Yale University Press, 1969.

Americas, mapping of

MAPMAKING WAS known to many Native American peoples, from the Inuit of Greenland to the Aztecs of Mexico, long before Europeans came to the Americas. Native Americans carved maps in rock, bone, and ivory, and they drew maps on animal skins, tree bark, and paper made of plant fibers. A few precious examples of these maps have been preserved in museums, but much of what we know about Native American cartography comes from copies of Native American maps made by European explorers or settlers. Many Native American maps reflect more than just the geographic knowledge of their makers; the maps also reflect the history, culture, and spiritual beliefs of the indigenous peoples of the Americas. Maps were used to express information about kinship among families or groups, to celebrate sacred sites, and to record events in a community's history. As far as is known, however, the Native American mapmakers did not systematically set out to map the Americas; their maps were local or regional.

The Vikings were the first Europeans to arrive in America, reaching the coast of what is now Canada around A.D. 1000; however, no genuine Viking maps showing North America are known. A 1425 map by Danish cartographer Claudius Clausson shows Greenland, the large island off the eastern coast of North America that had been colonized by the Vikings, but Clausson made Greenland a part of the Eurasian landmass.

The European mapping of the Americas began with Christopher Columbus's voyage of 1492–93, about 500 years after the Vikings had settled in Greenland. One of Columbus's own maps, a sketch of part of Hispaniola's coastline, survives. Several other maps

The name America *was first used on this 1507 map by Martin Waldseemüller—in the lower left, near the southern tip of South America.*

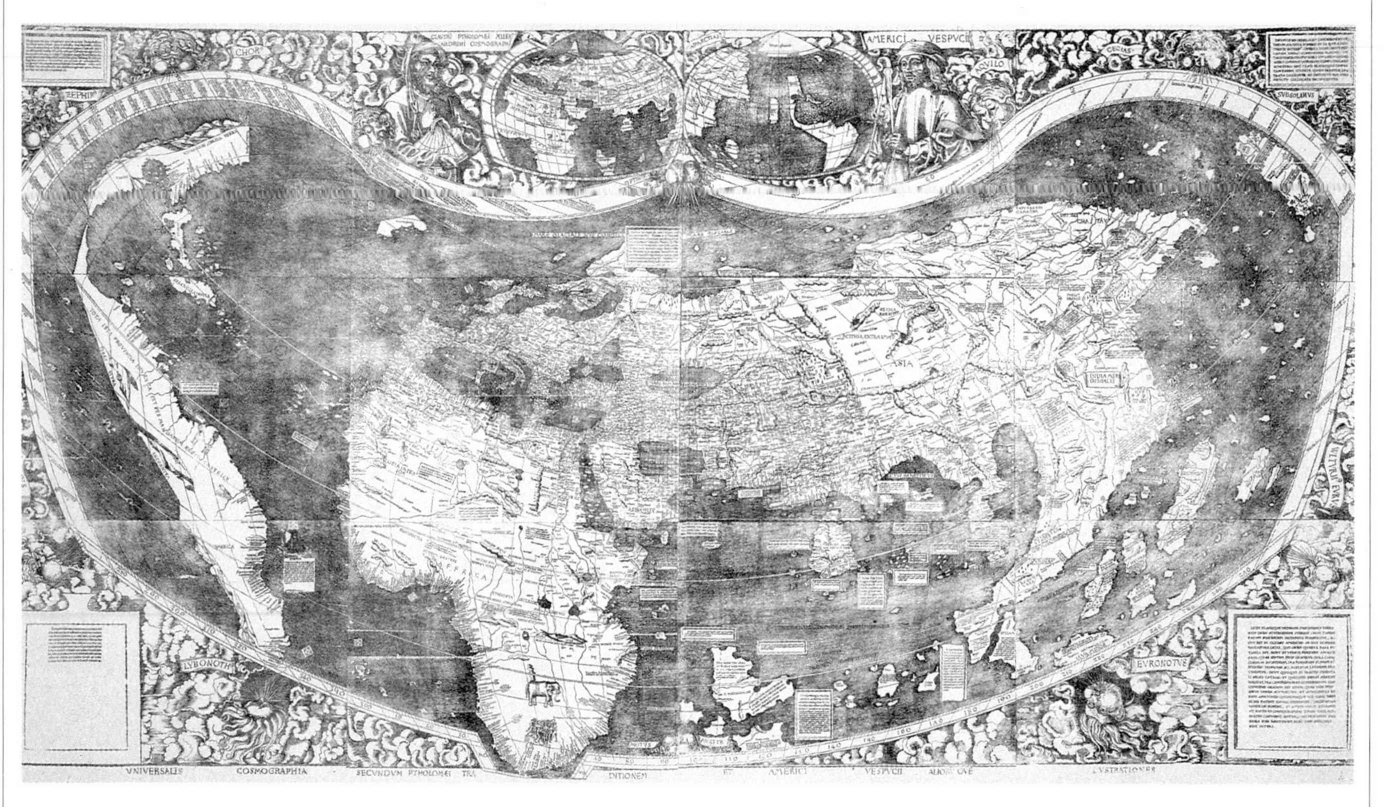

are said to have been drawn by Columbus's brother Bartholomew, but they may be copies of Bartholomew Columbus's maps drawn by Italian cartographer Alessandro Zorsi, an early 16th-century Italian collector of explorers' reports and maps. The oldest known world map to show Columbus's voyage is the La Cosa map of 1500. The Cantino map of 1502 shows even more of the Americas. Both of these maps are hand-drawn. The first printed map to show the Americas was Italian cartographer Giovanni Contarini's 1506 map of the world, which features a landmass in the sea northwest of Europe and another to the southwest: North and South America, separated from each other by a wide band of ocean.

The name America first appeared in 1507 on a world map made by Martin Waldseemüller, a German mapmaker. By that time Europeans had begun to realize that the lands Columbus had explored were not, after all, part of Asia, as he believed. They were a *Mundus Novus,* or New World—and what should they be called? Waldseemüller put the name America on South America in honor of the Italian navigator Amerigo Vespucci, who made several voyages to South America in the first years of the 16th century. Waldseemüller pointed out that in his accounts of his voyages, Vespucci had added "a fourth part" to the world. The name America was quickly adopted by other cartographers; it appears on a 1520 woodblock map by the German scholar Peter Apian.

The Americas appeared on dozens of world maps in the 16th century. One of the most widely circulated of these maps was Sebastian Münster's 1540 map of the world. The first map of the Americas alone appeared in 1556, in a book called *Voyages,* a collection of explorers' tales prepared by Giambattista Ramusio. This map was probably made by Giacomo Gastaldi, the leading Italian cartographer of the day. The first map of North America alone was made in 1566 by Bolognini Zaltieri, another Italian mapmaker. The Zaltieri map is notorious among geographers because it perpetrated a piece of fictional geography: a mythical waterway called the Strait of Anian that flowed into the Pacific from North America. This misleading strait had a long and mischievous life. It was copied by other mapmakers for several centuries and helped reinforce the entirely erroneous idea that a Northwest Passage ran clear across the north part of the continent.

A map called "The North Part of America," made in 1625 by British mapmaker Henry Briggs, promoted another long-lived geographic confusion. Briggs showed California as an island. Although some 18th-century French maps correctly showed California as part of the continent, Briggs's mistake was repeated on a number of Dutch, German, and British maps up to 1747, when King Ferdinand VI of Spain announced that California was officially not an island.

Regional maps of the Americas began appearing as soon as Europeans started establishing colonies there. One of the most noteworthy of these local maps was made in the 1580s by John White, a member of Sir Walter Raleigh's ill-fated Virginia colony. When White returned to England in 1587, his map came into the hands of Théodore de Bry, a German engraver and publisher, who published it in 1590. The map's poignancy comes from the fact that it shows "Roanoac," the English colony from which all the settlers mysteriously disappeared between 1587 and 1590.

By the end of the 16th century, map publishers were beginning to issue the series of decorative atlases that to many people represent the "golden age of cartography." Maps of the Americas appeared in the atlases of Abraham Ortelius (1570), Gerardus Mercator (1595), John Speed (1627), and other leading map publishers of the 16th and 17th centuries. Many of these maps were ornamented with pictures of Indian villages, native American plants and animals, cannibals (on maps of South America, which had acquired a rather gruesome reputation from the reports of a few early explorers), and other picturesque scenes.

Caribbean region

The Bahamas and other Caribbean islands were the first part of the Americas encountered by Columbus and those who followed him, and as a result these islands served as the center or starting point for many later maps of the hemisphere. The islands were also mapped separately, beginning in the mid-16th century. As European nations hastened to stake their claims to the islands, maps reflected national interests. French atlases included maps of French colonies such as Montserrat and Martinique; British atlases were more likely to have maps of British colonies such as Jamaica and Barbados. The British mapmaker Thomas Jefferys produced the first separate atlas of the region, called the *West Indian Atlas,* in 1775.

South America

South America's general outline was fairly well known to geographers and mapmakers by the end of the 16th century. Ferdinand Magellan rounded the continent's southern tip in 1520; for years afterward the name *Magellanica* was used for the island of Tierra del Fuego, or sometimes for the southern part of the continental landmass, the region that is now divided between the nations of Argentina and Chile. In the decades after Magellan's voyage, both the eastern and western coasts of the continent were extensively explored—mostly by the Spanish and Portuguese, who laid claim to South America, but also by the Dutch and British.

On a 1611 Dutch map of the Americas, Lower California is mistakenly shown as an island, and the northern and western coasts of North America trail off into the unknown.

In the early 17th century, map publishers began producing maps of the individual regions of South America. Dutch mapmakers Jodocus Hondius and Willem Janszoon Blaeu and British mapmaker John Ogilby all issued maps of Venezuela, Chile, Peru, and other areas; some of these were printed from the same copper plates, which passed from owner to owner. Aside from an increase in the amount of ornamentation, the major trend in maps of the 17th and 18th centuries was better charting of the coastlines. The interior of the continent was often filled with detail on these maps, but much of this information was conjectural or badly distorted. Reliable maps of the interior, especially of the Amazon River Basin, did not appear until the 19th century, when traveling naturalists such as Alexander von Humboldt began to survey South America's mountains, rivers, and rain forests.

North America

The mapping of North America was quite unlike that of South America. The

South American coastline was fully mapped long before the interior of the continent was known. But the overall shape of North America was not known until the mid-19th century, when the Canadian Arctic was explored in detail. By that time, most of the continent's interior had already been mapped, at least in broad outline.

The 17th-century mapping of North America culminated in a large-scale map on two sheets made by Italian cartographer Vincenzo Coronelli in 1696. Decorated with scenes from the works of Théodore de Bry, this map showed the Great Lakes and the St. Lawrence River correctly because Coronelli had access to the reports of French explorers such as Jacques Marquette and René-Robert La Salle. French explorers had penetrated west of the Great Lakes by the 1690s, and by the middle of the 18th century the representatives of several British fur trading firms were beginning to probe the plains of central Canada and the tundra of the north.

In the 18th century, European mapmakers were greatly concerned with showing their countries' territorial claims in North America. In 1715 Herman Moll, a Dutch cartographer who lived and worked in England, made "A New and Exact Map of the Dominions of the King of Great Britain on ye Continent of North America." The map features insets with smaller, regional maps and one detailed picture of beavers busily building a dam; the beaver was a source of much interest to Europeans because of the thriving trade in its waterproof fur. Similarly, French mapmaker Guillaume De L'Isle marked out France's claims on his 1703 "Map of Canada or New France," on which Detroit, a French fort, appears for the first time. As the 18th century progressed, maps of North America took on a practical, military flavor. The conflict between the French and the British, and later between the British and the American colonists, created a demand for accurate, up-to-date charts and maps of disputed territory, troop movements, and battle sites. Many maps were made by army and navy engineers; these appeared in magazines and also in broadsheets—posters or flyers announcing the latest news. During the American Revolution, events happened so quickly that some maps were out of date by the time they appeared. For example, in November 1781 British cartographer William Faden published a map of Chesapeake Bay and Yorktown, but by the time the map reached the public, the British had lost the siege of Yorktown and surrendered to the Americans.

Several important atlases of America were issued in the late 18th century, including J. F. W. Des Barres's *Atlantic Neptune,* a series of sea charts of the Atlantic coast made for the British Admiralty.

Mapmaking by Americans began in 1608, when colonist Robert Tindall made a hand-drawn map of the Virginia colony. Other regional charts followed, and in 1784 Abel Buell produced the first map made entirely by American citizens. Matthew Carey's first American atlas was published in 1794, and in 1832 Henry Schenk Tanner's *New American Atlas* brought together the latest in American cartography.

In the 19th century, American mapmaking focused on the West. At the beginning of the century, the map was blank between the Mississippi River and the Pacific Ocean, except for some territory explored by the Spanish in the southwest. Explorers, wandering fur trappers, and army surveyors gradually filled in the blank space. One of the most significant maps of the century was William Clark's 1814 map of his journey across western North America with Meriwether Lewis in 1804–6. Stephen

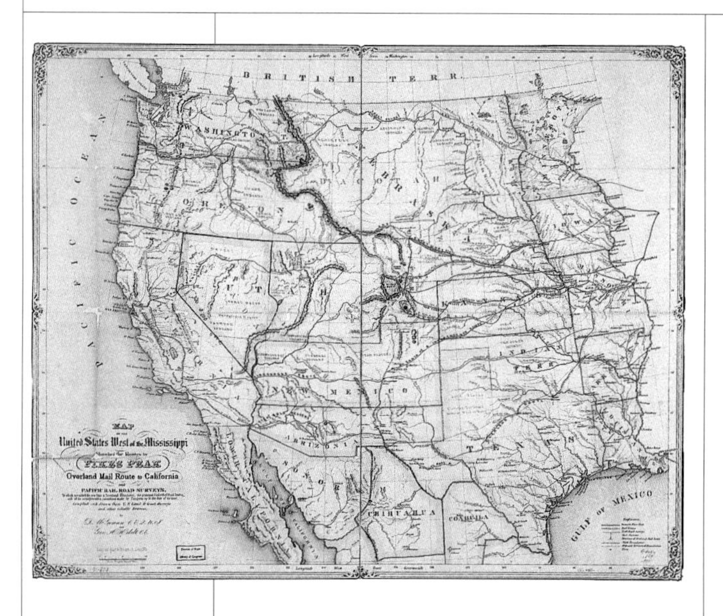

Government surveys of railroad routes in the 1850s led to increased knowledge of the American West, as this 1859 map shows.

Long explored a route along the Arkansas River into what is now Colorado and reported discouragingly—and incorrectly—that the American plains were really "a Great American Desert." This phantom appeared on maps for a few years. Army officers Zebulon Pike and John Charles Frémont, as well as Jedediah Smith and the other trappers and guides who came to be called the Mountain Men, filled in the Rocky and Sierra Nevada mountain ranges on the map.

In the 1850s, government-sponsored surveys reconnoitered the West looking for railway routes. From 1867 to 1878, Clarence King, Ferdinand Vandiveer Hayden, John Wesley Powell, and George Wheeler compiled detailed geologic and topographic data on the West in government-sponsored surveys that were forerunners of the U.S. Geological Survey. At the same time, polar expeditions were completing the mapping of the Arctic coast of North America. By the early years of the 20th century, the broad mapping of the Americas was complete, although remote sensors on orbiting satellites have located new islands in the Canadian Arctic as recently as the 1980s.

SEE ALSO

American mapmakers; Buell, Abel; Carey, Matthew; Clark, William; Clausson, Claudius; Columbus, Christopher; Contarini, Giovanni Matteo; Frémont, John Charles; Humboldt, Friedrich Wilhelm Heinrich Alexander von; La Cosa map; La Salle, René-Robert Cavelier, Sieur de; Magellan, Ferdinand; Marquette, Jacques; Native American mapmakers; Northwest Passage; Powell, John Wesley; U.S. Geological Survey; Vespucci, Amerigo; Viking explorers; Waldseemüller, Martin

FURTHER READING

Goetzmann, William H. *Exploration and Empire: The Explorer and the Scientist in the Winning of the American West.* New York: Norton, 1978.

Goetzmann, William H., and Glyndwr Williams. *The Atlas of North American Exploration.* Englewood Cliffs, N.J.: Prentice-Hall, 1992.

Goss, John. *The Mapping of North America: Three Centuries of Map-making, 1500–1860.* Secaucus, N.J.: Wellfleet, 1990.

Morison, Samuel Eliot. *The Great Explorers: The European Discovery of America.* New York: Oxford University Press, 1978.

Schwarz, Seymour I., and Ralph E. Ehrenberg. *The Mapping of America.* New York: Abrams, 1980.

Suárez, Thomas. *Shedding the Veil: Mapping the European Discovery of America and the World.* River Edge, N.J.: World Scientific, 1992.

Tooley, Ronald V. "The Americas." In *Landmarks of Mapmaking,* 191–244. 1976. Reprint. New York: Dorset Press, 1989.

Viola, Herman J. *Exploring the West.* Washington, D.C.: Smithsonian Books, 1987.

Amundsen, Roald

POLAR EXPLORER

- *Born: July 16, 1872, Borge, Norway*
- *Died: June 1928, in the Arctic*

ROALD AMUNDSEN studied medicine in Norway before turning to the sea. He served as a mate on the *Belgica,* a Belgian ship that carried a scientific expedition to

Antarctica in 1897. This was Amundsen's first taste of polar exploration, to which he devoted the rest of his life. The *Belgica* spent the winter in Antarctica, and Amundsen and Frederick Cook, an American doctor aboard the ship, used their medical knowledge to save the crew from a serious bout of scurvy, a vitamin-deficiency disease.

Amundsen's major contributions to geographic knowledge began in 1903, when he led a crew of six aboard the *Gjoa* in an attempt to sail through the Northwest Passage, the ice-choked waterway that winds among the islands of Arctic Canada, connecting the Atlantic Ocean with the Pacific. Amundsen almost failed to get the voyage started: he had to sail secretly to escape people to whom he owed money. The voyage was a success, however. The *Gjoa* emerged from the Arctic ice pack in 1906, making Amundsen the first navigator to sail the Northwest Passage. He also determined anew the exact position of the north magnetic pole; because the magnetic pole shifts over time as a result of variations in the earth's magnetic field, it had moved some distance away from its last known location.

Amundsen next turned his attention to the other end of the world: to Antarctica. Ambitious to become the first person to set foot on the south pole, Amundsen mounted an expedition that reached the pole on December 14, 1911, after an arduous two-month journey by skis and dogsleds to the center of the frozen continent. Amundsen beat a rival British expedition, led by Robert Falcon Scott, to the pole by 35 days. He became an international hero.

Amundsen succeeded where other explorers had failed for two reasons: First, he studied the customs and equipment of Arctic peoples such as the Inuit of Greenland and adapted their methods—the use of sled dogs, for example—to his own expeditions. Second, his approach to exploration was thoughtful rather than merely heroic. His expeditions were models of careful planning and good management.

Roald Amundsen set records at both ends of the earth: he was the first man to sail the Northwest Passage and the first to reach the south pole.

In 1918–20 Amundsen led an expedition through the Northeast Passage, sailing east from Norway to Alaska along the northern coast of Asia. As in his earlier journeys, he took frequent compass and sextant readings, gathering geographic information that was used to make new maps. His final success was a trip by dirigible across the north pole in 1926. Two years later Amundsen disappeared while trying to rescue another Arctic explorer, Umberto Nobile. He had written in his autobiography, "The victory of human kind over Nature is not that of brute force alone, but also that of the spirit."

SEE ALSO

Antarctica, mapping of; Arctic, mapping of; Cook, Frederick; Northeast Passage; Northwest Passage; Scott, Robert Falcon

FURTHER READING

Flaherty, Leo, and William H. Goetzmann. *Roald Amundsen and the Quest for the South Pole.* New York: Chelsea House, 1993.

Huntford, Roland. *Scott and Amundsen: The Race to the South Pole.* New York: Putnam, 1980.

Maxtone-Graham, John. *Safe Return Doubtful: The Heroic Age of Polar Exploration.* New York: Scribners, 1988.

Anaximander

GREEK ASTRONOMER AND GEOGRAPHER

- *Active: 6th century B.C., in Miletus, a Greek colony in Turkey*

ANAXIMANDER IS the first philosopher known to have developed a cosmography, a theory of the earth and its relation to the universe. According to ancient historians, he made and published a map of the known world that remained in use for centuries, although no copies of it exist today. Anaximander believed that the earth is spherical and that it floats in space. He was an associate of Thales of Miletus, another early cosmographer.

SEE ALSO
Ancient and medieval mapmakers

Ancient and medieval mapmakers

ANTHROPOLOGISTS BELIEVE that all human cultures have developed some type of cartography—not formal, elaborate maps in all cases, but some form of tools for finding routes from one place to another and for representing space in pictures or diagrams. A few of the cave paintings created during the last Ice Age (37,000 to 12,000 years ago) by prehistoric cultures in France, Spain, and elsewhere are thought to be maps, although their true meaning will probably never be known for certain. Some archaeologists think, however, that a large carved stone, or petroglyph, found in Jordan in 1978 may be the world's oldest stone map. It seems to show a region of more than a hundred villages and may be as much as 5,000 years old.

All of the oldest surviving maps come from the ancient Mesopotamian kingdoms of the Middle East. Mapmaking appears to have developed there as a way of keeping track of the land on which people owed taxes. A clay tablet found at Nuzi, in present-day Iraq, has been dated to about 2300 B.C. It shows a stream or river running through a valley between two ranges of hills or mountains, with settlements and farm plots marked on it. Similar tablets dating from around 2000 B.C. contain surveying notes, by which property lines were established.

Maps that show property lines and identify the owners of land are called cadastral maps. The term *cadastral* comes from the Latin word *capistratum,* a list of taxpayers. Cadastral maps, used to keep track of property owners for tax purposes, were made in many parts of the ancient Middle East. The Egyptians were making maps of their communities on papyrus as early as the 14th century B.C.; the Turin papyrus, the earliest surviving Egyptian map, dates from about 1320 B.C.

Local maps, town plans, and cadastral maps apparently were used long before mapmakers attempted to depict the larger world. The oldest known "world map" dates from Babylonia in the 6th century B.C. It is a clay tablet that shows the habitable world as it was envisioned by Babylonians of that time: the kingdom of Babylonia is in the center, with Assyria on the east, Chaldea on the west, and a great range of mountains in the north; the whole thing is surrounded by a circular ocean dotted with mythical islands.

The ancient Greeks made maps and even globes, although none of these has

Made in the Babylonian city-state of Nippur around 1500 B.C., this clay map shows fields separated by straight irrigation canals. The deeply curved channel in the center is a river or stream.

survived. What we know about Greek maps has been pieced together from descriptions of them in ancient and medieval texts. Miletus, a Greek colony on the coast of what is now Turkey, was a center of geographic study and mapmaking in the 6th century B.C. Thales of Miletus was a dealer in olive oil who made a fortune and then devoted the rest of his life to philosophy and science. His pupil Anaximander may have been the first person to publish a map of world geography. Hecateus, another Miletan scholar, wrote a two-volume geography book called *Travels Around the World* in about 500 B.C. The book survives only in fragments. Apparently, Hecateus viewed Greece as the center of the earth, which was surrounded by a ring of ocean. Like Anaximander, Hecateus is thought to have made maps on metal plates, perhaps tablets of bronze or silver. Although such maps would have been far more durable than paper or parchment ones, none has survived.

Herodotus, a Greek historian of the 5th century B.C., told of how one such map was used. Around 495 B.C. Aristagoras, ruler of Miletus, tried to form an alliance with Cleomenes, ruler of the Greek city-state of Sparta, to invade Persia (present-day Iran). Aristagoras took to Sparta a bronze tablet with a map of the world and pointed to all the various Persian realms, reminding Cleomenes how rich each one was and how much loot could be won there. The map made a favorable impression on Cleomenes—until he asked how long it would take to follow the route that Aristagoras was tracing. When he was told that the map route was a three-month march, Cleomenes angrily ordered Aristagoras and his map out of the city by sundown. He felt that the map had tricked him into thinking Persia was closer than it really was.

Herodotus himself traveled widely. He visited Libya, Persia, Egypt, and most Greek settlements in the Mediterranean world. His writings contain many geographic concepts that were later used by mapmakers. For example, Herodotus mentions the use of a grid to locate places on maps, and he suggested that the map of the world could be broken up into sections, as later cartographers were to do. His own geographic knowledge was advanced for his time. Herodotus knew that the Atlantic Ocean is connected to the Indian Ocean, for he reported that Phoenicians had sailed around Africa in the 6th century B.C. He also knew of the existence of the Indus River in India.

Mapmaking became part of Greek culture. In the 5th century B.C., it even appeared in a play—*The Clouds,* by Aristophanes. In the play, a world map is brought onstage. Characters point to Athens on the map and speak of measuring "the whole world." By the 3rd century B.C., Greek writers were regularly mentioning the maps and globes they owned.

The ancient Greeks knew that the earth is round. The philosopher and mathematician Pythagoras is credited with being the first person to describe the world as a sphere. The later philosophers Aristotle and Plato, whose works were influential for centuries, shared this conclusion, planting the idea of a round earth firmly in the tradition of classical learning.

The next great step forward in mapmaking was to measure the size of this spherical earth. Around 200 B.C. Eratosthenes, the head librarian of Alexandria, a Greek city in Egypt, succeeded in measuring the circumference of the world as 25,000 miles (40,233 kilometers)—a figure very close to the true circumference, which is 24,902 miles (40,075 kilometers). However, the Greek astronomer Hipparchus (180–125 B.C.), who was the first to divide the equator into 360 degrees and is credited by some sources with inventing the astrolabe, criticized Eratosthenes for not taking enough measurements of latitude. He also proposed a method of determining longitude based on timing solar and lunar eclipses; this method, however, could not be tried until precise timekeepers were invented centuries later. Poseidonis of Rhodes, a 1st-century-B.C. philosopher who wrote a history of science, was also critical of Eratosthenes. According to Poseidonis, the distance around the earth was only about 18,000 miles. Many later geographers, including Ptolemy, adopted this figure, with the result that 15th- and 16th-century explorers who relied on Ptolemy's geography greatly underestimated the size of the world.

Much of what we know about Eratosthenes and the other ancient Greek cosmographers and mapmakers comes from the writings of Strabo, a geographer who studied at Alexandria. Around 20 B.C. Strabo wrote a book called *Geography,* in 17 volumes. It discusses the work of earlier thinkers and writers. In many cases Strabo's account is all that remains of those earlier works, for by the 4th century A.D. the library at Alexandria had been destroyed.

Meanwhile, the Romans were developing their own maps. But where the Greeks had tried to picture the whole world and to come up with theories about geography, cosmography, and cartography, the Romans were concerned with more practical matters. They had an empire to run, and maps were tools to help them run it. Many of their maps dealt with transportation. One of the most significant achievements of the Roman Empire was the network of roads that linked Rome with the rest of Italy, and Italy with the rest of Europe. Maps of these roads showed the connections between highways, the distances from place to place, and features such as temples and towns along the roads. A map called the Peutinger table, named for the German map collector who owned it in the 16th century, has survived to give us an idea of what the Roman maps were like. The Romans also made surveyors' sketches, town plans, and maps showing military garrisons, battlefields, and army routes.

In the 1st century A.D., the emperor Augustus had an administrative map made of the entire empire. Based on local and regional road maps and surveys, the map took 20 years to complete and cov-

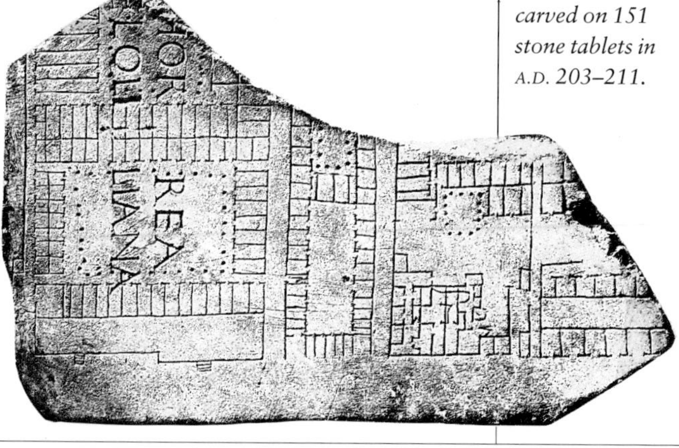

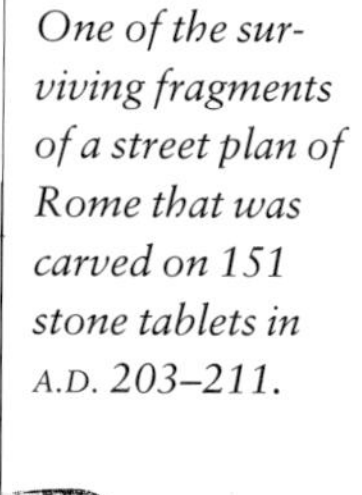

One of the surviving fragments of a street plan of Rome that was carved on 151 stone tablets in A.D. 203–211.

ered 50,000 miles (80,467 kilometers) of highways. It is said that a master copy of this map was carved into a marble wall near the Forum, the large public square at the heart of ancient Rome, but neither the marble map nor any copies of it have survived.

The most important figure in the early history of mapmaking lived and worked in Alexandria during the 2nd century A.D. He was a Greek geographer and astronomer named Claudius Ptolemaeus, called Ptolemy. Around 150 he published an eight-volume book called *Geography,* in which he set forth the principles that shaped the science of cartography during the following centuries. Ptolemy discussed ways of showing, or projecting, the earth's spherical surface on the flat plane of a piece of paper. He also introduced the terms *latitude* and *longitude* and showed mapmakers how to use a grid of parallels and meridians to locate places precisely in relation to one another. During the Middle Ages (roughly 400 to 1450), Ptolemy's work was neglected and forgotten in Europe, but it was preserved by the Islamic geographers and mapmakers of the Middle East and North Africa. In the 15th century, Europeans who had contact with the Arab world rediscovered Ptolemy's *Geography,* which had a profound influence on Western exploration and mapmaking for several centuries thereafter.

Medieval mapmaking

European mapmaking during the Middle Ages was shaped by Christianity, which introduced its own cosmography onto the maps of the period. For example, according to Christian dogma Jerusalem was the center of the earth. It was therefore placed at the center of most maps. Sometimes the Christian cosmographers rejected what had been learned by earlier geographers, as in the case of Cosmas Indicopleustes, whose name means "Cosmas, who traveled to India." Cosmas was a 6th-century merchant traveler turned monk who formed a picture of the world based entirely on the Bible. To Cosmas, the earth was a flat surface like the bottom of a box, and the stars were on the inside of the box's lid.

A small mappa mundi, *or world map, that was part of a 13th-century English prayer book. Jerusalem is in the center and east is at the top.*

One popular type of medieval map was similar to maps used in the late Roman Empire and by some Islamic scholars. Called T-and-O maps, these portrayed the earth as a flat disk (the O) divided into continents by intersecting bars (the T) that represented the Mediterranean Sea, the Nile River, and sometimes the Black Sea or the Don River in Russia. This simple framework, often little more than a schematic sketch, gradually developed into another type of medieval map, the *mappa mundi,* or world map. Large, elaborate, and generally circular, these maps were often painted in bright colors and ornamented with pictures. Some examples that survived into modern times, including the Hereford map and the Ebstorf map, are thought to have served as church decorations, celebrating how far Christianity had spread in the world.

Many makers of *mappae mundi* did strive to incorporate all the geographic knowledge they could, but they also strayed into the realms of myth, legend, and dogma. They drew biblical places such as the Garden of Eden and Paradise side by side with well-known cities such as Paris and Antioch. Not wanting large blank spaces to detract from the beauty of their creations, they filled deserts and unknown regions with monsters, imaginary countries, legendary kings such as Prester John, and fabulous beasts such as dragons, sea monsters, and hippogriffs, which roamed the world along with the almost equally fabulous elephants and giraffes. Also popular for decorative purposes were the bizarre races of humans described by the 3rd-century Roman historian Solinus, whose dog-headed men and other fanciful creations kept popping up on maps as late as the 17th century.

Although these illustrations have captivated map lovers for centuries, some modern historians of cartography suggest that they may have contributed to a warped, Eurocentric view of the world. Generations of maps that filled the outlying portions of the world with monsters and savages may have reinforced the notion that Europeans were the only civilized people and that the rest of the world was a savage place. When Europeans began venturing into Asia, Africa, and the Americas, they were all too ready to dismiss the peoples and cultures they encountered as primitive or even evil.

The tradition of the *mappa mundi* led to two of the finest maps that survive from the Middle Ages: the Catalan atlas of 1375 and the world map completed by Fra Mauro in 1459. Although these maps contain much that is inaccurate or simply fictional, they reflect a growing knowledge of world geography as well.

Road maps and sea charts

The world maps of the Middle Ages drew heavily upon theory and imagination, but another tradition of mapmaking was at work at the same time. This tradition produced maps that were more practical and more accurate than the *mappae mundi*. Among these were road maps, or itineraries, descended from the highway maps made by the Romans. The medieval version of the road map was the pilgrim's guidebook. Many such guidebooks were written from the 4th century on, and some of them included maps to guide religious pilgrims on their way to holy shrines or cities. These maps did not show scale and location accurately, but they gave fairly detailed accounts of the distances between points along the way and of the roads, mountain passes, inns, and churches that the traveler would encounter. Matthew

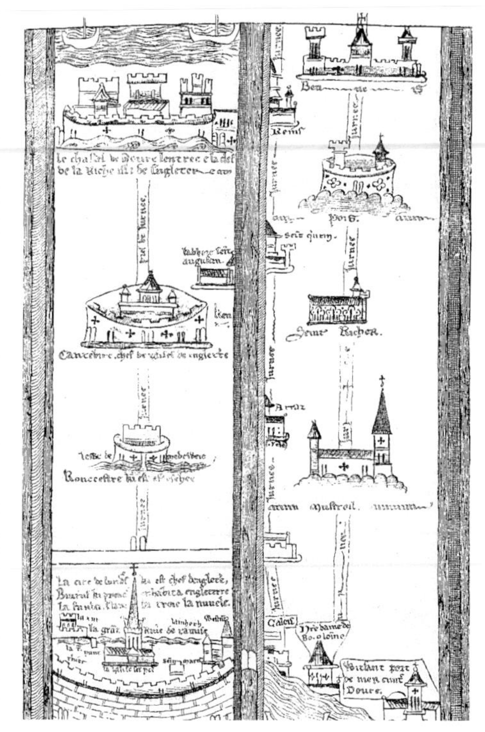

Matthew Paris's 1250 itinerary begins in London, at the lower left. A day's travel is shown as a "jurnee" (from the French journée, *or "day"). Boats at the upper left and lower right mark the crossing of the English Channel.*

Paris's 1250 map of Britain is in this tradition; it shows English pilgrims how to get to Dover, from which ships would carry them on their way to Rome and elsewhere.

Another type of local, practical map that underwent considerable development in the Middle Ages is the sea chart. Mariners, whose lives depended upon safe navigation, had been making coastal charts since ancient times. These came to be called portolans. The earliest portolans covered the Mediterranean and Black seas; as time went on, they were extended to cover the Red Sea, the Persian Gulf, the northern coasts of Europe, and the Atlantic coast of Africa.

Portolans were based on firsthand observation and, after about 1200, on the use of the compass. They were intended for shipboard use and were plain and functional, designed to convey information about prevailing winds and coastal hazards. The interiors of the landmasses were left blank. Gradually, however, some portolans became more elaborate and more decorative. While the seas and coasts were still meticulously drawn for accuracy, the continents were filled in, as on the *mappae mundi,* with a mixture of geography and imagination.

The Catalan atlas shows how the various traditions of medieval mapmaking merged at the close of the Middle Ages. It combines the accurate coastal detail and sailing information of a portolan with the expansive world view of a *mappa mundi.* As the work of Ptolemy was rediscovered in the West and the Middle Ages gave way to an era of European exploration and expansion, the two traditions continued to fuse. Sea charts covered an increasingly wider part of the world, and world maps became more accurate, reflecting the new discoveries of mariners and the careful position-finding of surveyors and geographers.

SEE ALSO

Anaximander; Beatus maps; Cartography; Catalan atlas; Mauro, Fra; Islamic geographers and mapmakers; Mappa mundi; Paris, Matthew; Peutinger table; Portolan; Prester John; Ptolemy; Strabo; T-and-O map; Turin papyrus

FURTHER READING

Boorstin, Daniel J. *The Discoverers.* New York: Random House, 1983.

Brown, Lloyd A. *The Story of Maps.* 1949. Reprint. New York: Dover, 1979.

Dilke, O. A. *Greek and Roman Maps.* Ithaca, N.Y.: Cornell University Press, 1985.

Harley, J. B., and David Woodward, eds. *The History of Cartography.* Vol. 1, *Cartography in Prehistoric, Ancient, and Medieval Europe and the Mediterranean.* Chicago: University of Chicago Press, 1987.

Thubron, Colin, and Time-Life editors. *The Ancient Mariners.* Alexandria, Va.: Time-Life Books, 1981.

Tooley, Ronald V. *Landmarks of Mapmaking.* 1976. Reprint. New York: Dorset, 1989.

Wilford, John Noble. *The Mapmakers.* New York: Knopf, 1981.

Andrade, Antonio de

ITALIAN MISSIONARY EXPLORER

- *Born: 1580, Italy*
- *Died: 1634, Goa, India*

ANTONIO DE ANDRADE became a member of the Society of Jesus, or Jesuits, a Catholic religious order that operated missions in areas of the world that had been visited by European explorers. He rose to a position of leadership among the Jesuits in India. In the early years of the 17th century, Andrade heard rumors of Christian communities living in exile in Tibet, a remote kingdom in the Himalaya Mountains that few Europeans had penetrated.

Determined to find out whether these rumors were true—and, if so, to help the Christians—in 1624 Andrade disguised himself as a Hindu pilgrim and traveled with a caravan along the age-old, winding footpaths of the Himalayas. He later recalled crossing mountain gorges on "bridges formed by frozen masses of snow." After a difficult and dangerous crossing of a high pass, he reached Tsaparang in Tibet and founded a mission there. Although he did not discover the mythical Tibetan Christians, Jesuits under his leadership remained in Tsaparang until 1631. Andrade spent the last four years of his life at Goa, a Portuguese colony on the coast of India. He had made three journeys through the Himalayas, and he had also sent his fellow missionaries on several exploring trips. For two centuries, Andrade's travel reports and those of his colleagues Francesco de Azevado and João Cabral remained the best geographic descriptions of the Himalayan region, although they were sketchy at best. Many European mapmakers based their maps of Tibet and the Himalayas on these accounts.

SEE ALSO

Asia, mapping of

Antarctica, mapping of

ANTARCTICA WAS discovered in the 19th century, although many older maps give the misleading impression that people in the 16th and 17th centuries knew of its existence. These maps show the Terra Australis, or Southern Land, a huge continent sweeping northward into the world's oceans from the south pole. The existence of this continent had been conjectured by Ptolemy in the 2nd century. Mapmakers continued to place the Terra Australis on their maps until voyages of exploration proved that it did not exist.

The Wilkes expedition visits an Antarctic ice island in 1840. Charles Wilkes made the first substantial map of Antarctica's coast.

The search for the Terra Australis and for seals to slaughter took ships into the high southern latitudes and led to the discovery of Antarctica. Captain James Cook probed the southern waters in the late 18th century. Other mariners landed on subantarctic islands, located in the southern seas near the Antarctic Circle. These islands—the South Shetlands, Kerguelen Island, and others—then began to appear on maps. The Russian explorer Fabian Bellingshausen circumnavigated the polar region in 1819–21 and was the first to see the Antarctic mainland, although doubt remained as to whether it was a continent or a smaller island. Other stretches of the Antarctic coastline had been sighted by 1835, mostly by whaling and sealing ships. Charles Wilkes of the Great U.S. Exploring Expedition (1838–42) was the first person to map an extended stretch of coastline. The French and British also discovered parts of the coast in the late 1830s and early 1840s. By mid-century, the outline of the continent was known in a general way, but no one had yet charted the coast in detail or tried to go inland.

A lull in exploration followed, until the Sixth International Geographical Congress of 1895 announced that Antarctica was the last great challenge of exploration. Nine nations sent 16 expeditions to Antarctica between 1897 and 1917. Some of these missions—such as the 1898–1900 expedition during which

Danish explorer Carsten Borchgrevink's crew became the first people to spend a winter ashore in Antarctica or the Norwegian explorer Roald Amundsen's conquest of the pole in 1912—were feats of endurance and patriotic flag planting; others were scientific research projects.

In the 1930s, American explorer Richard Byrd and Australian aviator Hubert Wilkins took airplanes to Antarctica and began the reconnaissance of the continent from the air. The U.S. Navy continued this effort with Operation High Jump, which began in 1946, sending photographers aloft in airplanes to take pictures that could be used as the basis for new maps. Mapping Antarctica proved more difficult than expected, however, because 99 percent of the continent is covered by a huge ice cap. Not only is the ice a hostile environment for ground exploration but it offers few features that can be used as fixed measuring points in aerial mapping. Antarctica posed a challenge to 20th-century scientific mapmakers: How could an ice-covered continent be mapped?

Mapping and exploration are no longer individual acts but, in most cases, team efforts that rely on technology and draw upon many different scientific disciplines. The final mapping of Antarctica is a good example. Twelve nations took part in research projects in Antarctica during the International Geophysical Year (IGY) of 1957–58, which launched a new flurry of scientific mapmaking in the polar region. But aerial photographs remained of little use for cartographic purposes without fixed marks on the ground to use as reference points. And the only way to create a network of such ground points was through triangulation—building two towers or markers and then taking careful sightings on a third point from each of them, so that the position of the third point could be determined by geometric calculations. Triangulation surveys, however, were not easy to perform in the wind-blown, frigid Antarctic wastes.

In the early 1960s, the U.S. Geological Survey (USGS) carried out an ambitious mapping project in Antarctica using newly invented electronic distance-measuring machines for triangulation. From helicopters, these devices bounced a signal to two ground stations, thus creating a triangle for surveying purposes. Aerial mapping surged forward. The surveyors established a line of fixed reference points across the Transantarctic Mountains, and by the end of the 1960s the USGS and the U.S. National Science Foundation had produced highly accurate, detailed maps of West Antarctica. Yet much of the continent remained to be explored in detail, and major features such as mountain ranges jutting through the ice sheet were added to the map as late as the 1970s.

At the same time, another type of mapping was taking place—the mapping of the invisible Antarctic continent, the rock beneath the ice. At the time of the IGY, scientists were not yet sure whether Antarctica's ice sheet rested atop a single continental landmass or a string of islands. Using gravimeters—instruments that record gravity pull—they began to measure the thickness of the ice sheet. Once they had measured the thickness of the ice in many places, they could begin to piece together a profile of the hidden land beneath. Seismic probes—small explosions that send sound waves through the ice to bounce off the rock—also contributed to this profile. Echo sounding, which uses radio waves instead of seismic shock waves, was introduced in the 1960s. From 1967 to 1975, the inventors of echo sounding, a team of British scientists from the Scott Polar Research Institute of Cambridge University, worked with American researchers from the navy and the National Science Foundation on a project to make a radio-echo map of the Antarctic continent.

By 1980 the rock continent had been mapped in all its unseen grandeur, including buried mountain ranges and immense deep basins that had lain unsuspected beneath the ice cap of East Antarctica. Satellites continue to provide data that are used both to improve the existing topographic and geologic maps of Antarctica and also to map new features, such as the hole in the ozone layer of earth's atmosphere that has appeared over Antarctica.

SEE ALSO

Amundsen, Roald; Bellingshausen, Fabian Gottlieb Benjamin von; Borchgrevink, Carsten; Byrd, Richard Evelyn; Cook, James; Great U.S. Exploring Expedition; Satellites; Scott, Robert Falcon; Terra Australis; Triangulation

FURTHER READING

Baughman, T. H. *Before the Heroes Came: Antarctica in the 1890s.* Lincoln: University of Nebraska Press, 1993.
Cameron, Ian. *Antarctica: The Last Continent.* London: Cassell, 1973.
Reader's Digest editors. *Antarctica: The Extraordinary History of Man's Conquest of the Frozen Continent.* 2nd ed. Pleasantville, N.Y.: Reader's Digest, 1990.
Stewart, John. *Antarctica: An Encyclopedia.* 2 vols. Jefferson, N.C.: McFarland, 1990.
Wilford, John Noble. *The Mapmakers.* New York: Knopf, 1981.

Antipodes

ANTIPODES ARE any two points that are directly opposite one another on the earth. The two points must be 180 degrees of longitude apart, and one must be as far north of the equator as the other is south. In a more general sense, the term *antipodes* has been used to refer to a place "on the other side of the earth." For example, people in Great Britain have called Australia and New Zealand the Antipodes, although this use of the word has fallen out of fashion. Today, the term is rarely used in either popular or scientific discussions of geography.

Antique maps

THE TERM *antique* is a slippery one. How old does something have to be before it is considered an antique? The answer varies, depending upon the kind of object being considered. It also varies from person to person: one person's broken-down old chair may be a promising-looking antique in someone else's eyes. The U.S. customs laws typically define an antique as something that is at least 100 years old.

In general, maps made before A.D. 400 are considered ancient. Very few actual maps survive from ancient times; most ancient maps are known to us from copies made later. Maps from the Middle Ages (roughly A.D. 400 to the late 15th century) are called medieval. Antique maps date from 1477, when maps began to be printed instead of individually drawn, to about 1900. Some collectors and dealers of antique maps, however, are not interested in maps made after 1850 because those made after that date are generally more functional and less decorative than earlier maps. For the most part, maps dating from 1850 and later lack the ornamentation, illustrations, and cartographic curiosities that many people admire in older maps.

One of the challenges in dealing with antique maps is to determine when the map was actually made, and by whom. Most maps contain cartouches, or frames, that give the map's title, often its maker, and sometimes its date. But even if a map is dated, the map may not have been printed in that year. Sometimes, especially in the case of large atlases with many maps, plates were engraved years before the atlas was ready to be printed. For example, Christopher Saxton's atlas of the counties of England and Wales was published in 1579, but some of the maps are dated 1574.

In other cases, engraved plates were used over and over again for years, and a map printed in 1670 might have been made in 1610 or even earlier. Map publishers often removed the dates from their plates, or omitted them in the first place, so that the maps would not look out of date. But many maps are undated simply because the practice of dating them had not yet been widely adopted by cartographers.

Experts can generally figure out when an undated map was published. They pinpoint a map's date by examining references to the map or the mapmaker in historical works, the type of paper used, the time period during which particular engravers or publishers were active, and the accuracy of the geographic information that appears on the map. Most antique maps are well documented; historians of cartography have compiled dozens of reference books that list mapmakers and their works.

Antique maps are prized by a growing number of collectors. Some collectors try to acquire maps by a particular cartographer; other collectors are interested in maps from a certain period, or maps of particular parts of the world. Still other collectors specialize in certain kinds of maps: railway maps, for example, or military maps. But the fascination of maps—even antique maps—is not limited to those with money to invest in costly collections. Many books available in libraries or bookstores contain excellent reproductions of the classics of mapmaking.

FURTHER READING

Lister, Raymond. *Antique Maps and Their Cartographers.* Hamden, Conn.: Archon, 1970.

Potter, Jonathan. *Country Life Book of Antique Maps and How to Appreciate Them.* London: Hamlyn, 1988.

Shirley, Rodney W. *The Mapping of the World: Early Printed World Maps, 1472–1700.* London: Holland Press, 1983.

Anville, Jean Baptiste Bourguignon d'

FRENCH MAPMAKER

- *Born: 1697, France*
- *Died: 1782, France*

JEAN BAPTISTE Bourguignon d'Anville was one of the most important figures in cartography in the 18th century, a time when France was becoming a leader in mapmaking. He published more than 200 maps, beginning with a map of France in 1719.

Anville was a thoughtful mapmaker who rejected the practice of filling up the blank spaces on maps with guesswork or fantasy. He took a rigorously scientific approach to cartography and pored over travel and history books, as well as earlier maps, to verify facts. If he was not sure of something, he did not put it on a map. He wrote, "To destroy false notions, without even going any further, is one of the ways to advance knowledge." He put this philosophy into practice on his 1727 map of East Africa and his 1749 map of the entire African continent on two sheets. Anville, who was deeply interested in Africa and hoped to visit Timbuktu, dropped many mythical and uncertain elements from the map of Africa. His was one of the first major maps to have blank spaces. These admissions of geographic ignorance were revolutionary at the time, but Anville's were

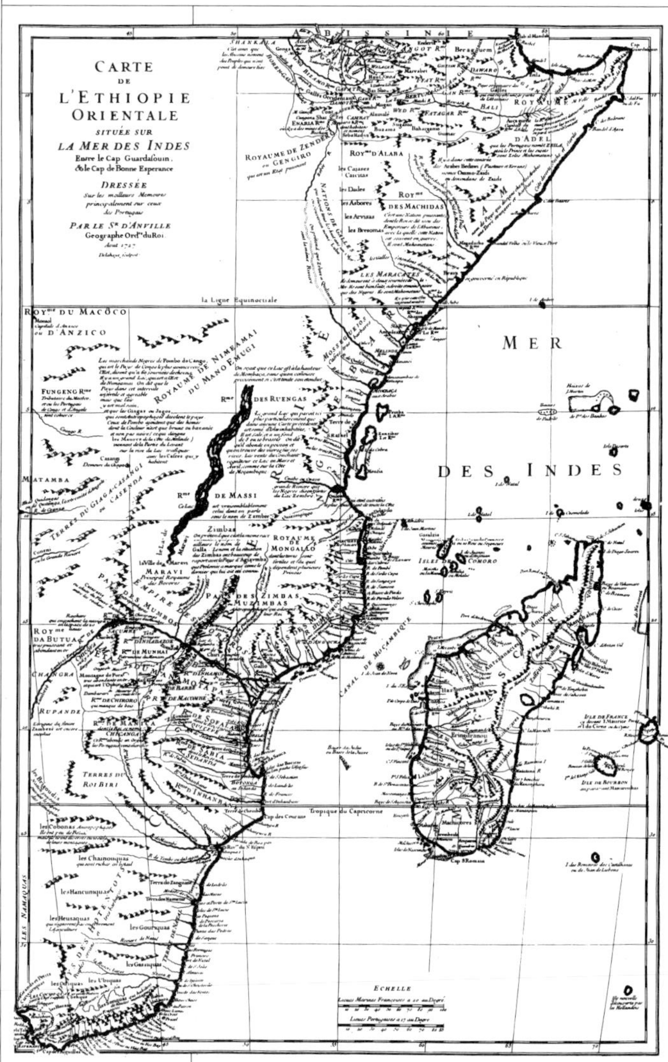

Stripped of myth and imagination, Anville's 1727 map of East Africa heralds a new era of scientific mapmaking.

among the most accurate maps of their period.

Anville published an atlas of China in 1737, using maps and surveys prepared by Jesuit missionaries and scholars in China. His atlas became the authoritative source of information on China; his maps of some parts of Tibet and Central Asia remained in use well into the 19th century. Another atlas, *Ancient Geography,* published in 1768, also remained in use until the mid-19th century. This influential work was a historical atlas in which Anville provided a geographic framework for ancient history.

Anville was also the author of 78 papers or books on geography and the owner of a remarkable map collection—more than 100,000 maps, now preserved in the French national library. In recognition of his contribution to geography and cartography, Anville was elected to the French Académie des Sciences in 1773. But if imitation is truly a form of flattery, he received an even greater honor from the many English, German, Dutch, and Italian mapmakers who copied, pirated, and imitated his maps for years after his death.

SEE ALSO

French mapmakers

FURTHER READING

Konvitz, Josef W. *Cartography in France, 1660–1848: Science, Engineering, and Statecraft.* Chicago: University of Chicago Press, 1987.

Apian, Peter

GERMAN ASTRONOMER AND GEOGRAPHER

- *Born: 1495, Leisnig, Saxony*
- *Died: 1552, Ingolstadt, Bavaria*

PETER APIAN (Peter Bienewitz) published a world map in Vienna in 1520. Four years later he published a book called *Introduction to Cosmography,* in which he described his geographic theories. A new edition of the book published in 1533 included an account of triangulation, a method of surveying, by Regnier Gemma Frisius. For many years thereafter, the combined Apian-Frisius book was *the* standard handbook for geographers and surveyors. Peter Apian's son Philip (1531–1589) used the principles of surveying described in his father's book when he made extensive surveys in Bavaria and produced maps of the region in 1563 and 1568.

Arctic, mapping of

THE ARCTIC is generally defined as the part of the globe north of the 75th parallel. It includes the northernmost part of Greenland and various clusters of islands north of Norway, Russia, and Canada. There is no land at the north pole, which is covered by a permanent ice pack.

The Arctic, an inhospitable region, was a realm of speculation for early mapmakers. It began to be mapped, although not very accurately, in the 16th century. Gerardus Mercator's 1569 world map includes a round panel showing his conception of the Arctic region, with a tall black rock of magnetic iron perched on the north pole, surrounded by a ring-shaped continent from which four rivers flowed into the Arctic Ocean. This attractively symmetrical vision was gradually replaced, in the century that followed, by the outlines of Spitsbergen (also called Svalbard), southern Greenland, Baffin Island, and other shores explored by Martin Frobisher, John Davis, Willem Barents, Henry Hudson, William Baffin, and other explorers and sealers.

In 1675 a mapmaker named Frederic de Wit made a handsome map of the north polar region, decorated with scenes from the whaling industry, that showed the state of geographic knowledge of the far north: at least half the map was blank, with only the northern coastlines of Europe, Russia, and eastern Canada filled in. In the following centuries, much mapping of the sea and land approaches to the Arctic was done by

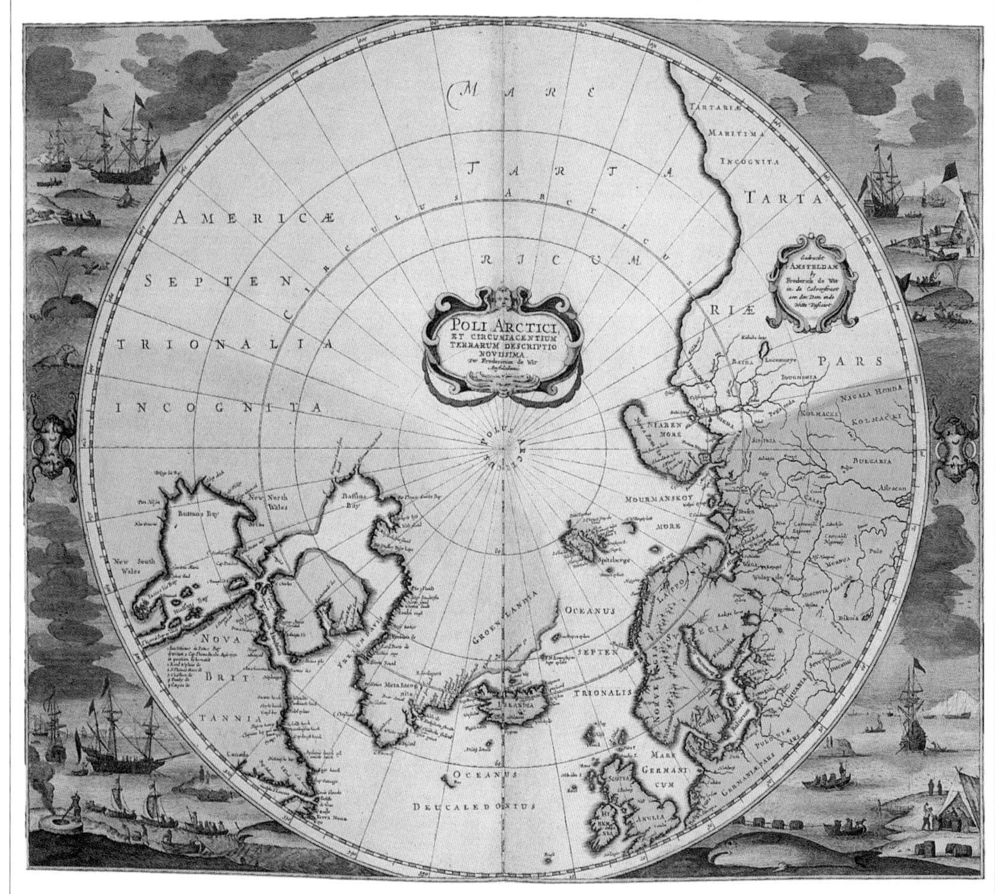

Frederic de Wit's 1675 map of the north pole reveals by its blankness how little was known of that enigmatic region. Scenes from the profitable whaling industry show why mariners were willing to venture into northern waters.

explorers who ventured north in search of the Northeast Passage, north of Asia, and the Northwest Passage, north of the Americas.

Although the ships that went north from North America always ran into impenetrable pack ice as they approached the Arctic, many 18th-century maps of the Arctic show what came to be called the "open polar sea." Reports of ice-free water around Spitsbergen (warmed by the Gulf Stream) led some cartographers to assume that the ice formed a narrow ring and that the inner Arctic Ocean was warm. William Parry, a British naval officer and explorer, thought otherwise. He rightly believed the pole to be ice-covered and tried in 1827 to reach it using boats mounted on runners like sleds. He got farther north than anyone had ever gone before but was forced to turn back after reaching the 82nd parallel.

In the mid-19th century, the search for the missing explorer Sir John Franklin resulted in a fairly thorough mapping of the Arctic Archipelago, the network of islands and channels north of mainland Canada. The long-sought Northwest Passage was found to be a daunting, commercially useless maze, not to be traversed by ship until Roald Amundsen did so in 1903–6. The High Arctic island clusters were explored and mapped between 1870 and 1900 by expeditions from Norway, Hungary, Great Britain, Italy, and the United States—except for Severnaya Zemlya, north of Russia, which was discovered by the Russians in 1900 and mapped by them between 1910 and 1935.

The symbolic conquest of the Arctic occurred when Robert Peary and his rival Frederick Cook claimed to have reached the north pole, Cook in 1908 and Peary in 1909. (Cook's claim was immediately discredited; Peary's has been questioned but is accepted by many experts.) The scientific exploration and mapping of the Arctic, however, were just beginning at the dawn of the 20th century. The 1920s and 1930s saw a host of scientific research projects in the Arctic, which all produced information that allowed new and better maps to be made. For example, the American explorer Louise Arner Boyd led several expeditions to the fjords of northeastern Greenland, sponsored by the American Geographical Society. Although her primary purpose was to study Arctic plants and rocks, she also gathered geographic data for maps.

Arctic exploration entered a new era in the 1930s, when the Russians introduced a revolutionary approach to polar studies: setting up research stations on large, drifting ice floes, from which large stretches of the Arctic Ocean could be observed. Ice stations continue to play a crucial role in polar studies. Since the 1970s, using new technological aids such as icebreaker ships, helicopters, satellites, and snowmobiles, scientists have produced many new maps of the Arctic—not just topographic maps but weather maps, wildlife maps, geologic maps, and more.

SEE ALSO

Amundsen, Roald; Baffin, William; Barents, Willem; Cook, Frederick; Davis, John; Franklin, John; Frobisher, Martin; Hudson, Henry; Northeast Passage; Northwest Passage; Parry, Sir William Edward; Peary, Robert Edwin

FURTHER READING

Berton, Pierre. *The Arctic Grail: The Quest for the North West Passage and the North Pole, 1818–1909.* New York: Penguin, 1988.

Brown, Warren. *The Search for the Northwest Passage.* New York: Chelsea House, 1991.

Maxtone-Graham, John. *Safe Return Doubtful: The Heroic Age of Polar Exploration.* New York: Scribners, 1988.

Scott, J. M. *Icebound: Journey to the Northwest Sea.* London: Gordon & Cremonesi, 1977.

Arrowsmith family

BRITISH ENGRAVERS, CARTOGRAPHERS, AND PUBLISHERS

THE ARROWSMITHS were the most prominent British mapmakers of the 19th century. Their engraving and map publishing business was founded by Aaron Arrowsmith (1750–1833), who was born in County Durham, Ireland, but established himself in London as a land surveyor around 1770. His surveys were used as the basis for an atlas published by two other mapmakers, William Faden and John Cary, in 1784. After that, Arrowsmith went into business for himself, publishing his first map in 1790.

Aaron Arrowsmith issued about 200 maps. Most of them were large-scale maps—that is, they showed individual regions, islands, or countries in great detail. Arrowsmith was more concerned with accuracy and clarity than with ornamentation and speculation, and although his maps look a bit austere next to the heavily decorated Dutch maps of the 17th century, they are elegant and clean-lined. He produced maps of Africa, Asia, and America that reflected Europeans' growing geographic knowledge of those continents; for example, his 1802 map of Africa in four sheets showed the discoveries of Scottish explorer Mungo Park along the Niger River in 1796. Today, however, Arrowsmith is best remembered for his many maps of the Pacific region, starting with a nine-sheet map of the whole ocean in 1798. His maps remain useful references for historians studying the history of 18th-century exploration in the Pacific.

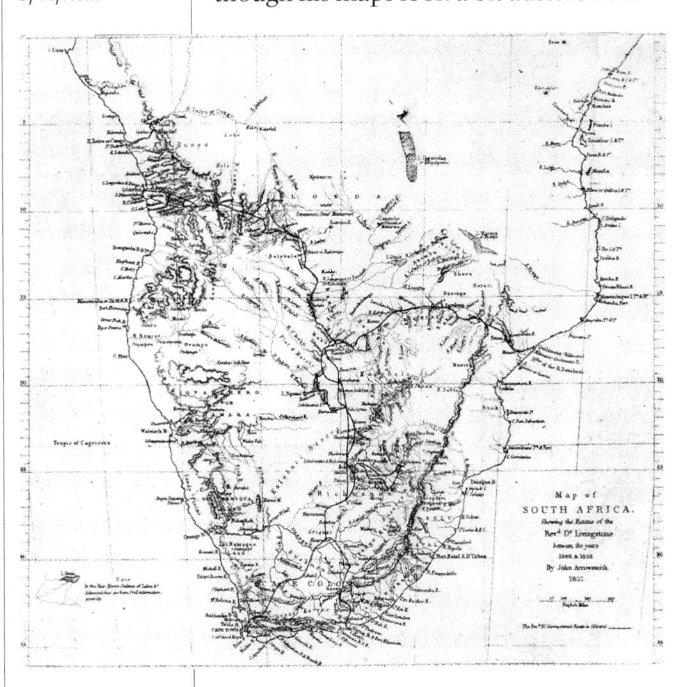

Carrying on a family tradition of mapping explorers' travels, John Arrowsmith made this map in 1857 for a book by David Livingstone, the Scottish missionary and explorer of Africa.

Aaron Arrowsmith was named Hydrographer to the King in 1820 and was universally regarded as a cartographic genius. His sons, Aaron, Jr., and Samuel, took over the business after his death. The last member of the dynasty was John Arrowsmith (1790–1873), nephew of Aaron, Sr., who published *The London Atlas of Universal Geography* in 1840. Containing a number of maps printed from plates made by Aaron, Sr., the atlas was revised and expanded many times to include new information about Australia and North America.

SEE ALSO

British mapmakers; Hydrography; Park, Mungo

Asia, mapping of

THE WORLD'S oldest known maps were made in Asia, in the kingdom of Babylonia, now known as Iraq. Various Asian cultures developed their own styles of mapmaking long before Europeans began to explore and map Asia. By the 7th century A.D., for example, Chinese cartographers were producing accurate,

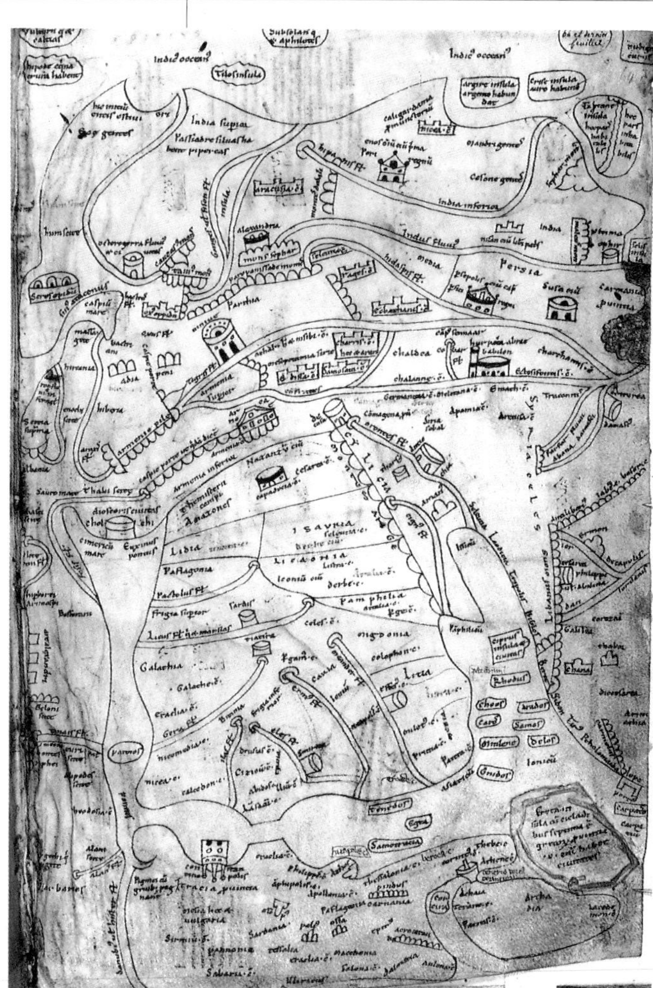

A 12th-century copy of an earlier map shows medieval Europeans' notions about Asia. East is at the top; the island-studded body of water at the lower right is the Mediterranean.

large-scale surveys of China's rivers and central provinces. By the 13th century, China had been mapped in great detail, although Chinese scholars gave little attention to mapping the world outside the borders of their empire. Japan developed its own cartographic tradition; like the Chinese cartographers, the Japanese mapmakers concentrated on mapping their homeland rather than on creating maps of the outside world. The earliest Japanese maps were sketches of estates and farms, created to help in recording land taxes. These local maps were later extended into detailed maps of Japan's provinces and roads.

A different mapmaking tradition arose in India, where the earliest known maps are religious or cosmographic, having to do with ideas about the structure of the universe rather than with geographic details. Images dating from before the 2nd century B.C. show the world divided into four continents like the petals of a lotus flower, with a sacred mountain called Meru at its center; later, some Indian cosmographers pictured the earth as having seven continents, each surrounded by a sea of some different substance, such as wine, milk, or butter. But a more practical kind of mapmaking may also have existed in India. Scholars believe that Indian merchants and navigators may have been skilled in making coastal charts and sailing manuals, although the only such document known today dates from 1664 and may have been influenced by European chart-making traditions.

The mapping of Asia in the European tradition began in the 2nd century A.D., when Ptolemy included 12 maps of Asia in his *Geography.* Most of these covered the parts of Asia that are closest to Europe and the Mediterranean world, such as Syria, Turkey, and Persia (now Iran). These areas were known to the Greeks by experience, while the remote parts of Asia were little more than rumors or legends. Later, the Romans incorporated western Asia—sometimes called Asia Minor—into their empire and their maps.

The Middle East

The Middle East appeared on most European maps made during the Middle Ages. These maps were shaped by Christianity and centered on Jerusalem. The Christian Holy Land of Palestine (present-day Israel and Jordan) was the most-mapped part of Asia. The first printed map of this region appeared in a 1482 edition of Ptolemy's *Geography,* and many others followed. Maps of the Holy Land were included in Bibles and history books, and later in atlases. Sixteenth-century Dutch cartographer

Abraham Ortelius is just one of the many leading mapmakers who made maps of the Holy Land; his was decorated with several dozen scenes from the Bible and is sometimes called the "Abraham's Travels" map. Town plans of Jerusalem were also included in all the major atlases of the 16th and 17th centuries.

Eastern Asia

Contact between Europe and China was stimulated in the 13th century by the rise of the Mongol Empire in Asia. The Mongols were well-disposed toward visitors from the West, and half a dozen or so missionaries and merchants from Europe made the long journey across steppes and mountains to the Mongol court in China. Although none of these travelers is known to have made maps, their travel stories—especially those of Marco Polo—impressed mapmakers. Scenes, place names, and topography from Polo's narrative began appearing on maps such as the Catalan atlas.

In the 16th century, Europeans began making maps of China based on firsthand knowledge. Jesuit scholar-missionaries such as Matteo Ricci began a two-way exchange of cartographic information: they made maps in the European style for the Chinese, and they made maps of China for the Europeans, often drawing upon traditional Chinese maps for geographic data. In 1655 another Jesuit, Martino Martini, compiled the first atlas devoted to China. It was replaced in 1737 by French cartographer Jean Baptiste d'Anville's *Atlas of China,* which contained a map of Siberia based on a manuscript by the explorer Vitus Bering, who had crossed northern Asia to the Pacific coast a decade earlier.

Europeans came later to Japan than to China. The first European maps of Japan appeared at the end of the 16th century, but they were not very accurate. For example, Ortelius showed five different shapes for Japan. Not until the mid-18th century did European maps show the Japanese islands in anything like their correct shapes and locations.

Southern Asia

India had existed on European maps of Asia from the time of Ptolemy, but its shape and size were badly distorted on all early maps. Not until the 16th century was it generally drawn accurately. Sri Lanka (called Taprobane in ancient times and Ceylon more recently) was even harder to pin down. On some maps it is east of India, on others west. Its size, too, was greatly variable.

Once the Dutch had established themselves in the East Indies trade, their maps of India improved steadily. However, the British East India Company gained a foothold in India in the early years of the 17th century, and soon the British were the European experts on

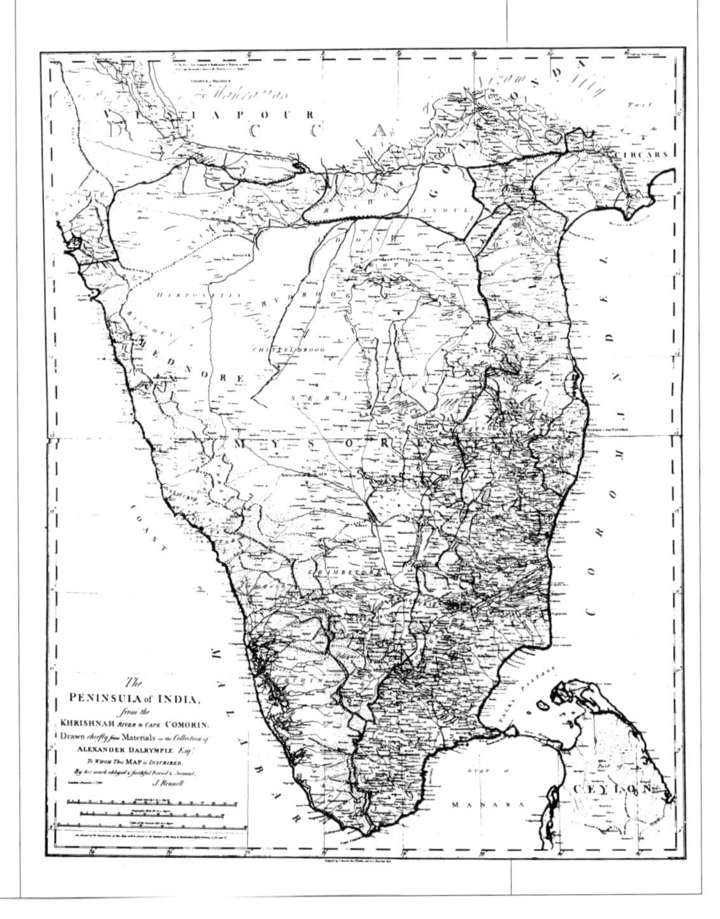

James Rennell's 1788 map of southern India, based on his 1782 map of India, shows the kingdoms and provinces of the tip of the subcontinent, as well as part of Ceylon (now Sri Lanka).

India. An Englishman, James Rennell, made the first detailed and generally accurate maps of India; his *Bengal Atlas* appeared in 1779 and his map of the whole subcontinent in 1782. The 19th-century mapping of India was dominated by the British, whose Great Trigonometrical Survey of their highly prized colony was the most ambitious cartographic project that had ever been undertaken.

Central Asia

Well into the 19th century, large tracts of Central Asia and the Tibetan plateau remained almost unexplored and unmapped by Europeans. British, German, and Russian explorers sought out these remaining "blank spaces" on their maps, partly out of geographic curiosity and partly for political reasons: Great Britain, Russia, and other nations were vying for influence and territory in Asia's heartland. These ventures, along with colonial expeditions into Southeast Asia, finally completed the mapping of Asia. In 1911 the India survey was joined to a similar survey of Central Asia that had been carried out by the Russians. For the first time, the outline and major features of Asia were accurately revealed on the map.

SEE ALSO

Chinese mapmakers; Great Trigonometrical Survey of India; Islamic geographers and mapmakers; Japanese mapmakers; Polo, Marco; Rennell, James; Ricci, Matteo

FURTHER READING

Harley, J. B., and David Woodward, eds. *The History of Cartography*. Vol. 2, *Cartography in Traditional Islamic and South Asian Societies*. Chicago: University of Chicago Press, 1992.

Severin, Timothy. *The Oriental Adventure: Explorers of the East*. Boston: Little, Brown, 1976.

Tooley, Ronald V. "Asia." In *Landmarks of Mapmaking*, 101–46. 1976. Reprint. New York: Dorset, 1989.

Astrolabe

AN ASTROLABE is an instrument that was developed in ancient times for use in making astronomical observations. The Greeks were using astrolabes by the 2nd century B.C., perhaps earlier. Arab astronomers refined the design and use of the astrolabe between the 7th and 17th centuries A.D.

The astrolabe measured the altitudes of stars, planets, the sun, and the moon. From these measurements, with the aid of certain mathematical tables and formulas, astronomers could figure out the latitude of the place where the measurement was taken. The astrolabe consisted of an alidade, a rotating bar that pivoted against a ring that was marked off in degrees. In the astronomer's astrolabe, behind the alidade was a star map. One or more interchangeable metal plates called tympans showed star maps for various latitudes. The mariner's astrolabe often lacked the tympan or star map; it generally consisted merely of the outer ring and the alidade, with the center part cut away so that the astrolabe would offer less resistance to the wind and would be easier to use at sea. Using the astrolabe, a navigator could determine the position of the sun above the horizon. Consulting tables, the navigator could then compare that reading to the sun's position at known latitudes on that same day, and from this information he could determine his own latitude.

The astrolabe allowed a navigator to determine latitude by measuring the sun's angle above the horizon.

The astrolabe was used in navigation until the 16th century, when it was replaced by the cross-staff, which was far easier to use.

SEE ALSO

Cross-staff; Navigation

Astronomical mapping

SEE Space mapping

Atlantic Ocean, mapping of

IN GREEK MYTHOLOGY, the Atlantic Ocean—the western ocean beyond the Pillars of Hercules—was a place of mythical wonders. Medieval mapmakers, too, filled the Atlantic Ocean with legendary places such as the island of Antillia (which later gave its name to the Antilles islands in the Caribbean).

The mapping of the Atlantic Ocean began in earnest in the late 13th century, when mariners from the Mediterranean nations began venturing out onto the ocean. They discovered the island clusters called the Canaries, Madeiras, and Azores. Surviving maps show that these islands were discovered within a short span of time. The Canaries first appear on a 1339 map made by Angelino Dulcert, and by the 1380s maps were showing all three groups. The islands played an important role in the history of exploration, for they served as the jumping-off points for later oceanic voyages, including those of Columbus and Magellan.

The North Atlantic was first explored by the Vikings, who reached Iceland by the 9th century, Greenland by the late 10th century, and the North American coast around the year 1000. No Viking maps survive, however; the so-called Vinland map, which appears to have been drawn in the 15th century and to show the Vikings' discoveries in North America, is now thought by most scholars to be a forgery.

The next recorded voyage across the Atlantic was that of Columbus in 1492–93. Other voyages soon followed. Sailing for the British, John Cabot reached Newfoundland, now part of Canada, in 1497. His son Sebastian explored the North American coast from Labrador

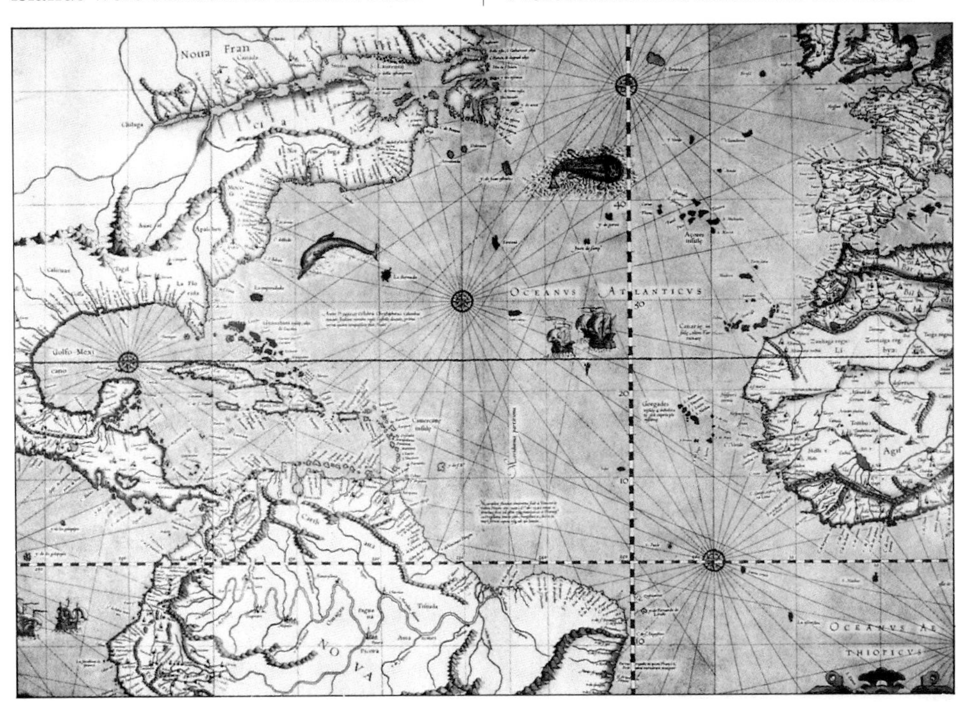

The Atlantic Ocean as it appeared on Gerardus Mercator's 1569 map of the world. The fantastical sea monsters of earlier maps have been replaced by a fairly realistic dolphin and a whale.

south to Chesapeake Bay in 1508–9. The Portuguese reached the coast of Labrador and Nova Scotia around 1500, and by 1506 Portuguese fishing fleets regularly visited the waters off Newfoundland. The North Atlantic was fairly well explored and mapped by the early 16th century, and from that time on explorers such as Martin Frobisher turned their attention to the search for the Northwest Passage.

Most of the South Atlantic was mapped by mariners who were trying to navigate past the southern tips of Africa and South America. Portuguese navigators Bartolomeu Dias, Vasco da Gama, and Pedro Álvars Cabral had used the wind system of the South Atlantic to round Africa by 1500. Ferdinand Magellan, sailing for Spain, pioneered the route around South America in 1520, becoming the first European to see Tierra del Fuego. Throughout the 16th century, Portuguese and Spanish explorers probed the coast of South America; their reports, combined with those of the Portuguese mariners who had been charting the west coast of Africa for generations, yielded a fairly reliable map of the South Atlantic by the end of that century. The Falkland Islands may have been discovered in the 16th century as well, but the first evidence of their discovery appears on an 18th-century map by Amédée-François Frézier. The mapping of the South Atlantic was largely complete after 1615, when the Dutch navigators Jakob le Maire and Willem van Schouten sailed around Tierra del Fuego, proving that it was an island and not part of a great southern continent.

SEE ALSO

Cabot, John; Cabot, Sebastian; Cabral, Pedro Álvars; Columbus, Christopher; Dias, Bartolomeu; Frézier, Amédée-François; Frobisher, Martin; Gama, Vasco da; Magellan, Ferdinand; Northwest Passage; Undersea mapping; Viking explorers; Vinland map

FURTHER READING

Morison, Samuel Eliot. *The Great Explorers: The European Discovery of America.* New York: Oxford University Press, 1978.

Morris, Roger. *Atlantic Seafaring: Ten Centuries of Exploration and Trade in the North Atlantic.* Camden, Maine: International Marine Publishing, 1992.

Atlas

AN ATLAS is a collection of maps bound into a single volume. Although maps and sea charts had long been collected in portfolios, the first bound atlases appeared in Italy in the mid-16th century. These were collections of individual maps bound together in volumes to suit the needs of each customer. (The Catalan atlas appeared much earlier, in 1375, but it was not a true atlas, because it was a single map on multiple sheets, rather than a number of separate maps bound together.)

The 16th-century Italian atlases are sometimes called Lafreri atlases after Antonio Lafreri, an engraver and publisher who produced many of these specially ordered collections, although other Italian cartographers and publishers did the same thing. About 50 Lafreri atlases survive. They are dated from 1556 to 1575. Some of them contain as many as 160 maps from a variety of different mapmakers. Lafreri put into many of these volumes a title page with a picture of Atlas, a giant from Greek mythology who in ancient times was said to have supported the world. This is the first known association of the word *atlas* with a map collection.

On the title page of Mercator's book of maps, the titan Atlas holds the world in his lap. Such volumes are now called atlases.

Because Lafreri collected existing maps rather than printing his own, and because his atlases were individual collections rather than standardized editions, credit for producing the first modern atlas is given to Dutch cartographer

Atlases come in all sizes. The Klencke Atlas in the British Library's map collection is a bound set of wall maps that belonged to King Charles II of England (1630–1685).

Abraham Ortelius. In 1570, Ortelius published an atlas called the *Theatrum Orbis Terrarum (Theater of the World),* which contained 70 maps, each re-engraved by Ortelius from the originals, with full credit going to the original cartographers. Ortelius did not use the word *atlas* to describe his volume of maps, however; the first person to do so was his friend and colleague Gerardus Mercator.

In 1585 Mercator published the first volume of a set of maps that he called *Atlas sive Cosmographicae meditationes de fabrica mundi et fabricata figuri* (*Atlas, or cosmographical meditations upon the creation of the universe and the universe as created*). Like Lafreri, Mercator prefaced his maps with a picture of the titan Atlas holding the world. The atlases of Ortelius and Mercator were printed and reprinted in many editions over the years and were far more widely circulated than the Lafreri collections had been. Other map publishers, seeing how popular Ortelius's and Mercator's atlases had become, produced their own volumes of maps. No longer mixed and matched to suit the whims of each customer, these were standardized volumes in which each buyer received the same maps. People began calling these volumes atlases after Mercator's great work.

World atlases were immensely popular, but mapmakers also began making national atlases, collections of maps of a single country. One of the first of these was Christopher Saxton's 1579 atlas of England and Wales. An atlas of France appeared in 1594, and Cornelis Wytfliet published an atlas of the Americas in 1597.

But, as French geographer Bruzen de la Martinière explained in 1734, there were two major problems with these splendid atlases. "The first is that they are priced so that many scholars are unable to afford them," wrote Martinière. "The second is that because of their grandeur . . . they are, so to speak, nailed up in the book-case" instead of being handled and used. Map publishers solved these problems by making pocket atlases—small, relatively inexpensive versions of the large ornamental atlases. The atlases of Ortelius and Mercator were published in pocket-size versions within a few years of their initial publication. English mapmaker John Speed produced a pocket atlas of England in 1627, using surveys made earlier by Christopher Saxton. Martinière himself published a pocket atlas of France "for the use of travelers and officers" in 1734.

Innumerable atlases have been published since the days of Lafreri, Ortelius, and Mercator. Each year sees the publication of many more. Some are world atlases, with maps of all parts of the world. These usually include several large maps of the entire world, followed by maps that show the continents, regions, countries, and even cities in greater detail. Some atlases cover only a particular continent or nation, or even a state. Some are historical, with maps showing empires, nations, trade routes, and battlefields from various times in history. Thematic atlases are sets of maps organized by the kind of information they display. Almost every imaginable subject, from archaeology to zoology, has been the theme of at least one atlas. For example, there are atlases of Islam, of Nazi Germany, and of nuclear energy. Some atlases are now published in the form of computer software, with disks and databases that can be updated almost instantly to reflect political changes or new, more accurate geographic surveys.

SEE ALSO

Catalan atlas; Mercator family; Ortelius, Abraham; Saxton, Christopher

FURTHER READING

Allen, Phillip. *The Atlas of Atlases.* New York: Abrams, 1992.

Makower, Joel, ed. *The Map Catalog.* 3rd ed. New York: Vintage, 1992.

Wilford, John Noble. *The Mapmakers.* New York: Knopf, 1981.

Australia, mapping of

THE ABORIGINAL peoples of Australia organized their lives and cultures around geographic concepts. To them, the land was traversed by what they called "song-lines," patterns of spiritual energy. Although the aborigines made no maps of the outside world, anthropologists believe that some of their paintings were actually maps: symbolic representations of their cosmography as well as their local topography.

In the Western sense—the fixing of places on paper—Australia began to be mapped centuries before it was actually discovered. Ptolemy's theory of a gigantic southern landmass called the Terra Australis caused mapmakers to draw a huge continent in the Southern Hemisphere long before any Europeans had actually seen this land. When European mariners began sighting bits of Australia's coastline, they thought it was part of the Terra Australis. The process of piecing together Australia's true shape went hand in hand with the gradual realization that the Terra Australis did not exist.

The first Europeans to see Australia were Dutch sailors in the East Indies trade who landed in northern Australia in 1605. The Dutch kept their discovery secret, however, and this landfall did not appear on a map until the 1630s. The Dutch discovery finally showed up on Henricus Hondius's world map of 1630 and on Willem Janszoon Blaeu's 1636 map of Southeast Asia. Then and for many years to come, Australia was called New Holland.

During the early 17th century, Dutch mariners charted pieces of New Holland's northern and western coastline. Then, in the middle of the century, the explorer Abel Tasman saw enough of the coast to be sure that New Holland was an island—a large one, but not the supercontinent called for by Ptolemaic geography. Dutch maps of the 17th century reflected Tasman's voyage and his improved knowledge of Australia's coastline, but on French maps of the same period New Holland's shape is greatly distorted: the northwestern coast is accurate, but the eastern part of the continent bulges far to the north and south. French cartographers made New Guinea and Tasmania part of Australia rather than showing them as the separate islands they are. This erroneous image was copied by Dutch mapmakers of the early 18th century.

Captain James Cook of Great Britain ushered in the next phase in the mapping of Australia. On his first Pacific voyage (1768–71) he filled in the map of the still-unknown east coast of the continent. His account of the voyage was published in 1773, and just 15 years later Britain established the first European settlement—a prison colony—near where the city of Sydney stands today.

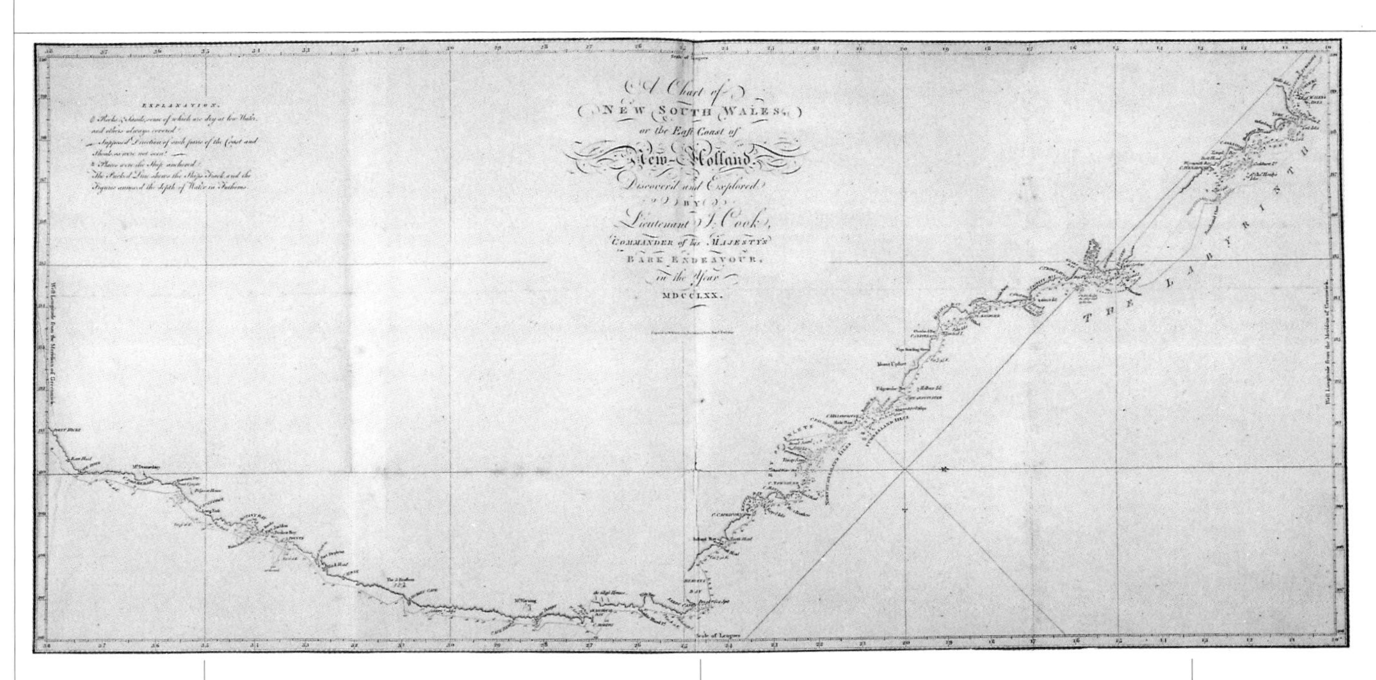

After nearly coming to grief when his ship struck the Great Barrier Reef, James Cook mapped the east coast of New Holland, as Australia was called in the 18th century.

The charting of the continent's coastline was completed in the early years of the 19th century by a handful of explorers, mostly British, including Matthew Flinders, who introduced the name Australia.

Up to this time, Australia had generally appeared only on world maps, maps of the Pacific Ocean, or maps of the East Indies. But at the beginning of the 19th century, mapmakers started adding separate maps of Australia to their atlases. The Arrowsmith family of British mapmakers, for example, produced about 70 different maps of Australia starting in 1800. Maps of the individual regions and states of Australia appeared in many 19th-century atlases. One of these atlases, published by the British mapmaker John Tallis in 1851, is particularly interesting because it was one of the last decorated atlases ever made. In the style of Dutch atlases from the 17th century, it featured vignettes of aborigines, kangaroos, and Australian towns.

Even in Tallis's time, however, most of Australia was still a blank space on the map. European exploration and settlement of the continent did not get far beyond the coast until the beginning of the 19th century. Around 1814 people began to cross the mountains that separated the coastal settlements from the interior, and the long, slow process of exploring and mapping inner Australia got under way. By mid-century the southeastern part of the continent was well mapped. By the third quarter of the century, explorers such as Charles Sturt and Robert O'Hara Burke had sketched out the interior, and the first comprehensive maps were being made. After 1901, when Australia's six colonies joined to form a single commonwealth, new maps and atlases were made to celebrate Australia's statehood.

SEE ALSO

Cook, James; Flinders, Matthew; Ptolemy; Tasman, Abel Janszoon; Terra Australis

FURTHER READING

Moorehead, Alan. *Cooper's Creek: The Opening of Australia.* New York: Harper & Row, 1963.

Tooley, Ronald V. "Australia." In *Landmarks of Mapmaking,* 245–68. 1976. Reprint. New York: Dorset, 1989.

———. *The Mapping of Australia.* Watchung, N.J.: Saifer, 1980.

Azimuthal projection

SEE Projections

Backstaff

SEE Cross-staff

Baffin, William

ENGLISH NAVIGATOR

- *Born: about 1584, probably in London, England*
- *Died: Jan. 23, 1622, in the Persian Gulf*

IN 1612 AND 1615, William Baffin served on English ships looking for the Northwest Passage. During these voyages he explored parts of the Greenland coast and Hudson Bay. Baffin was a pilot, responsible for setting the ship's course, and he had mastered the art of location finding as it was understood in the early 17th century.

In 1616 Baffin returned to the Canadian Arctic to investigate a large island and a gulf north of Hudson Bay; today these are called Baffin Island and Baffin Bay in his honor. Baffin turned his hand to mapmaking, producing an outline map of Baffin Island that was criticized by geographers. Two centuries later, however, explorers discovered that Baffin's map was fairly accurate.

Baffin drew this map of the Mogul Empire after talking with a British envoy who had visited Emperor Jahangir of northern India in 1615.

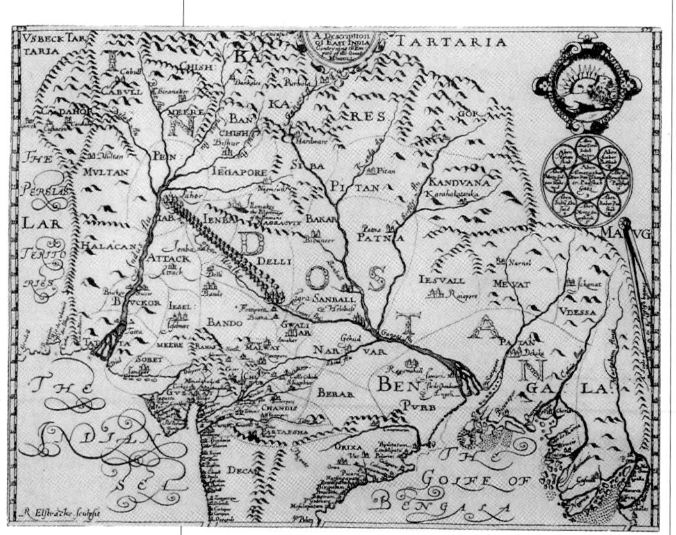

Baffin spent his final years in Asia, working as a sea captain for the British East India Company. In 1619 he drew a map called "The Empire of the Great Mogoll," showing the domains of the Mogul emperors in India and Central Asia. This map was published in a number of books about Asia.

SEE ALSO

Asia, mapping of; Northwest Passage

FURTHER READING

Brown, Warren. *The Search for the Northwest Passage.* New York: Chelsea House, 1991.

Lehane, Brendan, and Time-Life editors. *The Northwest Passage.* Alexandria, Va.: Time-Life Books, 1981.

Scott, J. M. *Icebound: Journey to the Northwest Sea.* London: Gordon & Cremonesi, 1977.

Baker, Samuel and Florence

EXPLORERS OF AFRICA

Samuel Baker

- *Born: June 8, 1821, London, England*
- *Died: Dec. 30, 1893, Sandford Orleigh, England*

Florence von Sass Baker

- *Born: Aug. 6, 1841, Hungary*
- *Died: 1916, Sandford Orleigh, England*

SAMUEL BAKER, a British adventurer and sportsman, spent time in Ceylon (now Sri Lanka) and Turkey before taking a boat trip down the Danube River in Europe. During this river voyage he visited a Turkish slave market, where he saw a young Hungarian woman named Florence von Sass being sold into slavery. He bought her to save her from this fate; soon they fell in love, and she later became his wife.

Samuel and Florence Baker, explorers of Africa.

Together they went to Egypt and traveled south up the Nile River. The search for the sources of the Nile was the biggest geographic puzzle of that age, and Baker and von Sass determined to join the search. In 1861–62 they explored the river's tributaries in what is now Ethiopia. They then continued upriver to the Sudanese city of Gondokoro, which at that time was the farthest point that had been reached by European explorers traveling south from Egypt.

At Gondokoro, Baker and von Sass met two British explorers, John Hanning Speke and James Augustus Grant, who had traveled north through the interior from Africa's east coast. Speke and Grant told them of rumors about a large lake lying somewhere off the route they had followed, and Baker and von Sass decided to locate this lake. They left Gondokoro and, after a dangerous and difficult journey, in 1864 they became the first Europeans to reach Lake Albert, between the modern nations of Uganda and Zaire. They made their way back to London and published an account of their trip, including a map of the Lake Albert region.

Baker estimated the size of the lake incorrectly; he thought it was bigger than it is. In fact, the geography of the lake country of East Africa was not fully understood until some years later. Nonetheless, their remarkable adventure won great popularity for Samuel Baker, who was knighted in 1866, and for Florence von Sass Baker as well.

After their marriage, the Bakers settled in England. Later, they made a second expedition to Gondokoro and the lands beyond, but they recorded no major new geographic finds.

SEE ALSO

Africa, mapping of; Speke, John Hanning

FURTHER READING

Baker, Samuel. *Albert N'Yanza: Great Basins of the Nile and Exploration of the Nile Source.* 1866. Reprint. Irvine, Calif.: Reprint Services, n.d.

———. *In the Heart of Africa.* 1884. Reprint. Westport, Conn.: Greenwood, 1970.

Hall, Richard. *Lovers on the Nile: The Incredible African Journeys of Sam and Florence Baker.* New York: Random House, 1980.

Middleton, Dorothy. *Baker of the Nile.* London: Falcon, 1949.

Stefoff, Rebecca. *Women of the World: Women Explorers and Travelers.* New York: Oxford University Press, 1992.

Balboa, Vasco Nuñez de

SPANISH CONQUISTADOR AND EXPLORER

- *Born: about 1475, Jerez de los Caballeros, Spain*
- *Died: Jan. 1519, Panama*

VASCO NUÑEZ DE BALBOA was the first European to see the Pacific Ocean from the American coast. In 1500, as the era of the conquistadors was beginning, he left Spain to seek his fortune in the New World. He went to Hispaniola, the island where Columbus had established Spain's first American colony (now divided between the nations of Haiti and the Dominican Republic). Soon, however, Balboa ran into debt. To escape from angry creditors, he hid on a ship bound for the Central American mainland. There, he helped establish a colony in what is now Panama.

Balboa added the South Sea to the map of the world.

In 1512 Balboa led an expedition southward, reaching the northern end of the Andes Mountains in Colombia,

South America. The next year, following reports from local Indians, he went west across Panama in search of a large sea. After crossing a series of knife-edged, jungle-covered mountain ridges, he reached the west coast of Panama on September 25, 1513, and gazed out at a body of water that he called the South Sea. He was correct in thinking that Asia lay on the other side, but he thought the sea was only a narrow channel, when in reality it is the largest ocean on earth.

Balboa claimed the South Sea, today called the Pacific Ocean, for Spain. Within a few years of Balboa's journey, the route he had discovered—overland across the Isthmus of Panama between the Caribbean Sea and the Pacific—began showing up on maps. For example, the land route to the "Mar del Sur," the South Sea, appears on a 1529 map of the American coast that was regarded as a top-secret document by the Spanish government. For two centuries, the overland route Balboa had pioneered continued to be used by Spanish explorers and traders, who preferred heat and mosquitoes to the long, hazardous sea route around South America. Balboa's reward for his discovery was the governorship of Panama. A rival conquistador plotted Balboa's downfall, however, and had him beheaded in 1519.

FURTHER READING

Garrison, Omar V. *Balboa: Conquistador.* New York: Carol Publishing, 1971.
Herrmann, Paul. *The Great Age of Discovery.* New York: Harper & Brothers, 1958.
Romoli, Kathleen. *Balboa of Darien, Discoverer of the Pacific.* Garden City, N.Y.: Doubleday, 1953.

Barents, Willem

DUTCH EXPLORER OF THE ARCTIC

- *Born: mid-16th century, Holland*
- *Died: June 20, 1597, in the Arctic*

WILLEM BARENTS was the chief pilot, or navigator, on three Dutch expeditions into the Arctic waters north of Europe.

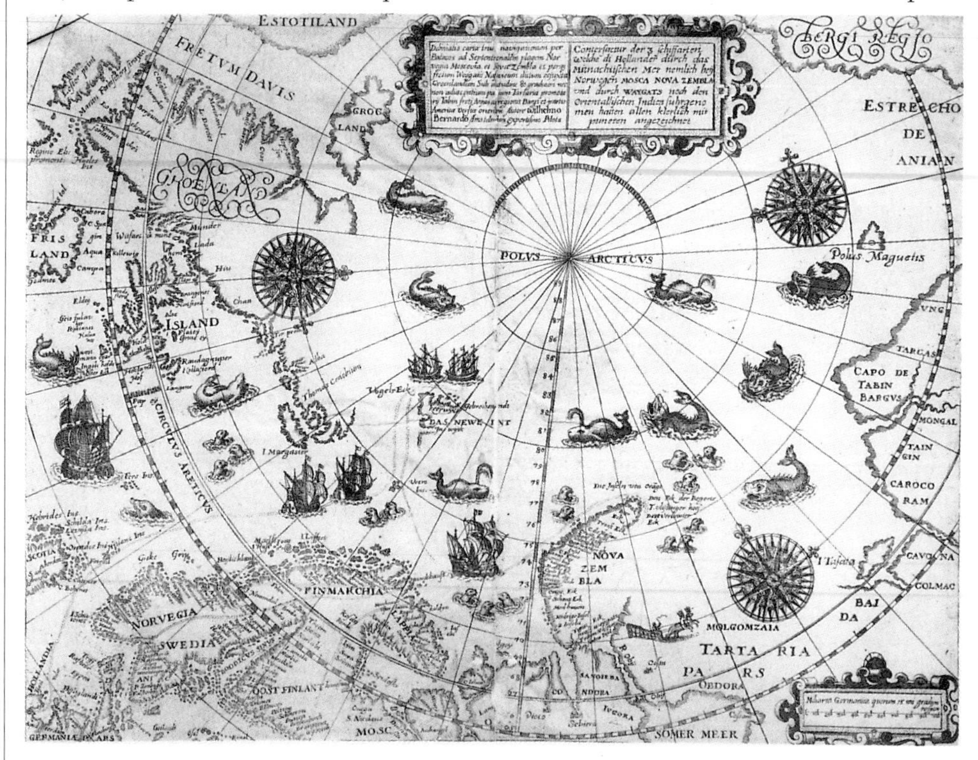

A 1598 Dutch map showing Barents's view of the Arctic. The Polus Magnetus in the upper right was a mythical mountain of iron, thought to be the source of magnetic attraction.

In 1594 and 1595 he crossed the part of the Arctic Ocean that lies between Norway and Russia; today it is called the Barents Sea in his honor. In 1596–97 he discovered Spitsbergen, a cluster of islands north of Norway, and then made the first known voyage around the north coast of Novaya Zemlya, a large island in the Russian Arctic. Barents died on the way home from this trip, but his shipmates told mapmakers of his discoveries, which were added to maps of the Arctic. A map of Barents's discoveries was engraved in Holland a year after his death.

SEE ALSO

Arctic, mapping of

Barrow, John

BRITISH GEOGRAPHER

- *Born: 1764*
- *Died: 1848*

JOHN BARROW served as secretary to a high-ranking British diplomat and accompanied his employer on missions to China and Africa, later describing these journeys in *Travels in China* (1804) and *Travels into Southern Africa* (1806). In 1806 he was made assistant secretary of the British Admiralty, and in the years that followed he used his position to promote voyages of exploration, including that of William Parry, who made three journeys into the Canadian Arctic between 1819 and 1825. Barrow was also one of the founders of the Royal Geographical Society, which sponsored exploring expeditions and published maps and geographic papers. In 1846, Barrow summed up the state of Arctic knowledge in *Voyages of Discovery and Research in the Arctic Regions.*

Barth, Heinrich

GERMAN EXPLORER OF NORTH AFRICA

- *Born: Feb. 16, 1821, Hamburg, Germany*
- *Died: Nov. 25, 1865, Berlin, Germany*

HEINRICH BARTH mastered Arabic before setting out for North Africa, where most people speak Arabic and follow the Islamic religion. He traveled on his own for some years, but in 1849 he joined a British expedition to cross the Sahara Desert from the north.

The British expedition leaders died in 1852, leaving Barth in charge. He explored the region around Lake Chad, settling several geographic questions about the region's rivers. Then he went west to the city of Timbuktu, a center of caravan trade that few Europeans had visited. Along the way he gathered an enormous amount of information about the land, people, and languages of sub-Saharan Africa. From Timbuktu he went north, crossing the Sahara a second time and mapping the age-old routes of the desert nomads.

Barth returned to Europe in 1855 and published a five-volume account of his travels. His maps were meticulous and detailed, based on thousands of measurements and notes. Because he devoted himself to filling in the blanks on the map with in-depth knowledge of people and places, he is regarded today as a master of scholarly exploration. Unfortunately, Barth did not become as well known as some other 19th-century explorers because just as his book appeared, public interest turned to East Africa and the search for the sources of the Nile.

SEE ALSO

Africa, mapping of

FURTHER READING

Herrmann, Paul. *The Great Age of Discovery.* New York: Harper & Brothers, 1958.
Pennington, Piers. *The Great Explorers.* London: Bloomsbury Books, 1979.
Stefoff, Rebecca. *Scientific Explorers: Travels in Search of Knowledge.* New York: Oxford University Press, 1992.

Base map

A BASE MAP is one from which other maps are compiled. A cartographer who creates a map from existing maps rather than from fresh survey data is using base maps. Most commercially produced maps are made from base maps.

Bathymetric maps

BATHYMETRIC MAPS show the topography, or shape, of the bottom of a body of water. They are relief maps of lake, sea, and river beds. Contour lines called isobaths, which connect points of equal depth, are used to show elevations and depressions on the bed.

The term *bathymetric* comes from the Greek words *bathys,* meaning "deep," and *metron,* meaning "measure." The first bathymetric map, a chart of the floor of the English Channel, was made in 1752 by French cartographer Philippe Buache. During the 19th century, the United States, Great Britain, and other nations launched ambitious projects to map the ocean floors. These efforts continue today; the newest bathymetric maps are generated by computers and based on millions of bits of information from undersea scanning devices and satellite monitors.

SEE ALSO

Contour lines; Hydrography; Isogram; Relief; Undersea mapping

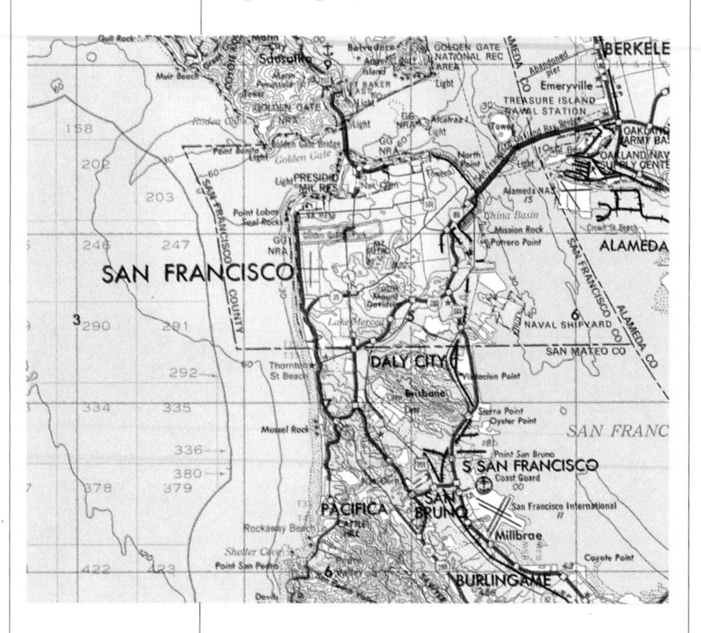

A bathymetric map showing the seafloor around San Francisco. The narrow, wriggling black lines are isobaths connecting points of equal depth at intervals of 200 feet, or about 30 meters.

Beatus maps

DURING THE Middle Ages, the most accurate maps were those made by the Arabic-speaking geographers of the Islamic Empire. European maps, for the most part, were as much about Christian dogma as they were about geography. Medieval Christian cartographers based their maps on the Bible and sprinkled them with places known only from the Bible: the Garden of Eden, for example, or the kingdom of the giants Gog and Magog.

The Beatus maps were somewhat more accurate than many other medieval maps. Several Beatus maps exist. They were made between the 10th and 13th centuries and are copies of a map made by a Spanish monk named Beatus of Valcavado to illustrate his book *Commentary on the Apocalypse,* which was published in A.D. 776. All but one of the Beatus maps have the east at the top;

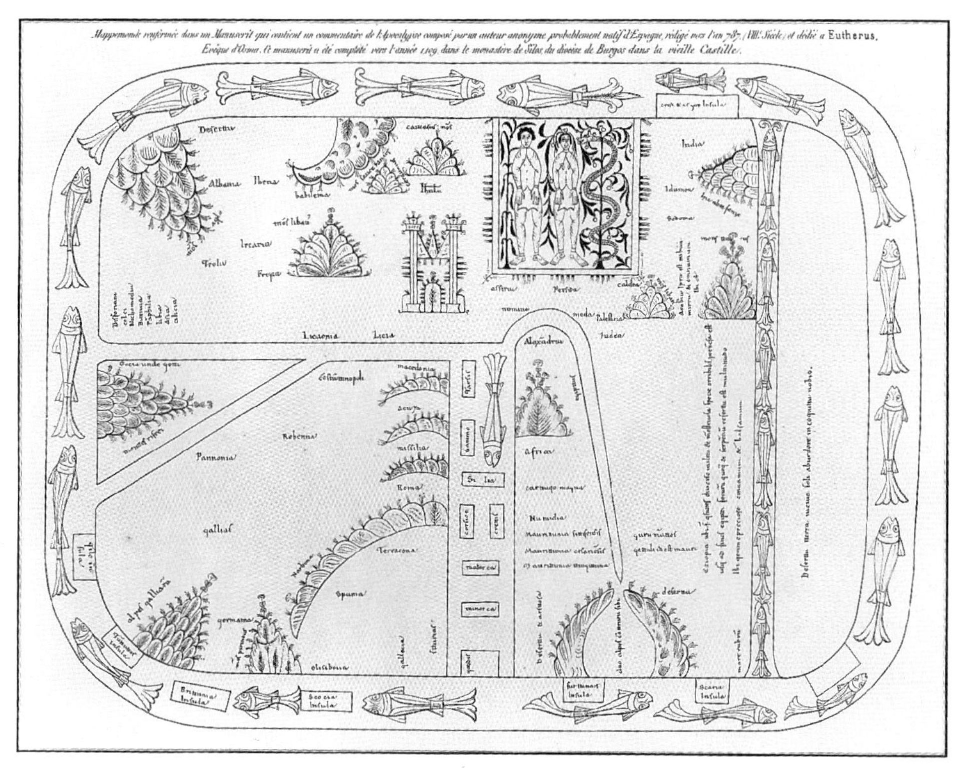

A 12th-century Beatus map with the Garden of Eden at the top. The long curving line right of the center is the Nile River; Europe is at the lower left.

Jerusalem, the scene of Christ's life, is at the center. The Beatus maps contain some fanciful biblical details, such as pictures of Adam and Eve coyly clutching their fig leaves as the snake whispers in Eve's ear. But the best of them also contain many real places, presenting a thorough picture of the world as it was known to Europeans of the Middle Ages.

SEE ALSO

Ancient and medieval mapmakers

Behaim, Martin

GERMAN GEOGRAPHER

- *Born: between 1436 and 1459, Nürnberg, Germany*
- *Died: Aug. 8, 1507, Lisbon, Portugal*

MARTIN BEHAIM came from a well-off family and was destined for a career in commerce; he also acquired the knowledge and skills of a geographer and navigator. Sometime around 1484 he went to Portugal, then a center of maritime activity and exploration, and found favor with the Portuguese court. He was made a member of the council that advised the king on nautical matters, he married a woman from an aristocratic family, and he was knighted.

Behaim visited the Azores and the Canaries, two island groups in the Atlantic, and he made one voyage to the West African coast, which had been the subject of Portuguese exploration for decades. His path may have crossed that of Christopher Columbus, for both men were in Portugal at the same time and both were deeply interested in navigation and geography, but there is no record of their having met.

Today, Behaim is remembered as the maker of the oldest surviving globe. In 1490 he visited his hometown of Nürnberg, Germany.

Martin Behaim made his "earth apple," the oldest surviving globe, in 1492.

Impressed by Behaim's knowledge of the world, Nürnberg's leading citizens asked him to make a globe showing all the recent discoveries of explorers from Portugal and elsewhere. Behaim designed a map that included more than 1,100 place names. It was painted on parchment in six colors by an artist named B. Glockenthon, who adorned it with 111 tiny pictures of everything from kings on thrones to elephants and mermaids. The parchment was then carefully fitted onto a wooden sphere. The globe was completed in 1492, around the time Columbus was getting ready to set sail on his first voyage to the Americas. Pleased with their globe, the Nürnbergers called it Behaim's Erdapfel (earth apple).

The Behaim globe is valuable not only because it is the oldest terrestrial globe in existence but because it shows the view of the world that was shared by many people in Columbus's time. Most important, it disproves the tired old myth that people before Columbus believed the earth to be flat.

Behaim's globe combines certain inaccurate geographic theories with up-to-date geographic discoveries. For example, Behaim followed the tradition of medieval mapmaking by placing mythical islands such as Antillia in the middle of the Atlantic Ocean, but he also showed recent discoveries such as the rounding of southern Africa by the Portuguese explorer Bartolomeu Dias in 1487–88. Perhaps the most important feature of Behaim's world view is the great eastward extent of Asia. Like Columbus, Behaim believed that the Atlantic Ocean separated the west coast of Europe from the eastern edge of Asia. He thought that the width of the ocean could not be very great because the portion of the earth covered by Asia was so large. On Behaim's globe, Japan—which no European had yet seen, but of which Marco Polo had heard tales in China—seems only a few days' or weeks' sail away from the Azores Islands.

Behaim's globe is preserved in the German National Museum in Nürnberg. Some of the inscriptions on it have faded or been rubbed off over the centuries, but we know what they were because the historian Johann Gabriel Doppelmayer copied the globe onto a two-dimensional map in 1730.

SEE ALSO
Columbus, Christopher; Globe

Bellingshausen, Fabian Gottlieb Benjamin von

RUSSIAN EXPLORER

- *Born: Aug. 18, 1779, Osel, Russia*
- *Died: Jan. 13, 1852, Kronstadt, Russia*

FABIAN GOTTLIEB Benjamin von Bellingshausen (sometimes referred to as Thaddeus von Bellingshausen because his Russian name Faddei can be translated either way) led a Russian exploring expedition into the South Atlantic Ocean in 1819. In January 1820, at the height of summer in the Southern Hemisphere, he sailed close to a region that is now called the Princess Martha Coast of Antarctica, near the Weddell Sea. Earlier explorers had probed the edges of Antarctica's pack ice and its outlying small islands, but Bellingshausen is considered the first person to have set eyes on the continent itself, although he did not know it was a continent. Bellingshausen is commemorated on modern maps of Antarctica by the Bellingshausen Sea, an inlet of the South Pacific Ocean that runs along the Antarctic coast.

Like his British counterpart Captain

James Cook, Bellingshausen was an excellent chart maker. When his expedition paused at South Georgia Island in the South Atlantic Ocean, he produced in only three days a survey of the island's dangerous south coast that remained in use until the mid-20th century. He also cleared up some cartographic confusion during his voyage. In the 1770s, Cook had sighted land south of South Georgia Island. He had named it Sandwich Land after the Earl of Sandwich, but he had not explored it. Some mapmakers drew this unknown landmass as part of a continent; others depicted it as a small island. Bellingshausen proved that it was an archipelago, or cluster of small islands. He made careful surveys of each island and also took astronomical sightings that enabled him to place each one accurately on the map of the Atlantic. When the account of his voyage was published, mapmakers of all nations updated their maps to include the South Sandwich Islands, which soon became a port of call for sealers and whalers.

SEE ALSO
Antarctica, mapping of

Bench mark

A BENCH MARK is a permanent object whose exact elevation, latitude, and longitude are known. It serves as a starting point for measurements of other points. Some bench marks are natural objects such as rock outcroppings. Others are erected by surveyors and may take the form of metal pipes, plates embedded in the ground, or concrete blocks or pillars. In any case, the actual bench mark is indicated by a small metal plate set in place by a surveying team. In the United States, bench marks bear the insignia of the U.S. Geological Survey, the Army Corps of Engineers, and other such organizations.

Bering, Vitus

DANISH EXPLORER OF SIBERIA

- *Born: 1680, Horsens, Jutland, Denmark*
- *Died: Dec. 19, 1741, Bering Island*

VITUS BERING was a Danish navigator who served in the Russian navy under Czar Peter the Great. In 1724 he led an expedition eastward across the breadth of Siberia to determine whether Asia and North America were separate continents or were linked by a land bridge across the North Pacific Ocean. The crossing of Siberia alone was a mighty effort, but once he had reached the Pacific coast Bering built a ship and explored the waters between Siberia and North America—waters that today are called the Bering Sea. In 1728 he discovered the Bering Strait, which separates the two continents, ending centuries of geographic uncertainty by proving that Asia and North America are not joined. (Modern scientists have found, however, that the two continents have been joined in the past by a land bridge that has appeared in the Bering Sea during ice ages, when the level of water in the world's oceans dropped sharply.)

Bering made a second trip across Siberia in 1733, accompanied by a team of scientists. In the course of a decade, this Great Northern Expedition explored and mapped much of Siberia, including its major rivers and its Arctic and Pacific coasts. Bering and another commander

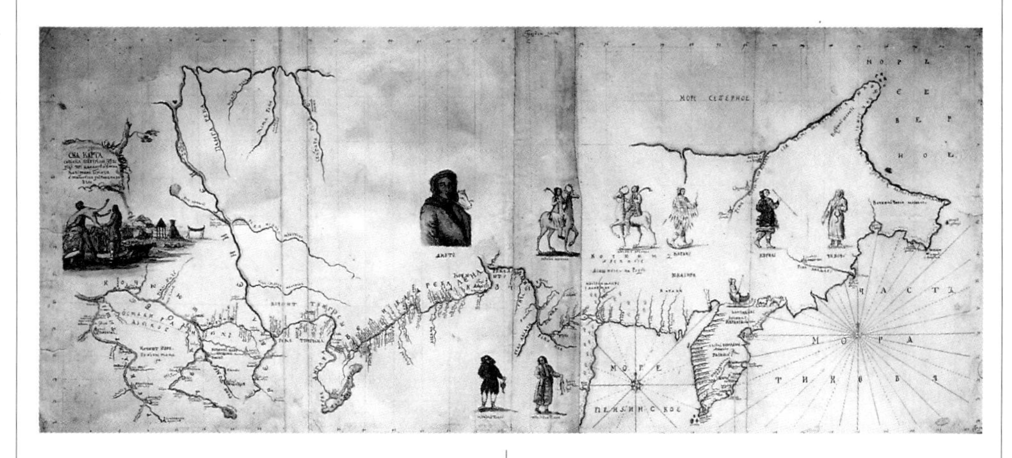

A map made after Bering's first expedition shows representatives of half a dozen native Siberian peoples; the Bering Strait is at the far right.

sailed as far east as Alaska, thus establishing Russia's claim to Alaska (Russian settlements would appear in Alaska before the end of the 18th century). On the way back to Siberia, Bering died of scurvy on an island in the Bering Sea, but the expedition returned to Russia with a wealth of geographic knowledge.

FURTHER READING

Muller, Gerhard F. *Bering's Voyages: The Reports from Russia.* Translated by Carol Urners. Fairbanks: University of Alaska Press, 1986.

Stefoff, Rebecca. *Scientific Explorers: Travels in Search of Knowledge.* New York: Oxford University Press, 1992.

Urners, Carol. *Bering's First Expedition: A Re-Examination based on Eighteenth-Century Books, Maps, and Mss.* New York: Garland, 1987.

Bianco, Andrea

ITALIAN MAPMAKER

• *Active: 1436–58, Venice, Italy*

ANDREA BIANCO was a navigator and cartographer who had practical experience at sea. His name appears in the records of the Venetian merchant fleet as a ship's officer or adviser, and he made several voyages to Holland and England. In the 1450s, he helped Fra Mauro of Venice make his large world map. He is known for two other surviving works: an atlas of 10 maps published in 1436 and a portolan chart made in 1448 that accurately records the recent discoveries of Portuguese explorers along the west coast of Africa.

SEE ALSO

Mauro, Fra

Blaeu family

DUTCH MAPMAKERS

WILLEM JANSZOON BLAEU and his son Joan Blaeu were among the foremost cartographers of Holland in the 17th century, the era of Dutch supremacy in mapmaking that has been called "the golden age of cartography."

Willem Janszoon Blaeu (1571–1638) was expected to enter his family's herring business in Amsterdam, but he hated the herring trade and took up astronomy instead, studying with a prominent Danish astronomer named Tycho Brahe. Blaeu then returned to Amsterdam to go into business as a maker of scientific instruments and globes. In 1604 he began making and publishing maps, first of individual countries and then of the continents and the world.

His world map was especially striking. It was published on 20 sheets, which, if joined, formed a map eight feet across. Sadly, only one battered copy of this impressive map survives.

In 1608 Blaeu produced his first sea atlas. He replaced this effort with a larger and more accurate set of sea charts in 1623. Using plates that he had bought from the Hondius family of mapmakers, Blaeu published his first land atlas in 1630, along with reprints of his own single-sheet maps. Willem Janszoon's son Joan Blaeu (1596–1673) assembled a larger atlas of 208 maps, published in two volumes in 1635. This Blaeu atlas went through many editions and was expanded with each new one. By the time it was reissued in several languages in 1662–63, it had grown to more than 600 maps in 9 to 12 volumes, depending upon the language. This final version was called the *Atlas Major* (*Grand Atlas*).

The *Atlas Major* was a superb work of cartography, the prized possession of kings and princes—and the most expensive printed book of the 17th century. Its maps were beautifully hand-colored and lavishly decorated with scenic views, vignettes from history and natural history, and other pictures. Some of the maps contain illustrations not found anywhere else, such as scenes of Dutch colonists and Indians in Brazil.

In general, the maps of the *Grand*

This 1630 world map by Willem Blaeu is perhaps the supreme artifact of the golden age of mapmaking. Vignettes at the top represent the moon and planets. At left are the elements; at right, the seasons. The seven wonders of the ancient world fill the bottom panels.

Atlas are as accurate as any maps of their day. When assembling material for their atlases, the Blaeus drew upon the best that was available. In the volume dealing with China, for example, Joan Blaeu based his maps on the work of the Jesuit missionary Martino Martini, who had surveyed China and who helped Blaeu make the map engravings in Amsterdam. The Blaeu-Martini maps of China were the first ones published in Europe on which Korea was correctly shown as a peninsula rather than as an island. Taken together, the maps that made up the *Grand Atlas* presented a comprehensive picture of the world as it was known to educated Europeans of the mid-17th century, as well as a portrait of commerce, science, and colonization in the far-flung corners of the world. In 1991 the Royal Geographical Society of London cooperated with a commercial publisher to issue a selection of 100 of the most interesting, influential, and beautiful of the Blaeu maps.

The Blaeu mapmaking business ended in 1672, when a fire in Joan Blaeu's warehouse destroyed most of his copper plates. He died the following year. The remaining plates and printed maps were auctioned to other mapmakers and remained in use for some years.

SEE ALSO

Dutch mapmakers

FURTHER READING

Goss, John, ed. *Blaeu's The Grand Atlas of the 17th-Century World*. New York: Rizzoli, 1991.

Bodega y Quadra, Juan Francisco

SPANISH CONQUISTADOR

- *Born: 1744*
- *Died: 1792*

IN 1775 Juan Francisco Bodega y Quadra was part of a Spanish expedition to explore California. He discovered Bodega Bay, north of present-day San Francisco. Later, he made a long voyage along the west coast of North America, surveying and fixing the locations of various landmarks. He made an important map of the coast from Mexico to Alaska that corrected many earlier errors.

Bodleian map

SEE Gough map

Borchgrevink, Carsten

NORWEGIAN ANTARCTIC EXPLORER

- *Born: Dec. 1, 1864, Oslo, Norway*
- *Died: Apr. 21, 1934, Oslo, Norway*

CARSTEN BORCHGREVINK left Norway for Australia as a young man. Australia was a natural starting point for voyages in the southern ocean, and in 1895 Borchgrevink found himself sailing in Antarctic waters aboard a whaling ship.

When he decided to become an explorer, Borchgrevink was sponsored by Great Britain. He led an expedition to Antarctica (1898–1900) in the ship *Southern Cross;* he and his men were the first people to spend the winter in Antarctica on land rather than aboard ship. They wintered at Cape Adare in Victoria Land, near the Ross Sea, and Borchgrevink surveyed and mapped the Ross Sea and the Ross Ice Shelf, which have been centers of Antarctic research activity ever since.

SEE ALSO

Antarctica, mapping of

FURTHER READING

Cameron, Ian. *Antarctica: The Last Continent.* London: Cassell, 1974.

Stewart, John. *Antarctica: An Encyclopedia.* 2 vols. Jefferson, N.C.: McFarland, 1990.

Bordone, Benedetto

ITALIAN MAPMAKER AND ENGRAVER

- *Born: 1460, Padua, Italy*
- *Died: 1531*

BENEDETTO BORDONE worked in Venice, a center of chart making during the 14th century. He is known to have made a globe and a map of Italy, although these are now lost. In 1528 he published *The Book of Benedetto Bordone,* more often known as the *Isolario.* An isolario, or "island book," was a book of directions for sailing around islands. Bordone's work was based on an earlier isolario that consisted of charts and sailing directions for the Greek islands. Bordone was the first person to make an isolario for all the islands in the known world, from the Americas to Indonesia. His book also included maps of Venice and Mexico City, which were considered "island cities" because they were built over water and threaded with canals.

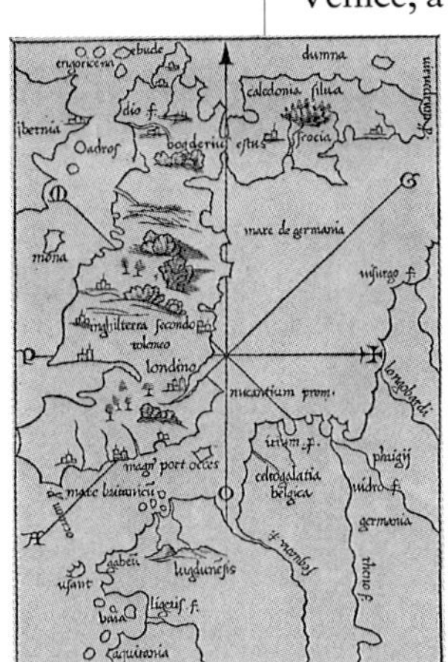

The British Isles as they appeared in Bordone's Isolario. *As on many early maps, Scotland is erroneously shown extending eastward into the North Sea.*

Bougainville, Louis-Antoine de

FRENCH DIPLOMAT AND EXPLORER

- *Born: Nov. 11, 1729, Paris, France*
- *Died: Aug. 31, 1811, Paris, France*

LOUIS-ANTOINE DE BOUGAINVILLE, a French aristocrat, served as a soldier in the war against England in Canada in the 1750s; he also saw service in Germany. He was chosen by King Louis XV to lead the first French expedition to circumnavigate the globe—that is, to sail around the world. Accompanied by several scientists, Bougainville made the voyage in 1767–69. Along the way he landed in Tahiti, which had received its first visit from Europeans—English mariners under the command of Captain Samuel Wallis—just a few months earlier. Bougainville then explored Espíritu Santo (later called the New Hebrides and now the nation of Vanuatu). He proved that Espíritu Santo was an island, not part of Terra Australis—the as-yet-undiscovered continental landmass that geographers expected to find in the southern ocean. Bougainville's expedition included the first woman known to have sailed around the world. She was the botanist's girlfriend, who sneaked aboard disguised as his manservant.

SEE ALSO

Terra Australis

FURTHER READING

Allen, Oliver E., and Time-Life editors. *The Pacific Navigators.* Alexandria, Va.: Time-Life Books, 1980.

Beaglehole, J. C. *The Exploration of the Pacific.* Stanford, Calif.: Stanford University Press, 1966.

Cameron, Ian. *Lost Paradise: The Exploration of the Pacific.* Topsfield, Mass.: Salem House, 1987.

Dunmore, John. *French Exploration in the Pacific.* 2 vols. Oxford, England: Clarendon Press, 1965–69.

Boundary lines

BOUNDARY LINES are lines on a map that reflect political boundaries in the world. They mark the borders of countries, states, and other political bodies. Some boundary lines can appear hard to follow on maps because they run on top of or next to other lines, such as parallels of latitude, meridians of longitude, or rivers. This is because many political borders have been established along natural boundaries such as rivers, and in other cases nations or territories have agreed to draw their mutual borders along a cartographic line. Much of the U.S.–Canada border, for example, runs along the 49th parallel north.

Bowen, Emmanuel

BRITISH MAPMAKER

- *Active: 1714–67, London, England*

EMMANUEL BOWEN should have died a prosperous man. He was part of the generation of British mapmakers who shook the dominance of the French cartographers and made England a center of mapmaking activity. Bowen produced a long string of maps, beginning with a map of Asia in 1714 and ending with an atlas in 1767, the year of his death. Among these maps were a world map (1717), a map of the Peruvian city of Cuzco (1720), an atlas in 1752 and another—called the *Royal English Atlas*—in 1762, and a map of the Americas in 1763. But despite his many publications and the fact that he was named royal engraver of maps by both King George II of England and King Louis XV of France, Bowen was penniless when he died. His son Thomas, also an engraver, fared even worse. After making maps for some of the most important books of the century, including Captain James Cook's *Voyages* (1773), Thomas Bowen died a pauper in a charity home in London.

Boym, Michal Piotr

POLISH MISSIONARY TO CHINA

- *Born: 1612*
- *Died: 1655*

AFTER JOINING the Society of Jesus, or Jesuits, Michal Boym went to China. Drawing upon the detailed surveys made in China by fellow Jesuits, Boym prepared an atlas of hand-drawn maps of China in 1653–55. He also made a map of China for printing; it was published in 1661, after his death. Boym's maps were part of the burst of Jesuit cartography that produced all of the important 17th-century maps of China. Like his fellow Jesuits Matteo Ricci and Martino Martini, Boym believed that accurate maps of China were needed to help the Christian missionary effort succeed there. But their maps, which remained for many years the only maps of China available to Europeans, found an eager audience that reached far beyond the Society of Jesus.

SEE ALSO
Asia, mapping of; Ricci, Matteo

Braun, Georg

GERMAN CHURCHMAN AND CARTOGRAPHER

- *Born: 1541*
- *Died: 1622*

GEORG BRAUN was the canon of Cologne Cathedral in Cologne, Germany. He was interested in maps and plans, and he compiled information and drawings from a wide variety of sources to prepare the *Civitates Orbis Terrarum*

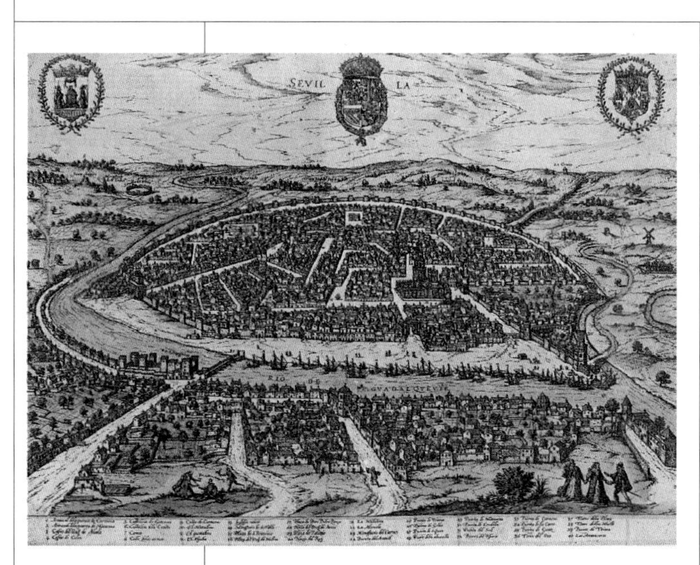

A bird's-eye view of Seville, Spain, from Braun's atlas of town plans.

(*Cities of the World*), an atlas of town plans that was published in six parts from 1572 to 1617. The *Civitates* contained more than 500 pictures or maps of cities. Most of the plans were of European cities, but cities in Africa, Asia, the Middle East, Peru, and Mexico were also included.

Drawings of the towns were made by an artist working under Braun's supervision, and then they were engraved on copper by engravers. There were three kinds of plans: panoramas, or scenic views, with details such as ships and landscape features; maps or charts that were similar to a surveyor's or builder's plan, with the layout of the city shown from directly above; and bird's-eye views, which combined the overview quality of the plan with the pictorial details of the panorama. The bird's-eye views were especially popular and were reprinted or imitated on scores of other maps. During the 17th century in particular, many maps were decorated with rows of town plans or panoramas in rectangular or oval frames along their edges. The plates used to print Braun's *Civitates* passed from one map publisher to another after his death and remained in use until at least 1708.

SEE ALSO
Town plans

British mapmakers

THE OLDEST known map made in Great Britain, known as the Anglo-Saxon map—in reference to the Anglo-Saxon culture of medieval Britain—is kept in the British Library. It dates from the late 10th century and shows the known world according to Ptolemy. One interesting feature of the Anglo-Saxon map is that, unlike most medieval maps, it does not have Jerusalem at its center.

The first significant British mapmaker was Matthew Paris, who worked in the 13th century and made the first detailed maps of Britain. John of Holywood (also known as Sacrobosco, the Latinized form of his name) was another influential Briton. Around 1230 he wrote a book on cosmography that included a world map. Although the map contained nothing original, it was

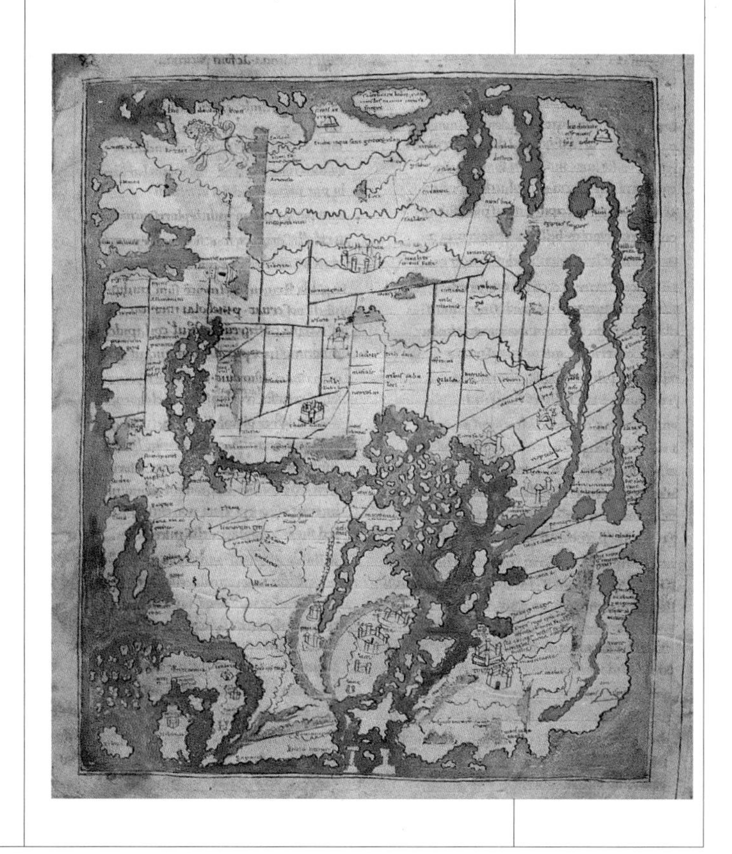

The Anglo-Saxon world map, with India at the top, Africa at the lower right, and Europe at the lower left. The columns at bottom center are the fabled Pillars of Hercules, gateway to the Atlantic Ocean.

circulated widely until the 15th century, especially among students. Other important early British maps are the Hereford *mappa mundi,* made around 1280, and the Gough map of Britain, made around 1335. Many maps of individual estates were also made, beginning in the 14th century.

British mapmaking produced an innovation in the late 16th century when Christopher Saxton published the first printed county maps. His maps of the counties of England and Wales were the first of many county surveys that would be made by British cartographers. Saxton also made a large-scale map of all of England and Wales that was reproduced in a pocket-size folding format in 1644. This was the first known map designed to be folded, the ancestor of the millions of portable folding maps that have since been published, right down to today's travel maps.

Also in the 16th century, British mapmakers and explorers alike began making notable world maps, although some of these maps—especially the manuscript charts of explorers such as Martin Frobisher and Sir Walter Raleigh—were never printed. In 1599 Edward Wright published a book called *Certaine Errors of Navigation,* which explained to mariners how to use the newly invented Mercator projection; the following year Wright published a world map that was copied by many of the 17th-century Dutch mapmakers.

The 17th century brought several landmarks in British cartography. In the early years of the century, John Speed published a series of new maps of the English counties, working in the tradition that had been established by Saxton. These maps were gathered into an atlas called *Theatre of the Empire of Great Britain* in 1611. Speed's maps were reprinted and copied dozens of times and are probably the best-known of all English county surveys. A few years later, Robert Dudley published the first sea atlas by a British chart maker. Later in the century, John Seller issued a stream of maps, charts, and atlases, including numerous collections of marine charts. Around the same time, map publisher Moses Pitt launched an ambitious venture that ended in disaster. He hoped to publish a 12-volume atlas in English to compete with the massive atlases being produced by the Blaeu family and other Dutch publishers, but he ran out of money after the first 4 volumes and was sent to prison for debt.

In the late 17th century, John Ogilby published the first road maps of England and Wales, thus inventing a new type of map publishing. After Ogilby's road maps were published, the makers of county maps began showing roads, something that had not been standard practice before. Greenville Collins's influential atlas of nautical charts for British waters—the first such charts produced by a Briton—appeared at the end of the century. Edmund Halley (1656–1742) is generally remembered today as an astronomer and the discoverer of Halley's Comet, but he also made some important contributions to cartography. He published the first meteorological chart, or weather map, in 1688 and the first map of the earth's magnetic fields in 1701. Another British first in scientific chart making came from geologist William Smith, who published the world's first geologic map in 1815; it showed the rock formations of England, Wales, and part of Scotland on 15 sheets.

Many 18th-century British mapmakers made atlases, all of which included maps of the British colonies in America. Emmanuel Bowen made two world atlases, each with numerous maps of America. John Senex's *English Atlas* (1714) and *New General Atlas of the World* (1721) were reprinted many times

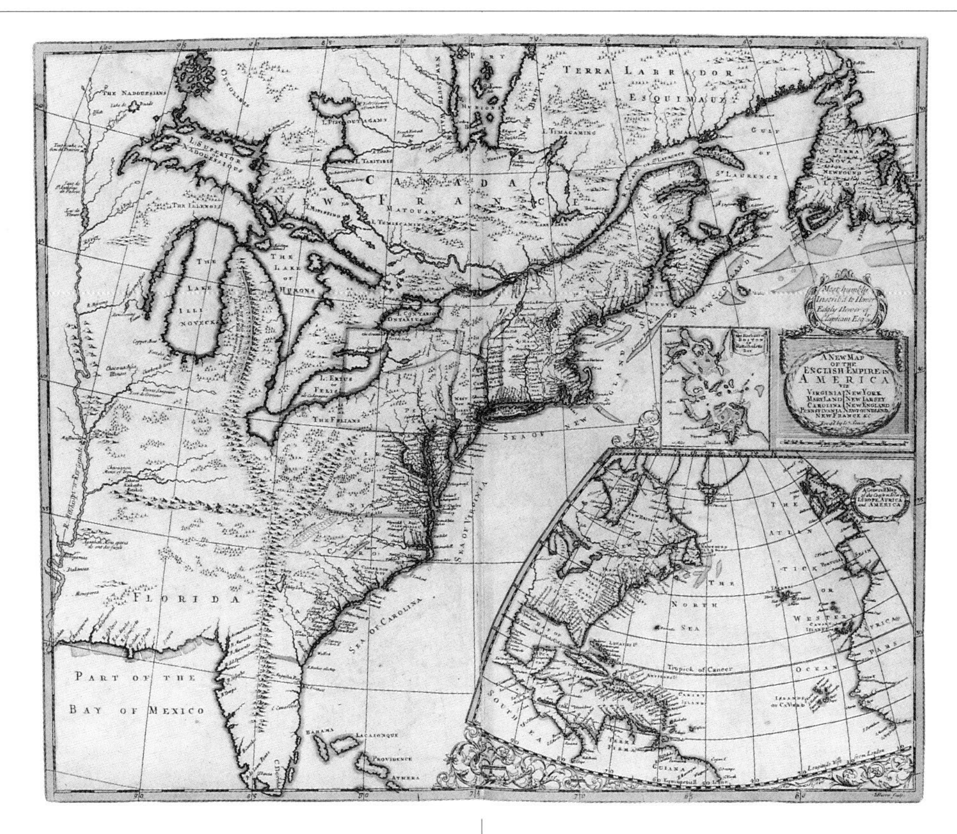

John Senex's map of "The English Empire in America" shows the five Great Lakes and the "Mitchisipi," or Mississippi, River.

until 1775. One of Senex's maps was called "The English Empire in America." It showed the British colonies in New England as well as other features of eastern North America, including a completely fictitious range of mountains running north through the middle of Michigan; this phantom range appeared on later maps that were copied from Senex.

The British colonies, and America in general, were popular subjects with British mapmakers. Cartographer Thomas Jefferys compiled an American atlas that was published after his death—it was issued in the fateful year 1776 and, of course, promptly required updating. Jefferys's business was taken over by William Faden, who produced many maps and globes in addition to his *North American Atlas,* which was not a standard edition but consisted of individual maps assembled to meet each customer's order. The tradition of the county survey was carried on by Thomas Kitchin and Emmanuel Bowen, who in 1755 published an atlas of 45 maps of the English counties, each map surrounded by text and illustrations.

The 18th century also saw the publication of Captain James Cook's outstanding maps. His maps of the Pacific Ocean, New Zealand, and the Australian coast are especially important because they clarified the true outline of the world's lands and seas for the first time. The first correct map of India was made around this time by James Rennell.

The leading figures in 19th-century English mapmaking were the Arrowsmith family and John Cary, a prolific cartographer whose maps were characterized by sound geography and up-to-date printing techniques. In addition to world maps, globes, and an atlas, he produced town plans and special-interest maps of railroads and canals.

Mapmaking underwent many changes in the 19th century, in Britain as elsewhere. The ornate, elaborately decorated maps and atlases of earlier years began to fall out of favor; an English atlas published in 1851 is one of the last to feature vignettes of people, animals, and landscapes on each map. The maps that were published with explorers' books became less fanciful and more earnest as scientific progress became the watchword of the era. In the age of the railroad and steamship, people wanted practical, functional maps. And with the dominance of the British Empire, the government played an increasing role in mapmaking. The Ordnance Survey, a government agency established in 1791 to map Great Britain, assumed responsibility for county surveys, and the Admiralty and Royal Navy began producing their own marine charts. As maps achieved a higher level of mathematical and geographic accuracy than ever before, they also became, in some ways, simpler and more standardized. The quirks of individual mapmakers disappeared, but high-quality mass-produced maps of all parts of the world became increasingly available to everyone.

SEE ALSO

Arrowsmith family; Bowen, Emmanuel; Collins, Greenville; Cook, James; Dudley, Robert; Gough map; Hereford map; Ogilby, John; Paris, Matthew; Rennell, James; Saxton, Christopher; Speed, John

FURTHER READING

Tooley, Ronald V. *Maps and Map-Makers.* 1949. Reprint. New York: Dorset Press, 1990.

Bruce, James

SCOTTISH EXPLORER OF ETHIOPIA

- *Born: Dec. 14, 1730, Larbert, Scotland*
- *Died: Apr. 27, 1794, Larbert, Scotland*

JAMES BRUCE was a wine merchant in London before the British government appointed him as its consul, or representative, in Algiers, an Arab city in North Africa. There he learned Arabic and became interested in African geography. At that time few Europeans had penetrated beyond the coast of Africa. Bruce was curious about the interior. He decided to explore Abyssinia (now Ethiopia), a mountainous land between the Nile River and the Red Sea that was almost unknown to Europeans. In 1786 he left Cairo, Egypt, and went by way of the Red Sea to Gondar, the capital of Ethiopia.

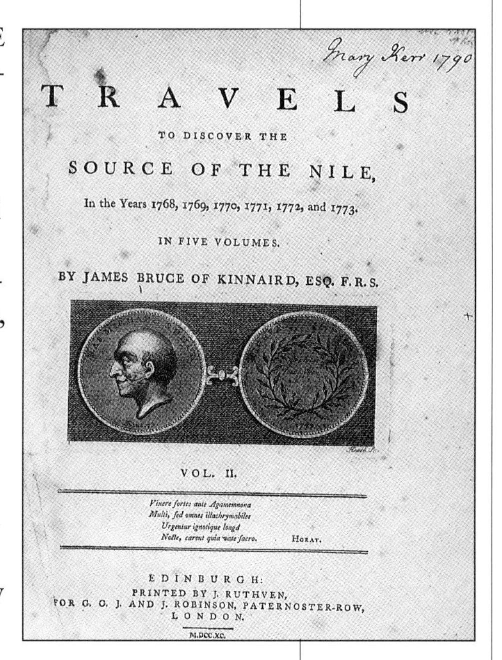

Mary Kerr 1790

TRAVELS

TO DISCOVER THE

SOURCE OF THE NILE,

In the Years 1768, 1769, 1770, 1771, 1772, and 1773.

IN FIVE VOLUMES.

BY JAMES BRUCE OF KINNAIRD, ESQ. F.R.S.

VOL. II.

Vixere fortes ante Agamemnona
Multi, sed omnes illachrymabiles
Urgentur ignotique longâ
Nocte, carent quia vate sacro. HORAT.

EDINBURGH:
PRINTED BY J. RUTHVEN,
FOR G. G. J. AND J. ROBINSON, PATERNOSTER-ROW,
LONDON.

M.DCC.XC.

One of the most controversial travel books of its time, Bruce's sober narrative earned its author the reputation of a liar.

Bruce arrived in Gondar in 1770. Ethiopia was then in the midst of civil wars, and he became a pawn in the turbulent politics of the country. Guarded closely by the emperor's soldiers, he was unable to leave for several years. He used the time to study the people and culture of the land, and to make geographic investigations—for example, he visited a spring that he thought was the source of the Blue Nile, one of the two Nile branches. He was wrong about the spring, but his maps of Ethiopia were accurate in many other respects.

Finally, in 1772, Bruce was allowed to leave Ethiopia. He followed the Blue Nile through mountain highlands, plains, and deserts—a region that was almost completely unexplored by Europeans—until he reached Egypt. Upon his return to London, Bruce found that nobody believed his story. He retired to Scotland in anger. In 1790 he published *Travels to Discover the Sources of the Nile,* which aroused a storm of ridicule and controversy. Readers were astounded by Bruce's report of life in Ethiopia, with its descriptions of steaks cut from living cattle, ancient Christian rites, and immense armies surging across the African landscape. Most readers dismissed Bruce as a liar and his journey as a tall tale. He died four years later without receiving credit for the accuracy of his account, which was confirmed by later travelers.

SEE ALSO

Africa, mapping of

FURTHER READING

Moorehead, Alan. *The Blue Nile.* New York: Harper & Row, 1962.

Bry, Théodore de

BELGIAN ENGRAVER AND PUBLISHER

- *Born: 1528, Liège, Belgium*
- *Died: Mar. 27, 1598, Frankfurt, Germany*

THÉODORE DE BRY set himself up in the engraving and publishing trade in Frankfurt, Germany. Around 1587 he visited England and met Richard Hakluyt, an English geographer who inspired in him an interest in travel books. Bry began collecting accounts of exploration, mostly in the Americas. He published these in a 25-part series beginning in 1590. Bry is important in the history of mapmaking not just because he put early travel narratives into print but because these narratives were illustrated with his engravings of explorers' drawings. Bry's pictures of Virginia, Florida, and South America were the first images that many people saw of the Americas, and they had a profound and long-lasting effect. Many people saw Bry's books, and in addition mapmakers used Bry's illustrations as vignettes—inset pictures or decorations—on countless maps in later years. For several generations, Bry's vision of the Americas helped shape the way Europeans pictured the American landscape and the Native American people.

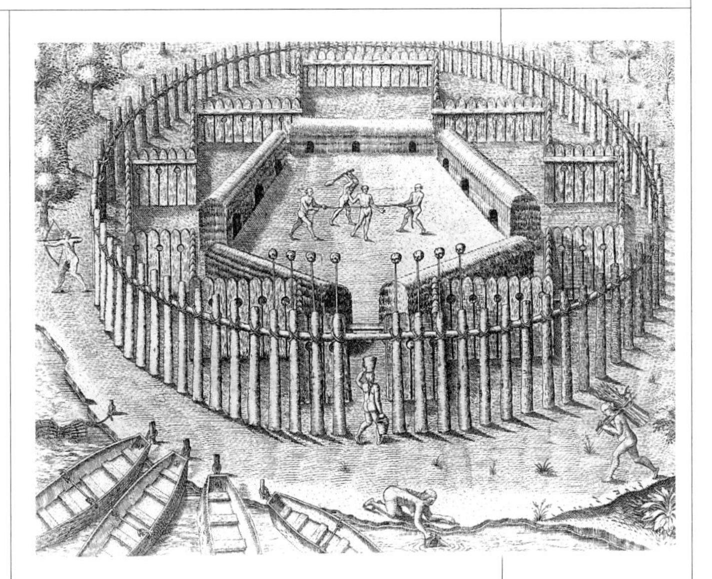

A Native American settlement in Virginia, engraved by Bry from a painting by John White, governor of the lost colony of Roanoke.

Buache, Philippe

FRENCH GEOGRAPHER

- *Born: 1700*
- *Died: 1773*

PHILIPPE BUACHE was the son-in-law of Guillaume De L'Isle, a celebrated French mapmaker. Carrying on the L'Isle family business, he published a number of atlases in the mid-18th century. Buache was the first person to try to map the undersea world. He made the first such bathymetric map in 1752; it was an accurate chart of the English Channel, showing the contours of the seafloor. Then Buache took a larger,

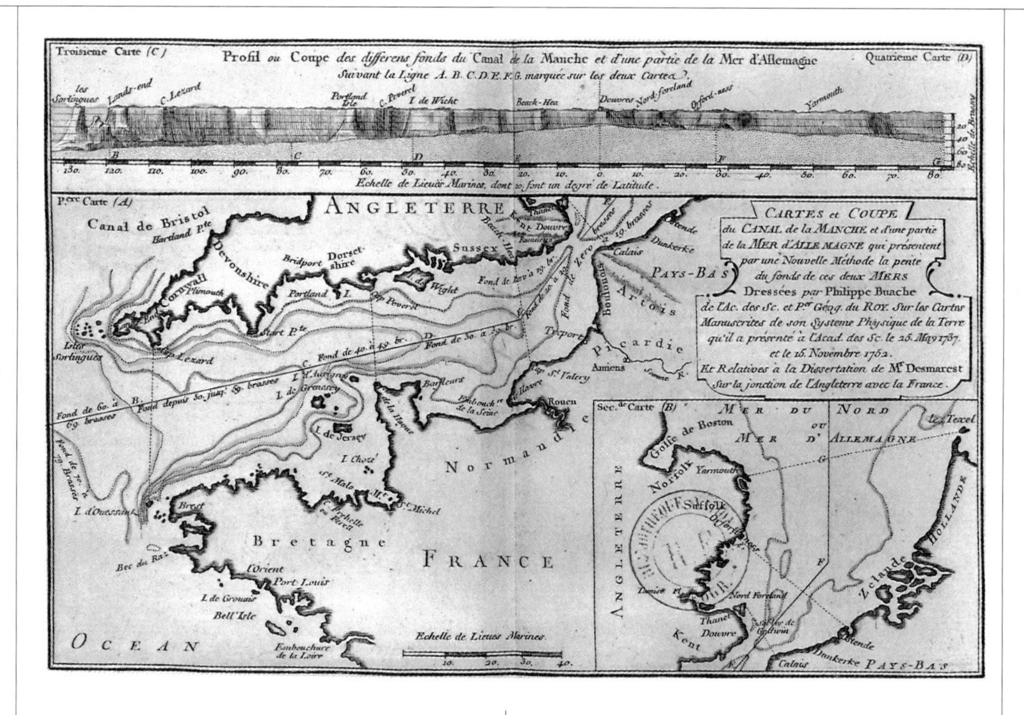

Buache's contour map of the seafloor between France and England. A pioneer of undersea mapping, Buache believed—correctly—that the world's seafloors and landmasses form a single geological system.

more theoretical approach. He concluded that the seas were crossed by chains of mountains, of which the existing islands were the tops. Although this theory is at least partly correct, the maps Buache created from it bear little relation to the true structure of the ocean floor. Nevertheless, he deserves credit for launching the art of undersea mapping.

SEE ALSO
Bathymetric maps; L'Isle family; Undersea mapping

Buell, Abel

AMERICAN SILVERSMITH AND ENGRAVER

- *Born: 1742*
- *Died: 1822*

ABEL BUELL was a silversmith who turned to engraving maps in 1770 and established himself in the Connecticut colony. On March 31, 1784, he announced in an advertisement in the *Connecticut Journal,* published in New Haven, that he had engraved a new map of the United States, calling it "the first ever compiled, engraved, and finished by one man, and an American." All earlier maps of the United States had involved the labor of more than one person, or had been produced in Europe or in the European colonies in North America, before the United States achieved independence. Buell's map, published the year after the Treaty of Paris ended the American Revolution and established the United States as a sovereign nation, was the first American map of America by an American. Its cartouche contains a picture of the American flag—the first time this flag appeared on a printed map.

Despite its historic significance, Buell's map is not very polished. Some scholars believe that its crudity may be due to haste and that Buell produced the map in a hurry so as to beat potential competitors. The geographic information contained in the map was probably copied from earlier maps by Lewis Evans and other cartographers. However, Buell extended American territory west to the

Mississippi River. Beyond that river, which formed the limit of geographic knowledge at the end of the 18th century, Buell identified a "River of the West," a mythical waterway thought to lead to the Pacific, and placed the words "Unbounded Plains Supposed to Extend to the South Sea."

FURTHER READING

Schwarz, Seymour I., and Ralph E. Ehrenberg. *The Mapping of America.* New York: Abrams, 1980.

Wheat, James C., and Christian F. Brun. *Maps and Charts Published in America before 1800.* New Haven, Conn.: Yale University Press, 1969.

Burckhardt, Johann Ludwig

SWISS EXPLORER

- *Born: Nov. 24, 1784, Lausanne, Switzerland*
- *Died: Oct. 17, 1817, Cairo, Egypt*

JOHANN LUDWIG BURCKHARDT went to England at the age of 22 to study biology and geography. Three years later, a British organization called the African Association picked Burckhardt to make a daring journey. He was to cross the Sahara Desert from north to south and report upon the tribespeople he met along the desert caravan routes.

Burckhardt's preparations for the Sahara trip were extensive. He went to Syria to learn Arabic; he also studied the Islamic religion because he planned to disguise himself as a Muslim. In 1812 he left Syria and traveled through the Middle East on his way to Egypt. Passing through what is now Jordan, he made a remarkable discovery. In a remote desert valley, he came upon Petra, a city that had been carved out of red sandstone cliffs centuries earlier.

The oldest caves at Petra date from the 6th century B.C. and were excavated by an ancient people called the Nabateans. For a long time, Petra flourished as a trading station where busy caravan routes met, but the routes across the Arabian Desert changed, and eventually Petra was abandoned and almost forgotten. Only a few isolated desert tribespeople were living in the rock-walled city when Burckhardt stumbled upon it. He placed Petra on the modern map, however. After his account of the city reached Europe, a growing trickle of travelers and historians made their way to it. The rock city was hauntingly described by the British poet John William Burgon, who in his 1845 poem "Petra" called it "a rose-red city half as old as time." Today Petra is familiar to moviegoers as the place where the final scenes of the movie *Indiana Jones and the Last Crusade* were filmed.

Upon arriving in Egypt, Burckhardt traveled some distance up the Nile River, where he located and described the ancient Egyptian temple of Abu Simbel. Postponing his Sahara trip, he went east to Arabia and visited Mecca, the holiest city of Islam. Mecca was forbidden to non-Muslims, but Burckhardt passed himself off as a Muslim pilgrim. Then he returned to Egypt, ready to tackle the Sahara at last, but he became ill and died in Cairo of dysentery before he could undertake his official mission.

In addition to his two important archaeological discoveries—Petra and Abu Simbel—Burckhardt made significant contributions to geography. His accurate surveys of the Dead Sea, the Sinai Peninsula, and parts of Arabia were reflected in many of the maps of the region that were made in the decades after his death.

SEE ALSO

African Association

Burke, Robert O'Hara

EXPLORER OF AUSTRALIA

- *Born: 1820, St. Cleram, Ireland*
- *Died: June 28, 1861, central Australia*

ROBERT O'HARA BURKE went to Australia at the age of 32 to serve in the police force of Victoria, the British colony in the southeastern corner of the Australian continent. A few years later, the Philosophical Institute of Victoria, when asked by a group of prominent local citizens to promote the exploration of the continent's interior, chose Burke and a young science student named William Wills to lead an expedition across Australia from south to north. Burke and Wills expected to be the first people ever to make this journey, although a rival party was preparing to set out at about the same time.

At the head of a large, well-equipped company, Burke and Wills left Melbourne in 1860. Quarrels among the men, combined with poor management of supplies and some rash decisions by the leaders, soon weakened the expedition, which split into several groups. After great hardships, Burke, Wills, and a few others reached the Gulf of Carpentaria on Australia's north coast in February 1861. On the return trip south, Burke and Wills died of starvation. Only one expedition member lived to complete the crossing. The expedition's fragmentary records helped to fill in the map of the interior of eastern Australia, but the various search parties that set out to aid Burke and Wills finally gathered more geographic information than the original expedition had done.

SEE ALSO

Australia, mapping of

FURTHER READING

Moorehead, Alan. *Cooper's Creek: The Opening of Australia.* New York: Harper & Row, 1963.
Tooley, Ronald V. *The Mapping of Australia.* Watchung, N.J.: Saifer, 1980.

Burton, Sir Richard Francis

ENGLISH ADVENTURER AND EXPLORER

- *Born: Mar. 19, 1821, Torquay, England*
- *Died: Oct. 20, 1890, Trieste, Italy*

SIR RICHARD Francis Burton was one of the best-known and most flamboyant explorers of the 19th century. From childhood he possessed the ability to learn languages quickly; by the time of his death, he had mastered 29 tongues. Restless and eager for a life of action, he got himself expelled from college so that he could join the army in India, which was then a British colony.

Burton spent some years in India, studying the languages and customs of Asia. He also became interested in Asian religions. Skilled in the art of disguise, he easily passed as an Afghan warrior. In this guise he made a daring journey in 1853, visiting Mecca and Medina, the holy cities of Islam in Arabia, while pretending to be a Muslim on a pil-

A master of disguise, Burton prepares to spy on the Muslim holy city of Mecca.

grimage. Soon afterward the British government sent him on another undercover mission, to the Ethiopian city of Harar, which was off-limits to foreigners. One of his companions was John Hanning Speke, a fellow army officer. Both were seriously wounded in a skirmish with Somali fighters. Nonetheless, Burton succeeded in his mission: making secret maps of the Ethiopian highlands and the route from the East African coast.

Burton and Speke explored together again in 1857, this time in search of the East African lakes that were thought to be the source of the Nile. Despite severe illnesses and other hardships, they traveled from Zanzibar, an island off the east coast of Africa, into the interior of the continent as far as Lake Tanganyika. They were the first European explorers to place this lake on the map. Speke, in a northerly march made on his own, sighted another lake, which he named Lake Victoria after the British queen. Speke was certain that Lake Victoria was the source of the Nile. Burton disagreed, and the two explorers quarreled publicly after their return to England.

Speke led a second expedition to Lake Victoria in 1860–63. Afterward, he claimed to have proved that the lake was the source of the Nile, but Burton continued to argue that Speke's geographic evidence was shaky and that Lake Tanganyika was really the river's source. Unfortunately for Burton, although Speke's evidence was poor, Speke was right about Lake Victoria. It, and not Lake Tanganyika, is the source of the White Nile, one of the two main branches of the Nile River.

Burton made no more major journeys of exploration, although he did travel in West Africa, North America, and South America. He served as a diplomat and devoted his enormous energy to writing more than 50 books, including a translation of the Arabic *Book of a Thousand Nights and a Night,* which became known to the English-speaking world as the *Arabian Nights.*

SEE ALSO

Africa, mapping of; Speke, John Hanning

FURTHER READING

Brodie, Fawn. *The Devil Drives: A Life of Sir Richard Burton.* New York: Norton, 1984.
Moorehead, Alan. *The White Nile.* New York: Harper & Row, 1960.
Rice, Edward. *Captain Sir Richard Francis Burton.* New York: Scribners, 1990.

Byrd, Richard Evelyn

AMERICAN AVIATOR AND POLAR EXPLORER

- *Born: Oct. 25, 1888, Winchester, Virginia*
- *Died: Mar. 11, 1957, Boston, Massachusetts*

RICHARD EVELYN BYRD was a born traveler—at the age of 12 he went around the world on his own, crossing the United States from Boston to San Francisco by train, taking a ship across the Pacific to visit a family friend in the Philippines, and returning by sea through the Indian and Atlantic oceans to Boston. This trip gave him a lifelong passion for the sea and made him determined to join the navy. Byrd began his military career as a naval officer, but after learning to fly an airplane in 1917 he concentrated on aviation, winning recognition as a pioneer of flight. He went on to create programs for the military and civilian use of airplanes and dirigibles.

Using aircraft, Byrd opened the polar regions to scientific mapmaking.

Byrd's career in exploration began in 1924, when he commanded naval flights over Greenland and the Canadian Arctic. In 1926, in a private expedition, he and a pilot flew his plane across the north pole. A year later he flew across the Atlantic Ocean. These successes made him a popular hero, and he was able to get funds for an expedition to Antarctica. In 1928 he established an Antarctic base called Little America. In November 1929 he and three companions flew over the south pole, making Byrd the first person to reach both poles by air. In this and three later expeditions (1933–35, 1939–41, and 1946–47), Byrd explored the region west of the Antarctic Peninsula; he named it Marie Byrd Land after his wife. Using airplanes to survey the territory, he produced new, more accurate maps of the Antarctic continent.

Byrd's books *Skyward* (1928), *Little America* (1930), *Discovery* (1935), and *Alone* (1938), as well as his articles for *National Geographic* and other publications, introduced millions of readers to the marvels of flight and to the remote and thrilling landscape of Antarctica. Acknowledged worldwide as the leader of Antarctic exploration, at the time of his death he was working on plans for the International Geophysical Year and for an international Antarctic treaty.

SEE ALSO

Antarctica, mapping of; International Geophysical Year (IGY)

FURTHER READING

Byrd, Richard E. *Alone.* 1938. Reprint. Los Angeles: Tarcher, n.d.

———. *Discovery: The Story of the Second Byrd Antarctic Expedition.* 1935. Reprint. Irvine, Cal.: Reprint Services, 1991.

Hoyt, Edwin P. *The Last Explorer: The Adventures of Admiral Byrd.* New York: John Day, 1968.

Rodgers, Eugene. *Beyond the Barrier: The Story of Byrd's First Expedition to Antarctica.* Annapolis, Md.: Naval Institute Press, 1990.

Rose, Lisle A. *Assault on Eternity: Richard E. Byrd and the Exploration of Antarctica, 1946–1947.* Annapolis, Md.: Naval Institute Press, 1980.

Steinberg, Alfred. *Admiral Richard E. Byrd.* New York: Putnam, 1960.

Cabot, John

ITALIAN NAVIGATOR

- *Born: about 1450, Genoa, Italy*
- *Died: disappears from historical record after 1498–99*

ALTHOUGH HE was born Giovanni Caboto in Italy, John Cabot is generally known by the English version of his name because in 1493 he moved to Bristol, the center of England's seafaring activity. Having heard of Columbus's voyage westward and the land he had found in the western Atlantic, Cabot persuaded the merchants of Bristol and King Henry VII to back him in his own venture of exploration.

In 1497 he sailed west across the Atlantic Ocean and reached Newfoundland, Canada, which he believed to be the easternmost edge of Asia. He explored part of the coast, finding traces of human habitation, before returning to England, where he reported that he had reached the coast of Cathay, as Europeans of the time called China. In 1498 he set out again with five ships. Two of the ships turned back, and three were lost at sea. Nothing certain is known of Cabot's fate. Most historians believe that he was aboard one of the ships that sank. Old court records state that he was paid a pension by the British government in 1499, which would mean that he had returned safely from the 1498 voyage, but the pension might have been paid to his widow if he had indeed been lost at sea. It is unlikely that the mystery of John Cabot's death will ever be solved, but his geographic mistake lived on. For years, geographers and navigators continued to think that the east coast of Canada was really China, their long-sought goal.

SEE ALSO

Cabot, Sebastian

FURTHER READING

Brown, Warren. *The Search for the Northwest Passage*. New York: Chelsea House, 1991.
Burrage, Henry S., ed. *Early English and French Voyages, Chiefly from Hakluyt, 1534–1608*. 1908. Reprint. New York: Barnes & Noble, 1967.
Lehane, Brendan, and Time-Life editors. *The Northwest Passage*. Alexandria, Va.: Time-Life Books, 1981.
Morison, Samuel Eliot. *The Great Explorers: The European Discovery of America*. New York: Oxford University Press, 1978.
Morris, Roger. *Atlantic Seafaring: Ten Centuries of Exploration and Trade in the North Atlantic*. Camden, Maine: International Marine Publishing, 1992.
Scott, J. M. *Icebound: Journey to the Northwest Sea*. London: Gordon and Cremonesi, 1977.
Williamson, James A. *The Cabot Voyages and Bristol Exploration under Henry VII: With Cartography of the Voyages by R. A. Skelton*. London: Hakluyt Society, 1972.

Cabot, Sebastian

NAVIGATOR AND MAPMAKER

- *Born: 1474, Venice, Italy, or 1476, Bristol, England*
- *Died 1557, London, England*

THE SON of John Cabot, Sebastian Cabot was born either in Italy or in England. He was trained as a navigator

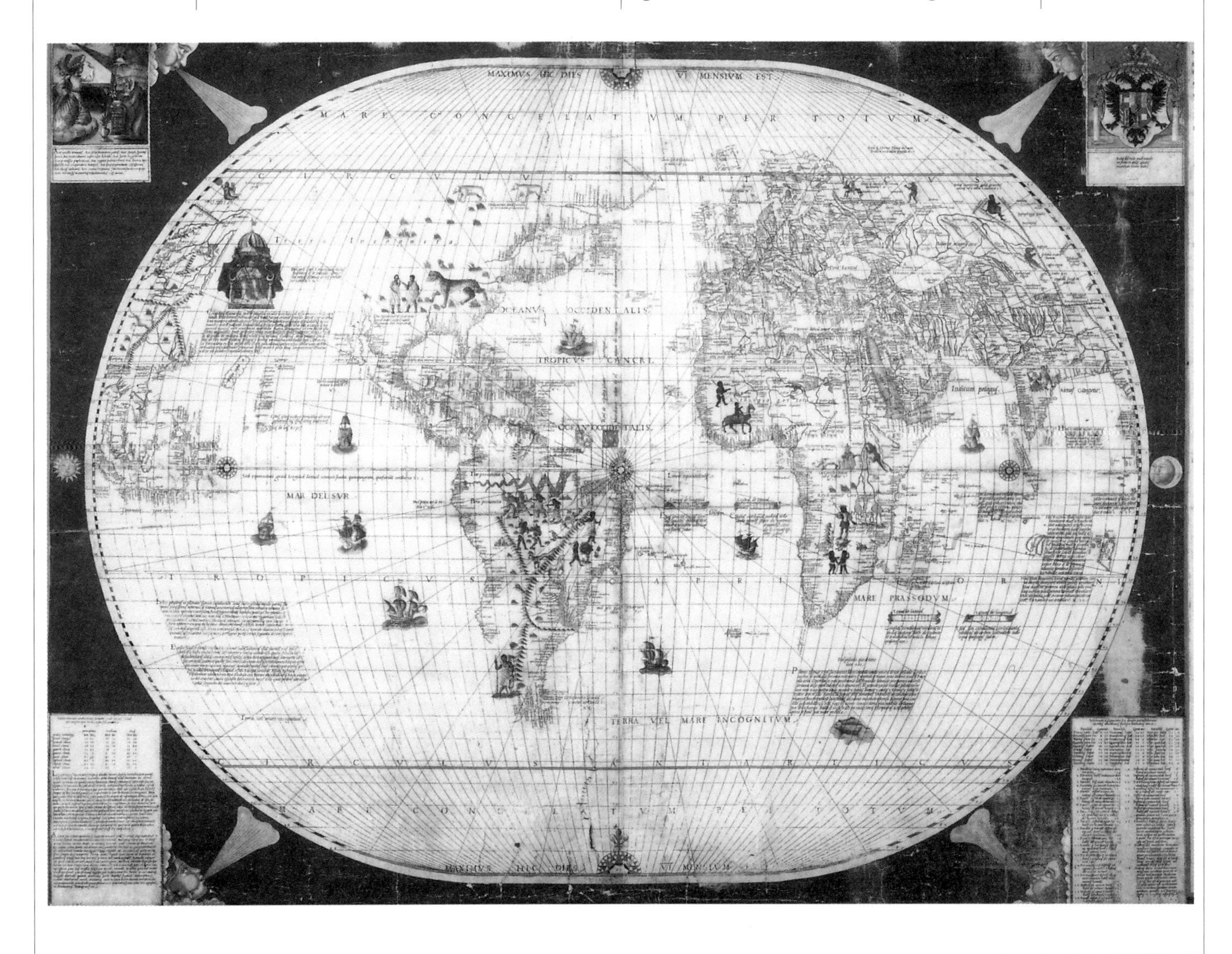

Sebastian Cabot's sole surviving map shows the world as he knew it. The throned monarch at the upper left suggests the wealth and majesty of Asia, the goal of many voyages of exploration.

and, like his father, explored the east coast of the Americas in the hope of finding a route to Asia.

In 1508–9 he sailed along part of the North American coast, taking measurements that he later used to make maps. Not all of his maps were of newly explored territory—in 1511 he mapped part of France for the British crown. He then entered the service of Spain and became the pilot major, or chief of pilots, of the Casa de la Contratación, the Spanish bureau that supervised exploration. In this job he received firsthand reports of all Spanish explorations in the New World. In 1526–30 Cabot led an expedition to the coast of Argentina, looking for silver or other riches, but he found so little reward that Spain banished him to Algeria for two years.

Cabot moved back to England in 1547; there he helped organize the Company of Merchant Adventurers to find a route to China through the Northeast Passage, which connects the Atlantic and Pacific Oceans north of Russia. He organized three expeditions along this route between 1553 and 1556; although these voyages did not reach China, they succeeded in opening up trade between Russia and England.

Cabot is thought to have made many maps during his career, but only one of them is known today, and only a single copy of it survives. Held in France's national archives, it shows the world as it was known in 1544, the year Cabot published it. North America is decorated with drawings of polar bears and Indians, and a point on the northeastern coast, where his father, John Cabot, made landfall in 1497, is labeled *prima terra vista*—the land first seen.

SEE ALSO

Cabot, John; Casa de la Contratación de las Indias; Northeast Passage

FURTHER READING

Brown, Warren. *The Search for the Northwest Passage.* New York: Chelsea House, 1991.

Burrage, Henry S., ed. *Early English and French Voyages, Chiefly from Hakluyt, 1534–1608.* 1908. Reprint. New York: Barnes & Noble, 1967.

Lehane, Brendan, and Time-Life editors. *The Northwest Passage.* Alexandria, Va.: Time-Life Books, 1981.

Morison, Samuel Eliot. *The Great Explorers: The European Discovery of America.* New York: Oxford University Press, 1978.

Morris, Roger. *Atlantic Seafaring: Ten Centuries of Exploration and Trade in the North Atlantic.* Camden, Maine: International Marine Publishing, 1992.

Scott, J. M. *Icebound: Journey to the Northwest Sea.* London: Gordon & Cremonesi, 1977.

Williamson, James A. *The Cabot Voyages and Bristol Exploration under Henry VII: With Cartography of the Voyages by R. A. Skelton.* London: Hakluyt Society, 1972.

Cabral, Pedro Álvars

PORTUGUESE NAVIGATOR

- *Born: about 1467, Belmonte, Portugal*
- *Died: about 1520, possibly in Santarem, Portugal*

IN 1500, on his way from Portugal to India around the southern tip of Africa, Pedro Álvars Cabral led his fleet far into the western Atlantic Ocean and unexpectedly found land. Claiming the land for Portugal, he named it Tierra da Vera Cruz (Land of the True Cross) and sailed on for India.

Cabral had been the first European to discover Brazil, a find that was recorded on the Cantino map of 1502. In Brazil, Cabral's men captured some lively, noisy parrots, and Cabral took a cage full of them back to Europe. The maker of the Cantino map adorned the Tierra da Vera Cruz with a sketch of three of these brightly colored birds.

Although Cabral never returned to Brazil, his landfall there had far-reaching effects. Because Cabral's was the first known landing in that part of South America, Portugal was able to claim the region, which later became the vast colony of Brazil, Portugal's largest overseas territory. Although Brazil has been an independent nation since 1822, Portuguese remains its national language.

SEE ALSO

Americas, mapping of; Cantino map

FURTHER READING

Smith, Anthony. *Explorers of the Amazon.* New York: Viking, 1990.
Stefoff, Rebecca. *Accidental Explorers: Surprises and Side Trips in the History of Exploration.* New York: Oxford University Press, 1992.

Ca da Mosto, Alvise da

ITALIAN TRAVELER

- *Born: 1432, Venice, Italy*
- *Died: 1488, Venice, Italy*

Alvise da Ca da Mosto was born in the trading port of Venice to a prominent family. He became interested in the explorations of West Africa that were being carried out by Portuguese expeditions, and he attached himself to the Portuguese court. In two voyages in 1455 and 1456, he explored the mouth of the Gambia River and made contact with outposts of the African empire of Mali. He also claimed to have discovered the Cape Verde Islands off the African coast, but the truth of this claim is uncertain.

Ca da Mosto's contribution to the history of exploration was not just the discovery of unknown lands. His true importance lies in the detailed and perceptive account he wrote about his voyages. This narrative not only gave information to mapmakers such as the Venetian cartographer Grazioso Benincasa, who based his 1468 map of Africa on Ca da Mosto's report, but also served as an early model for the study of new peoples and cultures.

SEE ALSO

Africa, mapping of; Henry, Prince of Portugal

Cadastral map

A CADASTRAL map is one that shows how land is subdivided, usually within a small area. It shows the boundaries between parcels of land; it may also show features such as streams and buildings. The term *cadastral* comes from the French *cadastre,* a register of property, based on the Latin *capitastrum,* or list of taxpayers. Cadastral maps have been used since ancient times to define individually owned properties and to record who owns them. Many of the oldest surviving maps from Babylonia, Egypt, and other civilizations are cadastral maps, used by the authorities to determine such things as property inheritance and land taxes.

Caillié, René-Auguste

FRENCH EXPLORER OF NORTH AFRICA

- *Born: Nov. 19, 1799, La Rochelle, France*
- *Died: Apr. 17, 1838, La Badère, France*

RENÉ-AUGUSTE CAILLIÉ was a young man living in Senegal, West Africa, and studying Arabic when he learned that a French geographic society had offered a cash prize to any explorer who could reach the Muslim city of Timbuktu and return safely. Located near the Niger River in present-day Mali, Timbuktu was inhospitable to non-Muslims. The only other European known to have visited the city had been murdered. Caillié was determined to win the prize—and the glory—for visiting Timbuktu and returning to tell about it.

Like most other 19th-century European explorers of the Arabic world, Caillié disguised himself as a Muslim to make the dangerous journey into the Niger River basin. He arrived in Timbuktu in 1828. From there he headed north with a caravan, crossing the Sahara Desert to Tangier on the Mediterranean coast; he is believed to have been the first European to complete the Sahara crossing, although European travelers had ventured far into the desert several centuries earlier. Caillié claimed the prize and published an account of his journey that added some useful details to the map of North Africa. It also contained maps and sketches he had drawn of Timbuktu.

SEE ALSO

Africa, mapping of

FURTHER READING

Welch, Galbraith. *The Unveiling of Timbuctoo.* 1939. Reprint. New York: Carroll & Graf, 1991.

Cantino map

THE CANTINO MAP was made in 1502. The cartographer who drew the map is unknown but was certainly Portuguese. The map was made for Alberto Cantino, an agent of the duke of Ferrara, in Italy. The duke was interested in the new discoveries being made by Spanish and Portuguese navigators, but his interest was not purely geographic. Like many other Italians, the duke of Ferrara feared that the new discoveries would threaten the traditional spice trade routes. For years, spices from southern Asia had entered Europe by way of the eastern Mediterranean Sea, passing through Italian ports such as Venice and bringing great wealth and prestige to Italian merchants. The Italians feared that their financial interests might be damaged if the Spanish, Portuguese, or English succeeded in finding a new way to reach India, where they could buy spices without the help of Italian middlemen.

Spanish and Portuguese authorities were very protective of the geographic information their mariners gathered. Captains' logs and charts were carefully guarded. Nonetheless, Cantino managed to have a map of the most recent Portuguese discoveries made and then to smuggle the map out of Portugal. The map is the earliest known Portuguese chart to depict discoveries in the Americas. It shows Vasco da Gama's 1497–99 voyage to India, although no report of that voyage was made public until 1506, four years after the Cantino map was made.

Cantino's map is also the first one known to show Brazil and Madagascar, both of which were first sighted by the Portuguese in 1500. Asian coastlines, including those of India and Sri Lanka, are shown more clearly on the Cantino map than ever before. But the anonymous cartographer still believed that the Americas were part of Asia; he called Columbus's Caribbean landfalls "islands recently discovered in parts of India."

The maker of the Cantino map carefully assigned the newly discovered territories to either Spain or Portugal, the two nations who were dividing the known world between them. The Carib-

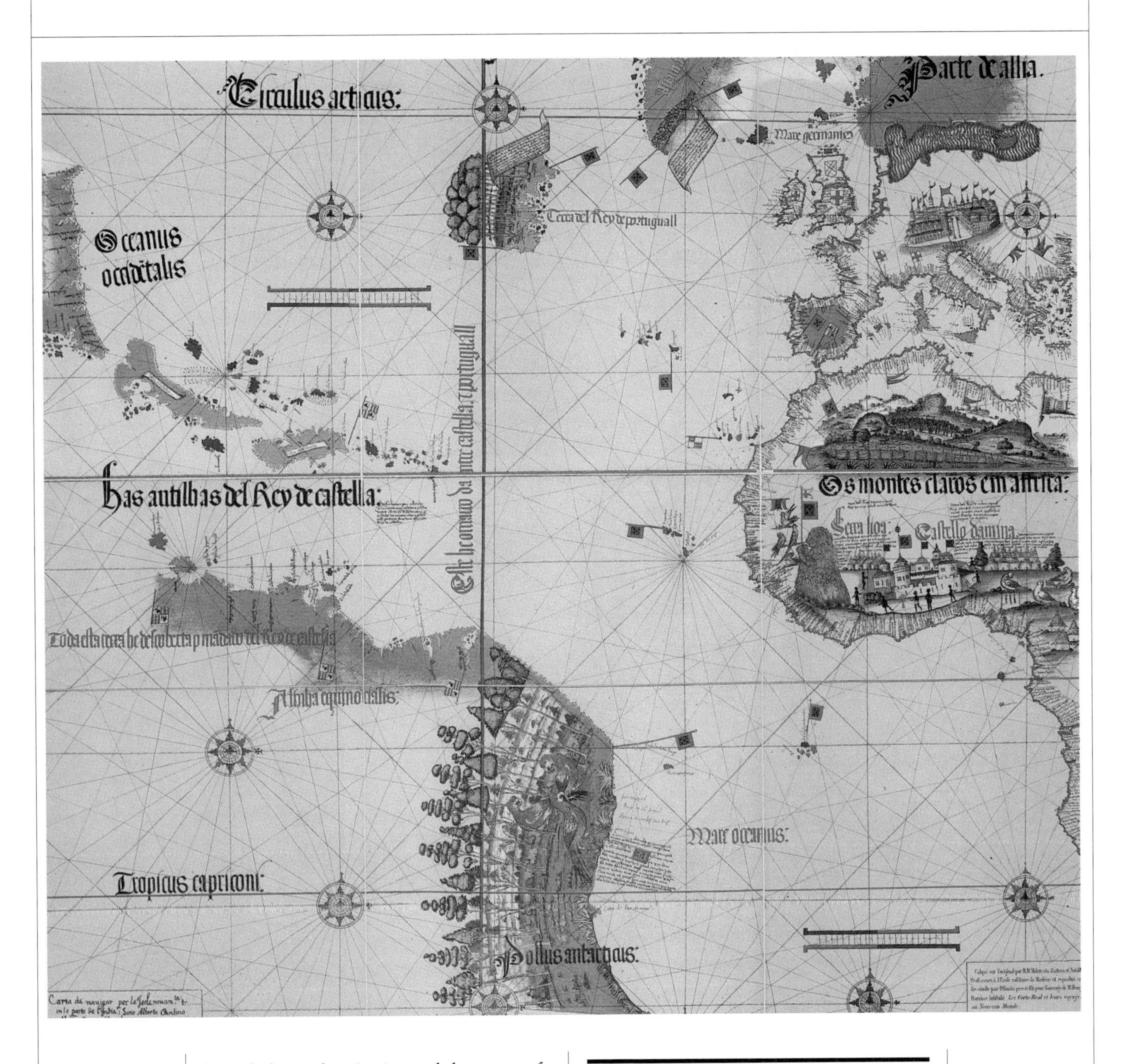

This portion of the Cantino map shows the line set by the Treaty of Tordesillas. Spain's New World territories are to the left of the line, Portugal's to the right.

bean belonged to Spain, and the coast of Brazil to Portugal. Neither of these countries would have been happy to see their secret geographic information fall into the hands of Italy. Today, the map is one of the treasures of the Biblioteca Estense in Modena, Italy.

SEE ALSO

Americas, mapping of

Cão, Diogo

PORTUGUESE NAVIGATOR

• *Active: about 1480*

DIOGO CÃO came from a seafaring family. In the 1480s, King João II of Portugal sent Cão on two voyages to extend the map of Africa's west coast. Cão succeeded in going farther south than any previous navigator. He explored the mouth of the Congo River and probed the African coastline as far as Walvis Bay in what is now Namibia. After Cão's voyages, European cartographers began adding the Congo region to their maps of Africa. On many 16th- and 17th-century maps, the kingdom of Prester John, a mythical Christian emperor, was located in the Congo, but Portuguese explorers sought for this elusive empire in vain.

SEE ALSO

Africa, mapping of; Prester John

FURTHER READING

Boxer, Charles R. *The Portuguese Seaborne Empire, 1415–1825.* New York: Knopf, 1965.

Divine, David. *The Opening of the World: The Great Age of Maritime Exploration.* New York: Putnam, 1973.

Forbath, Peter. *The River Congo.* New York: Harper & Row, 1977.

Stefoff, Rebecca. *Vasco da Gama and the Portuguese Explorers.* New York: Chelsea House, 1993.

Carey, Matthew

AMERICAN PUBLISHER

- *Born: 1760, Dublin, Ireland*
- *Died: 1839, Philadelphia, Pennsylvania*

MATTHEW CAREY was born in Ireland, but he left his homeland to escape oppression by the English. He had been apprenticed to a printer, and for a time he worked for Benjamin Franklin in Paris. He went to Philadelphia in 1784 with letters of introduction from Franklin and set himself up as a publisher of journals. Later, he turned to book publishing, a more profitable branch of the business. He reprinted books by European authors but was especially interested in encouraging American authors.

Carey owes his place in cartographic history to the *American Atlas,* which he began publishing in 1795. The plates from which the maps of Europe, Africa, and Asia were printed had to be imported from Europe, but the plates for the American maps in the atlas were engraved in the United States. Carey's was the first world atlas to be published in the United States, although an atlas of coastal charts called the *American Pilot* had appeared three years earlier.

SEE ALSO

American mapmakers

Cartier, Jacques

FRENCH EXPLORER OF NORTH AMERICA

- *Born: 1491, St. Malo, France*
- *Died: Sept. 1, 1557, St. Malo, France*

BORN A YEAR before Columbus's first voyage, Jacques Cartier studied in Dieppe, a port city that was the center of French maritime science, and became a navigator. He made at least one voyage to the Americas before 1534, possibly to Brazil. In 1534 King Francis I of France sent Cartier to North America to look for gold and other riches. Cartier explored a large gulf, which he named the Gulf of St. Lawrence, and landed on the coast of Quebec's Gaspé Peninsula. This voyage was the basis of France's claim to Canada. In 1535 Cartier sailed up the St. Lawrence River to where Quebec City stands today and then pushed on to the island of Montreal.

In 1541 Cartier returned to Canada, establishing a settlement near Quebec City. He gathered what he thought to be

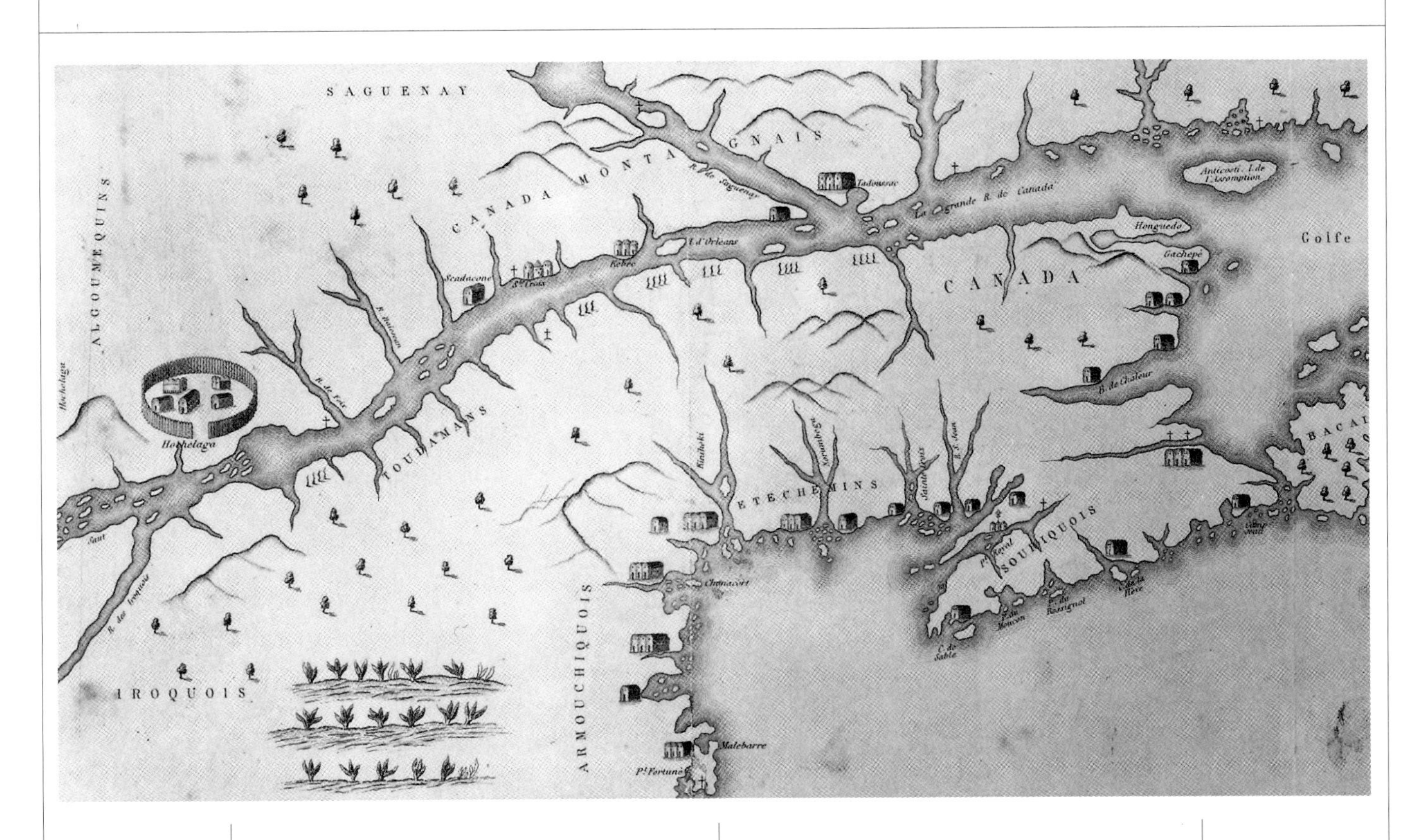

A 16th-century French map shows Cartier's discoveries along the St. Lawrence River. The settlements are those of Native Americans, whose tribes are named.

gold and diamonds; back in France, however, they turned out to be worthless rocks. Cartier's journeys established France's claim to Canada, but France did not press that claim for another half century. By that time, Cartier's discoveries had been reflected in French maps, such as one made by Pierre Desceliers in 1544 that gives the Native American name of Canada to the land north of the St. Lawrence River.

SEE ALSO

Americas, mapping of

FURTHER READING

Cartier, Jacques. *A Shorte & Briefe Narration of the Two Navigations to Neue France.* Translated by John Florio. 1966. Reprint. Norwood, N.J.: Walter J. Johnson, 1975.

Coulter, Tony. *Jacques Cartier, Samuel de Champlain, and the Exploration of Canada.* New York: Chelsea House, 1993.

Morison, Samuel Eliot. *The Great Explorers: The European Discovery of America.* New York: Oxford University Press, 1978.

Cartogram

A CARTOGRAM is a map that presents statistical information about a region, usually so that it can be compared with equivalent information from another region. Cartograms need not be geographically accurate; the sizes, shapes, and relative locations of countries may be distorted to reflect the particular information that is being communicated. For example, a cartogram could show the average annual rainfall of the countries of the world by making the wettest countries the biggest and the driest countries the smallest. The resulting map—although it would not be true to geography in terms of latitude and longitude and would not resemble the familiar image of the world—would present a true picture of comparative rainfall.

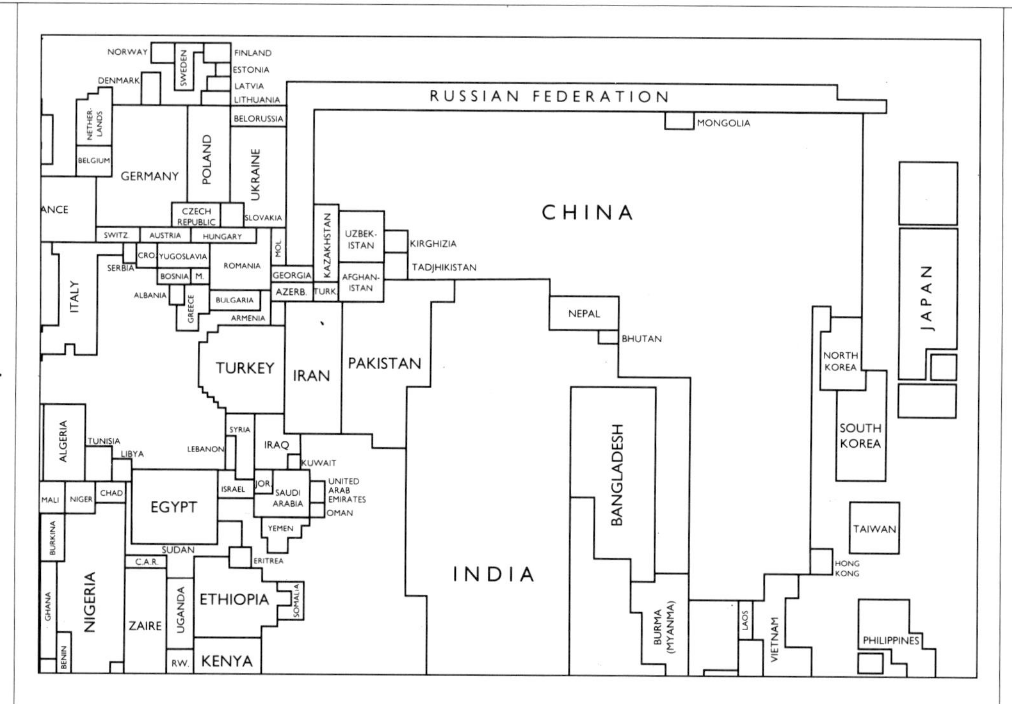

Cartograms distort the familiar outlines of nations and continents to show statistical information. In this cartogram from 1994, each country's size reflects its share of world population.

Today, cartograms are widely used to convey reams of social, economic, and environmental information. A world population cartogram might represent the different countries in terms of their populations, making China the biggest country on the map (Russia, the largest country in terms of land area, is the biggest country on a geographically accurate map). Or the cartogram could display the average family income in each country, in which case tiny, prosperous countries such as Luxembourg would vastly overshadow huge, poor countries such as Sudan. These and other cartograms can communicate a great deal of information quickly and in a direct, powerful way. New cartograms are being invented all the time. One notable collection of cartograms is *The New State of the World Atlas,* which presents many different types of information about the world on maps and cartograms.

FURTHER READING

Kidron, Michael, and Ronald Segal. *The New State of the World Atlas.* 4th ed. New York: Simon & Schuster, 1991.

Cartography

CARTOGRAPHY IS the art and science of drawing maps or charts; one who practices it is a cartographer. In its broadest sense, cartography includes all of the actions needed to make a map. Depending upon the area being mapped, these could consist of surveying the land, gathering information about place names and topographic features, compiling data from other maps, drawing the map, and printing the final version. In practice, however, the term *cartography* generally refers to the actual drawing of the map. The cartographer must decide what the map will include, what its scale will be, what information will be communicated by the map, and what colors and symbols will be used on it. Once the manuscript of the map has been drawn, the cartographer checks it for accuracy against survey data, aerial photographs, and maps that already exist; then the cartographer prepares it for printing.

Cartography has a long and rich history. No one will ever know when the first map was made. Maps are as old as human society, and the concept of mapmaking was probably invented many times over in many different communities. Almost every culture known has created some kind of map. Some scholars have suggested that certain prehistoric relics—carvings on ivory and paintings on cave walls—are really maps. Such speculations are fascinating but cannot be proved. The earliest known maps come from the Middle East and are thousands of years old.

From earliest times, maps have served two purposes. One type of map was local, practical, and based on firsthand observation. It depicted a village, a city, or a valley; it showed buildings and roads or perhaps outlined land ownership. Such maps were useful to governors, tax collectors, and other officials. Also practical and based on firsthand observation was the navigational chart. Made by seafarers, these charts showed coasts, harbors, currents, and perils such as submerged rocks.

The other kind of map covered a larger area and was not always realistic or intended for practical use. Maps of whole countries or regions, and especially maps of the entire world, included much that was based on myth, theory, hearsay, or legend—and often much that was inaccurate. Such maps often represented not firsthand observation but the cosmography or worldview of a particular culture or religion. They could also depict a philosopher's or scientist's ideas about the world, in which the borders of known lands were extended by guesswork or theory to the end of the earth.

Beginning in the late 15th century, when the nations of Europe began probing the rest of the world in a great burst of exploration, the two kinds of maps began to converge. As firsthand, reliable geographic knowledge became available for ever-wider regions, utilitarian maps, surveys, and sea charts covered more and more of the world. At the same time, the rise in geographic knowledge wiped many faulty theories and fanciful images from the larger maps, so that world maps gradually achieved a high level of accuracy and usefulness. By the end of the 19th century, most of the world had been mapped, although much detail remained to be filled in.

Cartography evolved rapidly and dramatically during the 20th century. New technologies, including photography, flight, and space flight, allowed mapmakers to gather more information faster than ever before. Computers came into use in cartography in a variety of ways: for processing, sorting, and storing information such as survey and climate data; for scanning photographs and satellite images and turning them into easily read maps; and for turning streams of digital information into maps that are stored on disk and databases and can blossom on computer screens at the touch of a button. Cartography will undoubtedly continue to change and grow, for each new advance in science, each new technological invention, and each new view of human society and history brings with it new kinds of knowledge to be mapped, as well as new ways of looking at old knowledge.

SEE ALSO

American mapmakers; Ancient and medieval mapmakers; British mapmakers; Chinese mapmakers; Computers in mapmaking; Cosmography; Dutch mapmakers; French mapmakers; Geography; Islamic geographers and mapmakers; Italian mapmakers; Japanese mapmakers; Map; Native American mapmakers; Pacific Islands mapmakers

FURTHER READING

Bagrow, Leo. *History of Cartography.* Revised and expanded by R. A. Skelton. Cambridge: Harvard University Press, 1964.

Boorstin, Daniel J. *The Discoverers.* New York: Random House, 1983.
Brown, Lloyd A. *The Story of Maps.* 1949. Reprint. New York: Dover, 1979.
Dilke, O. A. *Greek and Roman Maps.* Ithaca, N.Y.: Cornell University Press, 1985.
Hall, Stephen. *Mapping the Next Millennium: The Discovery of New Geographies.* New York: Random House, 1992.
Harley, J. B., and David Woodward, eds. *The History of Cartography.* Chicago: University of Chicago Press, 1987– .
Lister, Raymond. *Antique Maps and Their Cartographers.* Hamden, Conn.: Archon, 1970.
Nordenskiold, Adolf E. *Facsimile Atlas to the Early History of Cartography: Reproductions of the Most Important Maps Printed in the Fifteenth and Sixteenth Centuries.* New York: Dover, 1973.
Robinson, Arthur H. *Elements of Cartography.* New York: Wiley, 1993.
Skelton, R. A. *Explorers' Maps: Chapters in the Cartographic Record of Geographic Discovery.* London: Routledge & Kegan Paul, 1958.
Wilford, John Noble. *The Mapmakers.* New York: Knopf, 1981.

Cartouche

MANY MAPS contain cartouches—frames or scrolls that enclose the title of the map, the name of its maker, and perhaps other information, such as the date and place of publication. The cartouche is sometimes called the titlepiece.

The cartouches of maps made before the 19th century are often quite elaborate, decorated with ornamental engravings and poetic inscriptions; those of later maps tend to be plainer and more functional.

An aristocratic British hunting scene decorates the cartouche of a 17th-century road map by John Ogilby.

Casa da India

THE CASA da India (House of India) was the bureau of the Portuguese government that organized seagoing expeditions and controlled trade with India and other parts of Asia. During Portugal's great age of seaborne exploration and colonization in the 16th and early 17th centuries, the Casa da India used its powers to guard trade routes, navigational charts, and other information that might be commercially useful. Ship captains and navigators were supposed to turn over all charts and logs to the Casa, which kept them under lock and key, although new discoveries generally became known through word of mouth or security leaks. The archives of the Casa da India contained many reports, maps, and other records that would be priceless to modern scholars, but unfortunately many of these papers were destroyed when a major earthquake rocked Lisbon, the capital of Portugal, in 1755.

Casa de la Contratación de las Indias

THE CASA de la Contratación de las Indies (House of Trade with the Indies) was Spain's equivalent of Portugal's Casa da India: a government bureau that supervised voyages of exploration and trade and also controlled the maps and other records of Spain's ventures into the

Americas and Asia. Many mapmakers and explorers were employed by the Casa. Cartographer Diego Gutierrez worked there, and explorer and mapmaker Sebastian Cabot was hired by the Casa to teach navigation to ships' pilots.

Cassini family

FRENCH ASTRONOMERS AND GEODESISTS

THE CASSINIS were among Europe's foremost scientists in the 17th and 18th centuries. For four generations, they were in charge of the Paris Observatory. They took a leading part in the effort by scientists of the period to determine the exact size and shape of the earth. Geographers, navigators, and mapmakers all were hoping for new, more accurate measurements, and the Cassinis perfected survey methods that had a major effect on mapmaking.

The founder of the clan, Jean Dominique Cassini (1625–1712), was born Giovanni Domenico Cassini in Italy but changed his name after becoming a citizen of France. At the Paris Observatory, he discovered four moons of Saturn, as well as the divisions in that planet's rings. Cassini turned his attention to the earth when Jean-Baptiste Colbert, the minister of finance, asked for his help in preparing a complete, accurate topographic map of France. One of the great surveying problems of the age was the difficulty of determining longitude, which requires that an observer know the exact time at each of two widely separated points—not an easy task when timekeepers were unreliable and a mistake of even a second could ruin a calculation. Cassini developed a complex method of calculating longitude based on observations of Jupiter's moons; although it was too cumbersome to use at sea, requiring large telescopes and a perfectly steady surface, this method could be used to obtain fairly accurate measurements on land.

Jean Dominique Cassini at the Paris Observatory, with a globe and a rolled map. In the background is the long telescope he used for sighting the moons of Jupiter.

A team of surveyors headed by Jean Picard began taking sightings all over France using Cassini's method. Soon, the latitudes and longitudes of a number of points had been accurately determined. When a simple outline map was drawn around these points and placed over the existing map of France, it became clear that France's actual area was smaller than it appeared on the old, erroneous map. King Louis XIV saw this on a visit to the observatory and exclaimed to Cassini, "Your work has cost me half my estate!"

But the work continued. On the floor of a tower room, Cassini created a huge world map on which he recorded latitude and longitude observations sent to him from all over the world. This was an important moment in the history of mapmaking—the first time that relatively accurate latitude and longitude readings were systematically gathered and used to map the entire earth. Gradually, the continents and seas began to take on their proper proportions, as Cassini corrected errors that had been passed from map to map since the Middle Ages.

After Picard's death, Cassini had to devote more time to supervising the field survey of France. His son Jacques Cassini (1677–1756) helped and later took over the task. The Cassinis' goal was to determine the exact line of an arc, or part of a meridian of longitude, through the whole length of France. Once the arc was fixed, they could use triangulation to fill in the rest of the map by relating new points to points whose location was already established. Jacques Cassini and his son, César François Cassini de Thury (1714–1784), worked on the survey together. The framework of survey points was complete by 1740.

In 1744, César François began publishing the survey results that had been so laboriously accumulated by his family. The map he produced in eight sheets showed all of the triangulation points, but it was still only a framework: most of the long-awaited topographic details between these points still had to be filled in. The surveying of roads, rivers, towns, and other landmarks began in 1750 and was not finished when César François died. The task was passed on to his son, Jacques Dominique Cassini (1748–1845).

Jacques Dominique finally finished the great national atlas of France in 1791—but by that time the country had been convulsed by the French Revolution. The antimonarchist revolutionaries threw Cassini into jail because of his family's friendship with the royal family; Cassini spent nine months in jail and was lucky to leave with his head still attached to his neck (other accused royalists fared less well). He retired to the country and worked quietly on the *Topographic, Mineralogic, and Statistical Atlas of France,* which he published in 1818.

The original Cassini topographic atlas, the fruit of four generations of labor, was published in 1793 by the French Academy of Sciences. A landmark in cartography, it set the standard for all national mapping projects in the future. The Cassinis had proved that accurate topographic mapping over a large area was possible, and their work paved the way for other large-scale mapping projects such as the survey of India. The Cassini family was the cartographic ancestor of all modern mapping teams and projects.

SEE ALSO

Triangulation

FURTHER READING

Konvitz, Josef W. *Cartography in France, 1660 to 1848: Science, Engineering, and Statecraft.* Chicago: University of Chicago Press, 1987.
Wilford, John Noble. *The Mapmakers.* New York: Knopf, 1981.

Catalan atlas

THE CATALAN ATLAS is a world map on 12 sheets completed in 1375 by cartographer Abraham Cresques. Although the term *atlas* had not yet come into use, modern historians generally call Cresques's set of maps an atlas, even though it contains one-of-a-kind hand-drawn maps rather than the printed maps found in modern atlases. Cresques's set of maps is called the Catalan atlas because it was produced in Catalonia, a province of northeastern Spain that includes the port of Barcelona and the island of Majorca. Catalans, as

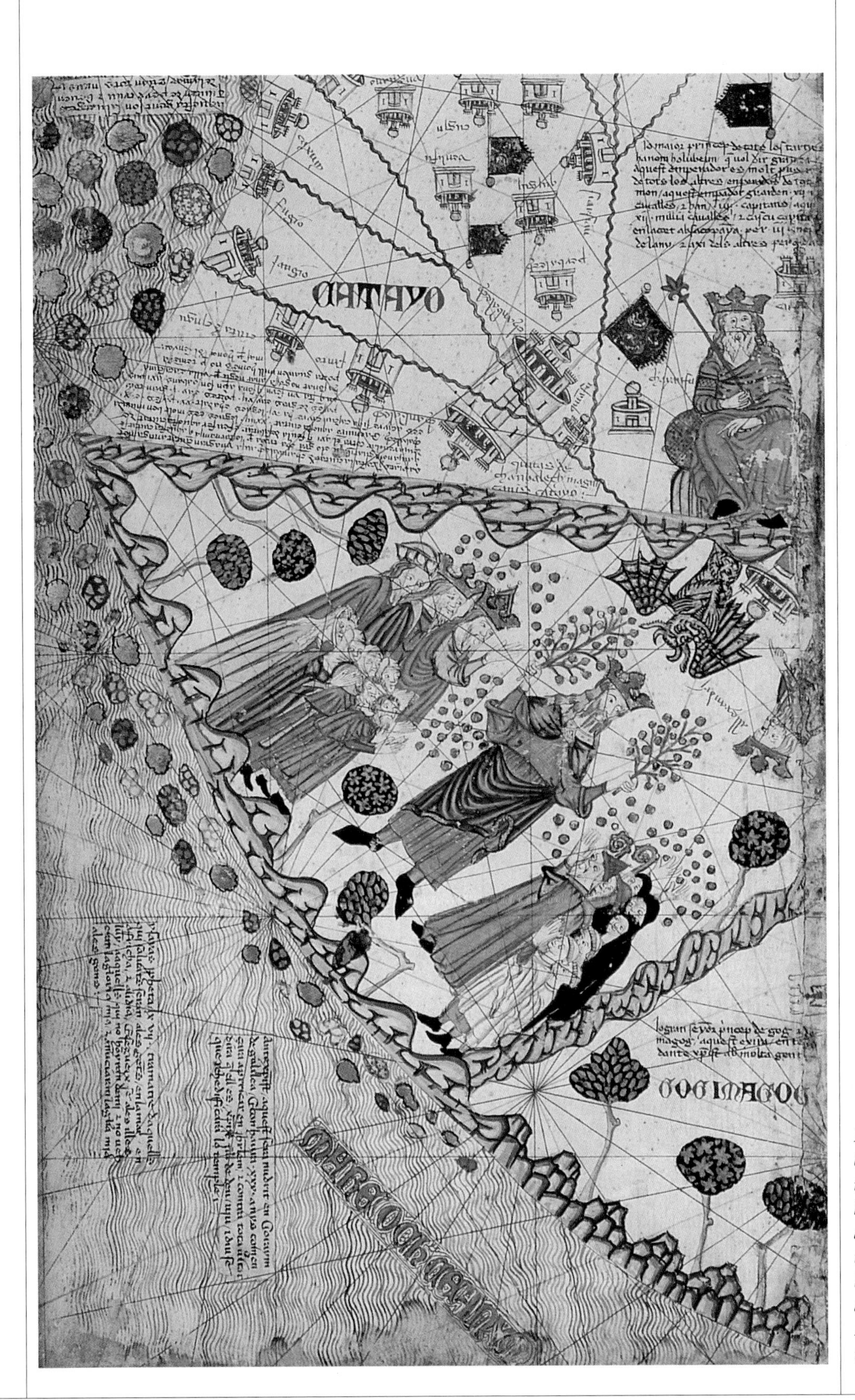

Marvels described by Marco Polo—the Great Khan in his green robes, and cities scattered across the landscape—appear on the Asian portion of the Catalan atlas.

the people of Catalonia are called, were among the leading navigators, geographers, and mapmakers of the Middle Ages. Their knowledge of the Mediterranean Sea, in particular, was unsurpassed. Among the Catalans were many Arabs and Jews, who brought an awareness of African and Middle Eastern geography.

The Catalan atlas is one of France's national treasures. It was made for King Charles V of France, who wanted a new map of the world. It was drawn on sheepskins, painted in brilliant colors, and adorned with gold leaf to make a magnificent, kingly display. But the Catalan atlas was more than just a showpiece. It was one of the best and most thorough maps of the medieval period.

The Catalan atlas combines features of two different kinds of maps: portolans, or sea charts used for navigation, and *mappae mundi,* or illustrated world maps. Coastlines are shown in accurate detail, as on a portolan. The interiors of the continents are decorated with many informative drawings and descriptive labels, as on a *mappa mundi.* Two panels show Catayo, or China, as described by Marco Polo, complete with illustrations of the marvels mentioned in his book; one drawing shows Polo's camel caravan wending its way across Asia with a troop of helmeted Mongol escorts.

Another of the maps shows the African kingdom of Mali and its ruler, Mansa Musa, called by the mapmaker "the richest and noblest king in the world." Cresques's illustrations showed cities, palaces, and treasures in the lands south of the Sahara Desert, as well as caravans crossing the desert. But although they are heavily illustrated, the Catalan maps contain fewer mythological and fictional places and creatures than most maps of the time. All in all, the Catalan atlas is considered a masterpiece of medieval European mapmaking, a collection of maps designed to be both useful and beautiful.

SEE ALSO
Cresques, Abraham; Mappa mundi; Polo, Marco; Portolan

Celestial maps

SEE Space mapping

Celestial navigation

CELESTIAL NAVIGATION means navigating by the heavens—that is, fixing locations and setting courses by the sun and stars. When you determine direction by reminding yourself that the sun rises in the east and sets in the west, you are practicing a simple form of celestial navigation. For centuries, celestial navigation was the only method available to ship pilots, many of whom possessed as much knowledge of the heavens—based on years of firsthand observations—as any astronomer. Although modern mariners possess sophisticated instruments that can pinpoint their location at the push of a button, celestial navigation is still taught for use in emergencies, or to those who want to pilot their boats the old-fashioned way. Pilots using celestial navigation can determine their latitude and longitude using a few simple tools: a sextant, a precise timekeeper that shows Greenwich time, and some astronomical and mathematical tables.

SEE ALSO
Navigation

Census maps

SEE Statistical maps

H.M.S. Challenger *plied the world's oceans while its scientists studied the world below the waves.*

Challenger Expedition

THE BRITISH Challenger Expedition was the first oceanographic exploring expedition. Beginning in 1872, five scientists spent three and a half years cruising the world's oceans aboard the square-rigged sailing ship *Challenger.* They covered 79,292 miles (127,607 kilometers) and gathered samples of 13,000 different species of plants and animals, many of them from depths that had been thought too great to support life. More than 4,400 of these specimens were of previously unknown species. The data from the Challenger Expedition—ocean temperatures, currents, depths, chemical composition of the water, and so on—filled 50 volumes and laid the foundation for oceanographic study.

SEE ALSO
Undersea mapping

Champlain, Samuel de

FRENCH EXPLORER OF NORTH AMERICA

- *Born: 1567, Brouage, France*
- *Died: Dec. 25, 1635, Quebec, Canada*

THE SON of a sea captain, Samuel de Champlain started his career as a soldier but soon turned to the sea. He sailed to the West Indies and Central America in the service of Spain. In 1603 he was appointed geographer of a French expedition to New France, the Canadian territory that was claimed by France. He explored the St. Lawrence River and the coast of North America as far south as present-day Cape Cod, Massachusetts; later he made highly accurate maps of Maine and New England. His map called "Description of the Coasts and Islands of New France" (1607) is surprisingly accurate, considering that he did not have the standard surveying tools of the time on his voyage. Modern archaeologists have verified the locations of Native American villages that Champlain showed on the map.

Called the "Father of New France," Champlain founded two colonies, one in Nova Scotia and another at Quebec. In his efforts to encourage the growth of the fur trade in New France, Champlain explored the region around the St. Lawrence River. On one of his journeys he visited the lake on the New York–Vermont border that now bears his name; it appears on his 1612 map, along with inviting drawings of Native Americans, fish, fur-bearing animals, and fruits to illustrate New France's wealth of resources. In 1615 Champlain traveled to the Great Lakes region. His map of this trip includes not only Lake Ontario and Lake Huron, which he saw, but other lakes that had been described to him by the Indians.

Even after he stopped exploring on his own, Champlain sponsored many expeditions into the American interior and collected information for numerous excellent maps. Thus, he was responsible for a great leap forward in geographers'

Champlain's hand-drawn map of New France, prepared in the winter of 1606–7 with the help of Native American informants.

knowledge of North America. And by encouraging merchants, trappers, and missionaries to venture into Canada, his maps promoted the colonization of French North America. Champlain served several times as governor of New France before dying there in 1635.

SEE ALSO

Americas, mapping of

FURTHER READING

Champlain, Samuel de. *Voyages of Samuel de Champlain, 1604–1618*. Irvine, Calif.: Reprint Services, 1991.

Coulter, Tony. *Jacques Cartier, Samuel de Champlain, and the Exploration of Canada*. New York: Chelsea House, 1993.

Morison, Samuel Eliot. *Samuel de Champlain, Father of New France*. Boston: Little, Brown, 1972.

Chart

ANY TYPE of map can be called a chart, but the term *chart* is often used in a more limited sense to refer to maps that are intended to guide route finders, especially navigators at sea or in the air. Nautical (sea) charts show the depth of the water, the buoys and other markers that guide mariners, and hazards to ships such as reefs and shallows. Aeronautical (air) charts show features that are recognizable from the air, with the emphasis on airports, major air traffic routes, and hazards such as high towers and bridges. Both types of charts are highly specialized, full of information that is vital to pilots and navigators. People who have not been trained in how to use them, however, may find them confusing and not very useful as maps. Like other kinds of maps, navigational charts are now available on computer, either on disks or from on-line databases.

Cheng Ho

SEE Chinese mapmakers

Chinese mapmakers

CHINA HAS a very long history of cartography. Not only did the Chinese develop mapmaking skills before any other Asian people, but they also mastered many techniques of mapmaking before those techniques were known in the West. For example, some scholars believe that the Chinese were the first to learn the art of taking sightings on the sun and stars to determine the latitudes of places on the earth's surface. The Chinese also used compasses and sundials before these instruments were used in the Mediterranean world.

References in Chinese literature suggest that two types of maps developed in early China. One type was an accurate, detailed local survey, with special emphasis on the courses of rivers and streams. The use of waterways for transportation and irrigation was vital to the growth of Chinese civilization, as were the draining of swampland for farming and the damming of rivers to prevent floods. All of these activities required a knowledge of the country's watercourses, a knowledge that was reflected on local charts.

The second type of map covered the entire country, or even the world, and was more speculative. Although no early Chinese world maps survive, a book called the *Shu Ching* (*Historical Classic*), dating from the 5th century B.C., describes the Chinese world view of the time: China, called the Middle Kingdom, was at the center of the world, surrounded by zones of "allied barbarians." Farther out from the center were the realms of the "uncivilized savages."

A map of China is thought to have been compiled by 1100 B.C., and many smaller regional maps existed by that time. We know that maps were common by the 7th century B.C.; that an official survey of the country, including maps, was issued by the government around 450 B.C.; that new maps were made in the second half of the 3rd century B.C. after the borders of China's provinces were redrawn; and that maps were prized as works of art in that era. None of these maps survives, however.

The oldest known Chinese maps were found in 1973 in a tomb in Hunan Province that dates from the 2nd century B.C. There are two maps, drawn on silk, depicting Hunan and the lands on its borders. One gives a fairly detailed and accurate picture of topography and settlement, with rivers, streams, mountains, roads, towns, and villages carefully marked. The other shows the province's military defenses: army camps, walled fortresses, watchtowers, and the like. The cartographer who made these maps used a set of standardized symbols to indicate towns and other features. These symbols had no doubt been in use for some time, which means that the art of cartography was already well developed in China by the 2nd century B.C. At that time, Chinese cartography was the most advanced in the world.

The Chinese invented paper at the end of the 2nd century B.C., and maps be-

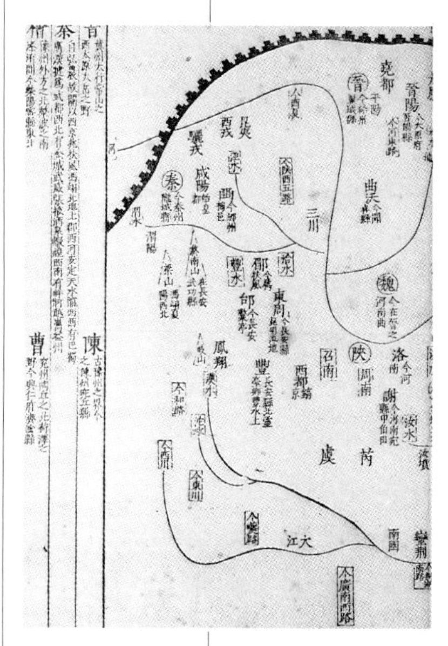

A printed map from the 12th century. The black structure at the top is the Great Wall of China; the circles contain the names of provinces.

gan to be drawn on this new material. Around the same time, a government department of mapmaking was created. Another technological innovation, the invention of printing, occurred in about A.D. 370—centuries before it was invented in the West. No one knows exactly when the Chinese began printing maps, but the oldest surviving printed map in the world is Chinese. It dates from the mid-12th century and shows three northern provinces, with the Great Wall snaking along their northern borders.

Chinese cartography took a great step forward in the 3rd century A.D., when Pei Hsiu issued a handbook for mapmakers. Rivers and their management continued to preoccupy cartographers, as a map of China carved on a slab of stone around A.D. 1100 demonstrates. The map is a careful grid—showing that the Chinese had adopted a system of coordinates—with rivers and the coastline carefully incised. Another landmark in Chinese mapmaking was the atlas of China issued by cartographer Chu Ssu-pen in the 14th century.

The Chinese had a well-developed sea trade, and sea charts were part of their cartographic tradition. Some were local charts, showing small strips of coastline with rocks and harbors marked. Others were larger in scope, showing the seas, islands, and coastlines of the South China Sea and the Indian Ocean. The best-known and farthest-ranging Chinese seafarer was Cheng Ho (1371–1435), a court official who commanded seven long voyages before 1433, taking Chinese imperial fleets to East Africa, Arabia, the Persian Gulf, India, and Southeast Asia. Although the Chinese were probably already aware of every place Cheng Ho visited, his voyages brought firmer geographic knowledge home to China's mapmakers.

By the end of the 15th century, the

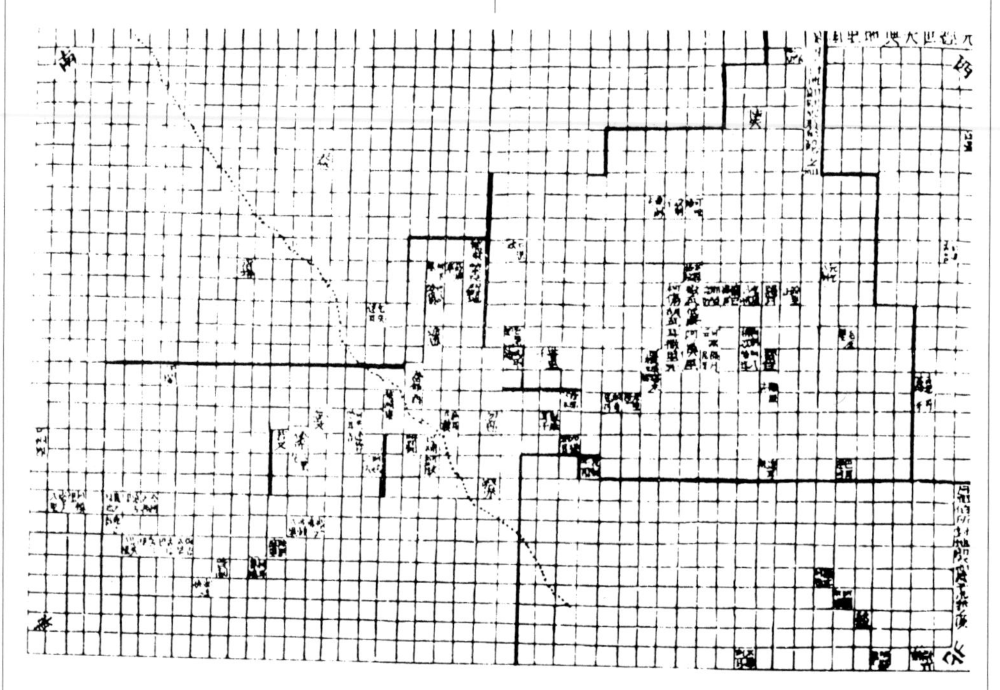

Beginning in the 14th century, the Mongol rulers of China used an abstract mapmaking style in which place names were positioned on a grid. This map shows northwest China; north is at the lower right corner.

Chinese had mapped in detail most of their own country, Southeast Asia, the east coast of India, and parts of the Philippines. Not well mapped, but known in general outline, were India and Central Asia, Africa, Arabia and the Middle East, and Madagascar. In the following century, Chinese cartography began to be influenced by Western missionaries to China, especially Matteo Ricci, who made a world map based on European knowledge and gave it to the Chinese. Jesuit missionaries thereafter were appointed the official mapmakers for the Chinese court, and in 1674 they were put in charge of the imperial observatory in Beijing. From that time on, Chinese cartography merged with the Western tradition of mapmaking.

SEE ALSO

Asia, mapping of; Boym, Michal Piotr; Chu Ssu-pen; Pei Hsiu; Ricci, Matteo

FURTHER READING

Bagrow, Leo. *History of Cartography.* Revised and expanded by R. A. Skelton. Cambridge: Harvard University Press, 1964.

Fernández-Armesto, Felipe, ed. *The Times Atlas of World Exploration: 3,000 Years of Exploring, Explorers, and Mapmaking.* New York: HarperCollins, 1991.

Lister, Raymond. *Antique Maps and Their Cartographers.* Hamden, Conn.: Archon, 1970.

Temple, Robert. *The Genius of China: 3,000 Years of Science, Discovery, and Invention.* New York: Simon & Schuster, 1986, pp. 30–33.

Tooley, Ronald V. *Landmarks of Mapmaking: An Illustrated History of Maps and Their Makers.* 1976. Reprint. New York: Dorset, 1989.

———. *Maps and Map-Makers.* 1949. Reprint. New York: Dorset, 1990.

Chronometer

A CHRONOMETER is a timepiece, or watch. In cartography, the term generally refers to a meticulously accurate timepiece used either to determine longitude or to time astronomical observations for surveying purposes. The development in the 18th century of a chronometer that would keep accurate time during long sea voyages was a technological breakthrough for mapmaking and navigation, allowing longitude to be reliably measured at sea for the first time.

SEE ALSO

Harrison, John; Longitude; Navigation

Chukei, Ino Tadataka

SEE Japanese mapmakers

Chu Ssu-pen

CHINESE MAPMAKER

- *Born: 1273*
- *Died: 1335*

CHU SSU-PEN was active during the Mongol dynasty, a time when China experienced increasing contact with Central Asia and the West and when ventures of exploration and mapmaking were supported by the imperial government. Around 1311–12 he prepared a set of maps of China in the form of an atlas, with maps of all the provinces. This atlas also covered the lands around China; it showed that the Chinese possessed accurate knowledge of the South China Sea and the Indian Ocean and that they knew at least the general outline of Africa and the location of the Nile and Congo Rivers.

Chu Ssu-pen's atlas was reprinted frequently; editions from 1579 and 1588 survive today. It served as the basis for

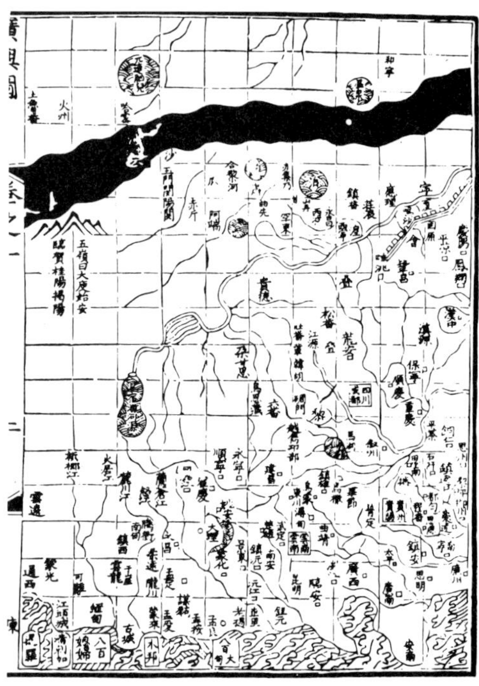

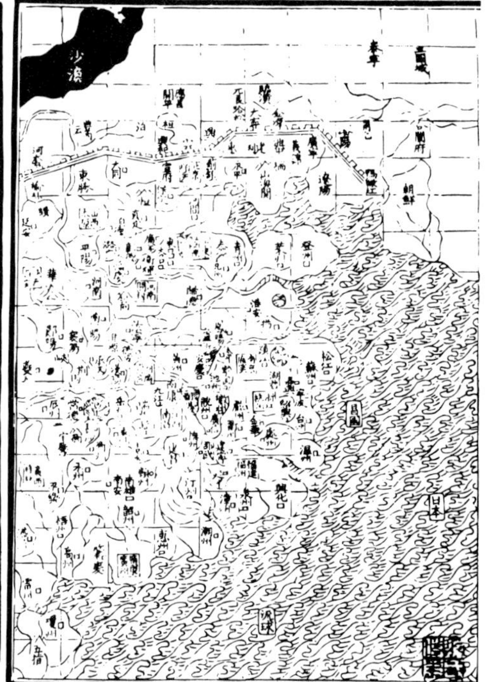

Part of Chu Ssu-pen's atlas. The Great Wall is shown by the zig-zag row of tiny squares just below the black band, which represents the Gobi Desert.

later maps of China, including those made by Jesuits Matteo Ricci and Martino Martini. Martini's map of China, based on Chu Ssu-pen's, was published in Europe in 1655 and was the standard European map of China until 1737.

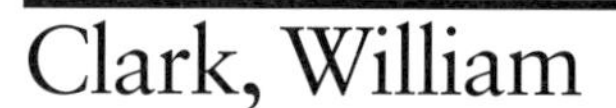

Clark, William

AMERICAN EXPLORER

- *Born: Aug. 1, 1770, Caroline County, Virginia*
- *Died: Sept. 1, 1838, St. Louis, Missouri*

WILLIAM CLARK was the younger brother of George Rogers Clark, the American militia commander who fought the British in the Ohio Valley during the American Revolution. When William Clark came of age, he joined the army and served in the wars against the Native Americans of the Ohio River region. During this time he met and befriended a fellow officer named Meriwether Lewis.

A few years later, President Thomas Jefferson selected Lewis to lead the first American exploring expedition into the land west of the Mississippi River. Lewis invited Clark to join the expedition, and the two explorers succeeded in their mission. They traveled from St. Louis, Missouri, to the Pacific Ocean coast of Oregon and back in the years 1804–6. Lewis and Clark returned from the West with a wealth of information about its inhabitants, geography, resources, and natural history.

Of the two men, Clark was the more skilled as a geographer and mapmaker. His map of western North America, published in 1814, is regarded as one of the most influential of all early American maps. It aroused widespread interest in westward migration, and it guided thousands of settlers across the Rocky Mountains into the Oregon Territory.

After returning from the West, Clark spent the rest of his life in government service on the frontier. He served first as the Louisiana Territory's Indian agent—the federal government's representative to the Native American peoples of the region. Later, he was governor of the Missouri Territory, and he ended his career as superintendent of Indian affairs for the upper Mississippi and Missouri rivers.

SEE ALSO

Americas, mapping of; Lewis and Clark Expedition

FURTHER READING

Cavan, Seamus. *Lewis and Clark and the Route to the Pacific*. New York: Chelsea House, 1991.

Duncan, Dayton. *Out West: An American Journey*. New York: Viking, 1987.

Lavender, David. *The Way to the Western Sea: Lewis and Clark Across the Continent*. New York: Harper & Row, 1988.

Otfinoski, Steven. *Lewis and Clark: Leading America West*. New York: Fawcett Columbine, 1992.

Steffen, Jerome O. *William Clark: Jeffersonian Man on the Frontier*. Norman: University of Oklahoma Press, 1977.

Clausson, Claudius

DANISH GEOGRAPHER

• *Active: early 15th century*

CLAUDIUS CLAUSSON (sometimes his name is given in its Latin form, Claudius Clavus) was the first geographer to show Greenland on a map. He is said to have voyaged to the Scandinavian colony in Greenland himself, although very little is known about his life.

In 1425 Clausson drew a map of northern Europe. It was intended to serve as an addition to Ptolemy's *Geography,* which did not include the northern regions. Clausson's map shows the Baltic Sea, Sweden, Denmark, Norway, Scotland, Iceland, and, squeezed into the northwest corner, part of the coast of "Greenland province." Although not very accurate by modern standards, this was the first known map of Scandinavia. It remained in use, copied by other mapmakers for their own editions of the *Geography,* until the early 16th century. The German cartographers Nicolaus Germanus and Henricus Martellus, among others, copied Clausson's map.

Climate maps

SEE Weather maps

Collins, Greenville

BRITISH NAVAL OFFICER AND HYDROGRAPHER

• *Active: 1669–98, England*

GREENVILLE COLLINS was a navy captain who in 1681 was appointed by Samuel Pepys, secretary of the British Admiralty, to survey the coasts of the British Isles and make new nautical charts. Great Britain had become a maritime power but produced few printed sea charts of its own. Even the *English Pilot,* published in 1671, had been printed from Dutch-engraved plates. British authorities wanted to produce their own sea atlas, both for practical reasons and for reasons of national pride.

Collins spent several years making his survey, then more years supervising the work of several engravers as they cut the plates from which the new charts would be printed. The collection of charts, called *Great Britain's Coasting Pylot,* was published in 1693 with a handsome title page that identified Collins as Hydrographer to the King. An elaborate illustration on the title page shows Neptune, the god of the sea, and

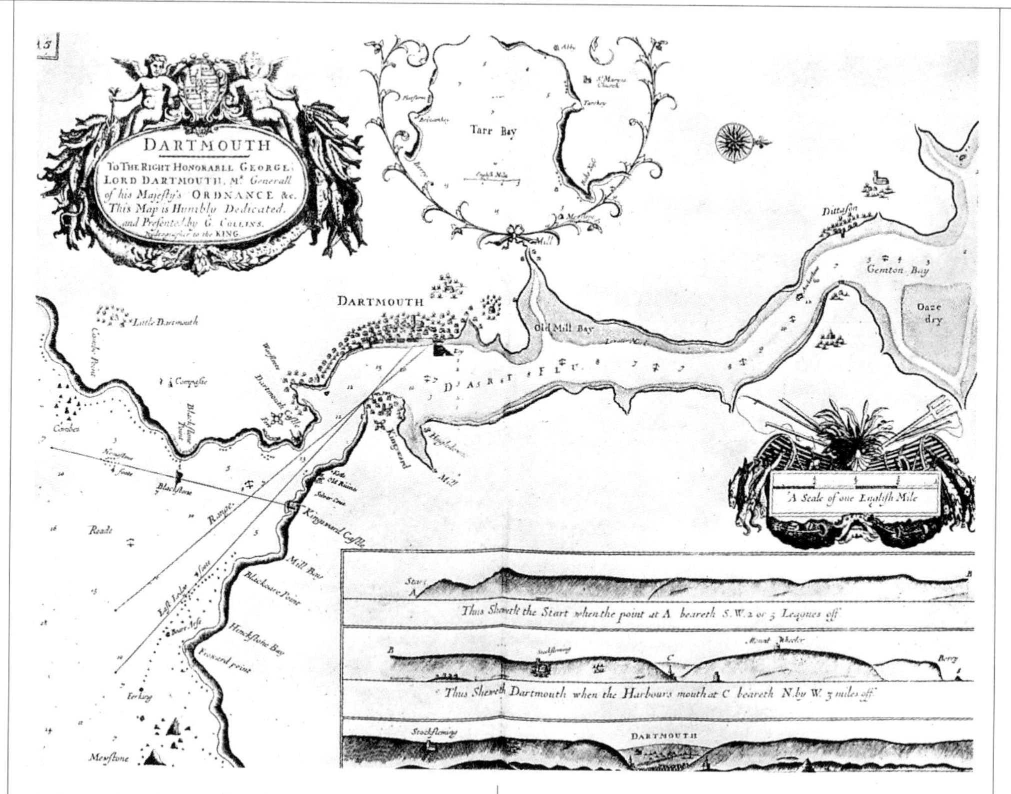

Collins's chart of Dartmouth, a port in southwestern England, shows the coastal rocks and reefs that had ripped the bottom out of many ships.

Britannia, the spirit of Britain, surmounted by the royal arms of Britain and a pair of cherubs blowing horns. Beneath all this, a mermaid and a merman unfold a miniature version of one of the charts to show the reader what awaits within.

Collins's survey was a landmark in British hydrography, and his charts were reprinted many times.

Columbus, Christopher

ITALIAN EXPLORER OF THE AMERICAS

- *Born: about 1451, Genoa, Italy*
- *Died: May 20, 1506, Valladolid, Spain*

CRISTOFORO COLOMBO—later known as Cristóbal Colón in Spanish and Christopher Columbus in English—received little formal education as a child. Later, he read widely and educated himself, with special emphasis on geographic subjects. Columbus became a seaman at an early age. By 1477 he was living in Lisbon, the center of Portuguese seaborne exploration. He sailed in the Mediterranean Sea and the Atlantic Ocean, and he probably made at least one voyage to Portuguese West Africa.

Portugal was engaged in a massive, long-term effort to find a sea route to "the Indies"—India and the other little-known lands of south and east Asia. Spices and other goods from this part of the world reached Europe only after a long, expensive passage through a network of traders in the Indian Ocean and the Middle East, making the nations of Europe eager to open direct trade with the Indies.

Like most well-read people of his time, Columbus knew that the world was round. He was familiar with the writings of Ptolemy and other geographers of the ancient world, and he had also read the travel narrative of Marco Polo, who had visited China in the 13th century and stimulated Europeans' interest in Asia. Along with most geographers of the 15th century, Columbus believed that it might be possible to reach the distant east by sailing west—that is, to reach Asia by crossing the Atlantic. Columbus calculated that Japan lay only 2,400 miles (3,862 kilometers) to the west and offered to make the journey. Neither Portugal, England, nor France was interested in backing Columbus's scheme. After years of delay, he finally won support from Spain and embarked in August 1492 with a crew of about 90 men in the *Niña,* the *Pinta,* and the *Santa Maria.*

Columbus had greatly misjudged the size of the world and the distance to Asia. Japan is more than four times as far away from Europe as he thought it was; furthermore, Asia is separated from Europe by the American continents. But although the Vikings had landed in North America 500 years earlier, knowledge of their discoveries had largely been lost. Columbus had no idea that the Americas existed. When he made landfall amid green semitropical islands on October 12, 1492, he believed he had reached the Indies, and that China and Japan lay just over the horizon. In reality, he was somewhere in the Bahamas.

Columbus held to his theory, however, for three more voyages, stubbornly insisting that he had reached Asia. In 1493 he set out with 17 ships and 1,000 men to establish a colony on Hispaniola, the island that today is occupied by Haiti and the Dominican Republic. He then spent nearly three years exploring Caribbean islands and coastlines, but he found no trace of the civilizations he expected to encounter in the Indies. In 1498 he tried again, this time sailing along part of the South American coast. He landed in Venezuela and saw the Orinoco River, whose size made him realize he had arrived at "a very great mainland," not an island. Still he believed that the Indies lay just past this obstruction.

Shipped back to Spain in disgrace because of grave charges that he had mismanaged the colony on Hispaniola, Columbus persuaded King Ferdinand and Queen Isabella to give him another chance. In 1502 he took four ships west on his fourth and final voyage to the Americas, hoping to find a channel that would lead him past the baffling landmass to the riches of the Indies. When at last he turned back, frustrated once again, he believed he was only 19 days' sail from the Ganges River in India. In reality, he was half a world away from it. Columbus returned to Spain in 1504, still convinced that he had landed on the fringes of Asia.

Columbus himself never acknowledged the full import of his voyages; he never admitted that he had reached a previously unknown part of the world instead of Asia. Other explorers and geographers, however, soon realized what he had done. As early as 1494 Peter Martyr, an Italian scholar who spread the news of Columbus's explorations throughout Europe, referred in a letter to

A woodcut resembling an imaginative map was published in 1494 with an account of Columbus's first voyage to the Americas.

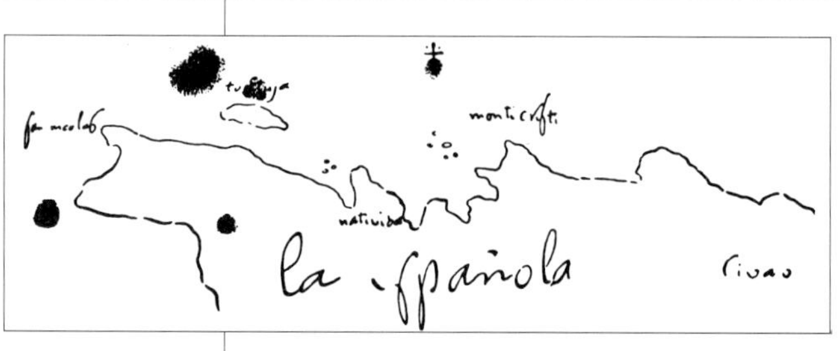

This sketch of "la española," or Hispaniola, is the only surviving map in Columbus's hand.

"the western hemisphere," which suggests that he suspected Columbus had found not a sea route to Asia but a new part of the world.

The only one of Columbus's own maps that survives is a sketch of the northwestern coast of Hispaniola, but his legacy soon began appearing on other cartographers' maps. Early 16th-century maps by Juan de la Cosa and Piri Reis showed the islands that Columbus had charted. Amerigo Vespucci popularized the idea that the lands Columbus had explored were "a new world," and within a few years of Columbus's death it was generally accepted that he had reached not Asia but a hemisphere that was ready to be measured, mapped, and conquered by Europeans. So, although he never achieved his goal of finding a trading route to the Indies, Columbus launched a centuries-long experiment in colonization that changed the face of the world.

SEE ALSO

Americas, mapping of; La Cosa map; Piri Reis map; Vespucci, Amerigo; Viking explorers

FURTHER READING

Bedini, Silvio, ed. *The Columbus Encyclopedia.* New York: Simon & Schuster, 1991.

Columbus, Christopher. *The Four Voyages.* Translated by J. M. Cohen. New York: Viking Penguin, 1992.

Cummins, John, ed. and trans. *The Voyage of Christopher Columbus: Columbus's Own Journal of Discovery.* New York: St. Martin's Press, 1992.

Dodge, Stephen C. *Christopher Columbus and the First Voyages to the New World.* New York: Chelsea House, 1991.

Fernández-Armesto, Felipe. *Columbus.* New York: Oxford University Press, 1991.

Phillips, William D., Jr., and Carla Phillips. *The Worlds of Christopher Columbus.* Cambridge: Cambridge University Press, 1991.

Russell, Jeffrey B. *Inventing the Flat Earth: Columbus and Modern Historians.* Westport, Conn.: Greenwood, 1991.

Sale, Kirkpatrick. *The Conquest of Paradise: Christopher Columbus and the Columbian Legacy.* New York: Knopf, 1990.

Wilford, John Noble. *The Mysterious History of Columbus: An Exploration of the Man, the Myth, the Legacy.* New York: Knopf, 1991.

Compass

THE COMPASS is our principal instrument for finding direction. Its origins are shrouded in mystery. Various peoples, including the Chinese, the Arabs, the Finns, the Vikings, the Greeks, and the Italians, have been identified as the first to use the compass. The first written record of the compass comes from China in the 11th century A.D., and the earliest European record of it dates from a century later. Although many scholars believe that the Chinese were the first to use the compass, it is possible that this useful device was invented independently in different parts of the world. At any rate, the compass—sometimes called the mariner's compass to distinguish it from a drafting tool called the compass—has been the principal aid to navigation since the 13th century.

The first compasses were probably nothing more than pieces of lodestone, a naturally occurring form of a mineral called magnetite, which aligns itself with the earth's magnetic field. At first, lodestone was prized just for its magnetic qualities—that is, for the way it drew iron to itself. Then people discovered that it possessed another virtue. If a bit of lodestone were dropped into a bowl of water, it would land pointing north. Iron and steel needles that had been rubbed with lodestone also pointed north. The mariner's compass developed from this discovery. The compass consisted of a magnetized needle stuck through a piece of wood, cork, or straw so that it would float in a bowl of water. The needle could move freely and always pointed toward the north.

To superstitious sailors, the compass needle's uncanny properties of showing direction and attracting other metals made it seem magical. Writing in 1600, a London physician and scientist named William Gilbert recounted some common beliefs about magnets: Contact with a diamond was thought to wipe out the magnetic power of a lodestone or magnetized needle. A huge mountain of lodestone, supposedly located somewhere in the far north, was powerful enough to draw the iron nails right out of ships' hulls, with dire results. Love potions made with lodestone were said to be able "to reconcile husbands to their wives, and recall brides to their husbands." And somewhere, it was thought, existed special kinds of magnets that would attract gold, silver, water, fish, and other substances. One of the strangest superstitions was the belief that onions and garlic could destroy magnetic power. Many sailors refused to eat these foods, fearing that their compasses would be rendered useless. Yet Gilbert spoke to several mariners who insisted quite firmly that they would sooner lose their lives than give up garlic and onions. Gilbert tested the notion himself and found that even after he belched garlic fumes on a compass needle and rubbed it with garlic juice, it pointed north as well as ever. Gilbert also found that—contrary to popular superstition—diamonds did not alter the needle's properties.

As time passed, the compass became more familiar to mariners. At some point, probably in the 14th century, the compass needle was mounted on the wind rose, a circle divided into segments to indicate the directions from which the winds blow. Wind roses had been used on charts for years, but one day a clever sailor or navigator drew a wind rose on a card or cut one from a chart, mounted the needle on a pivot, and stuck the pivot through the middle of the wind rose. Thus the compass rose—the circle of points that show the directions—was born. All a mariner had to do was let the needle settle and then move the card around so that "North" was under the needle. All of the directions could then be read from the card. Because the needle pointed north, north became the most important and elaborately decorated point on the compass rose; this may explain why most maps made after the Middle Ages are oriented with north at the top.

An ivory case with a cover (top) was made to house this 16th-century Italian compass. Compasses were highly prized and thought to have magical virtues.

In time, mariners and scientists observed that the compass needle did not point straight toward the north pole, as had long been thought. It pointed in a northern direction, but not true north. By the 16th century, people realized that the amount by which compass readings varied from true north depended upon the longitude at which the readings were taken—that is, the position of true north in relation to the north-pointing compass needle is different in different places.

The direction in which the needle pointed also seemed to change over the centuries. This is because the earth has two north poles, one geographic and one magnetic. The geographic north pole is fixed, but the magnetic north pole migrates slowly around the geographic pole. Navigators and others who used compasses gradually learned to allow for the difference between true north and magnetic north in compass readings (this difference is called the variation or declination). The first book listing compass variations for specific places was published in 1701. Today, the degree of declination or variation for a given region is often printed on maps of that region, especially on charts or topographic maps that will be used for route finding by compass. Anyone who learns to use a compass soon masters the task of allowing for variation.

Two basic kinds of compasses are used today: magnetic and gyroscopic. The magnetic compass uses a magnetized needle in a sealed compartment with a rotating base or frame—the compass card—that can be moved so that north is under the arrow. Some compass cards are divided into the points of direction, just like the wind roses of early charts, but others—called azimuth compasses—have dials that are divided into 360 degrees. A bearing from such a compass would be expressed as "245 degrees" rather than "West Southwest." Users of magnetic compasses must allow for variation, for such compasses are oriented to the magnetic north—currently located in the Canadian Arctic, 1,000 miles (1,609 kilometers) from the north pole.

The gyroscopic compass uses a rapidly spinning wheel that remains oriented to true north instead of magnetic north. Large and costly, gyroscopic compasses are used mainly in ships and aircraft.

SEE ALSO

Compass rose; Declination; Navigation

FURTHER READING

Jacobson, Cliff. *The Basic Elements of Map and Compass.* Merrillville, Ind.: ICS, 1988.

Kjellstrom, Bjorn. *Be Expert with Map and Compass: The Orienteering Handbook.* New York: Scribners, 1976.

Randall, Glenn. *The Outward Bound Map and Compass Handbook.* New York: Lyons & Burford, 1989.

Compass rose

COMPASS ROSES are the direction symbols printed on maps. On some maps, particularly old maps or ones designed to have an old-fashioned look, the arrows that indicate the principal directions—North, East, South, and West—are part of an elaborate design with 16 or even 32 segments that look like the petals of a rose. The compass roses of many maps, especially newer ones, feature only four or eight points. Others have a single north-pointing arrow to establish direction.

Compass roses were called wind roses in the age before compasses were used to establish direction. For centuries, mariners and geographers used the winds as symbols of direction. Ancient peoples—especially seafaring peoples, who relied on wind to power their vessels—spoke of direction in terms of winds, referring to the direction from which the winds blew. The directions and winds came to be associated with certain qualities. To the Mediterranean peoples, for example, the north wind symbolized cold and darkness because it blew from the cold northern lands where the days grow short during the winter, while the south wind, which crosses the sea from the burning sands of the Sahara Desert, meant warmth and pleasure.

By the time of Heraclitus, a Greek philosopher of the 5th century B.C., the sky was divided into four quarters: east, west, north, and south. Similarly, the Bible contains many references to "the four quarters of heaven" (Jeremiah 49:36) and "the four corners of the earth, holding the four winds of the earth" (Revelation 7:1). In the 4th century B.C. the Greek philosopher Aristotle, who laid the foundations of scientific knowledge in many disciplines, divided the four main winds into three parts each, creating a set of twelve directions.

These directions, or winds, had many different names in various times and places. One system that was widely followed for a long time was that of Eratosthenes, a third-century-B.C. Greek philosopher who condensed Aristotle's system into eight winds. These were represented by names and eventually were symbolized by mythological figures: the east wind was called Apeliotes, the southeast Eurus, the south Notus, the southwest Libs, the west Zephyrus, the northwest Argestes, the north Aparctias, and the northeast Boreas. Sometimes only four winds were identified, sometimes twelve. A remnant of this ancient system lived on in the figures of wind gods, cheeks puffed to blow forth their breezes, that adorn the margins of many maps made in Europe in the 15th and 16th centuries. The link with ancient mythology is strong: Boreas is often portrayed with ice or snow on his hair, blowing a frosty gale, while balmy Notus is curly-haired and smiling.

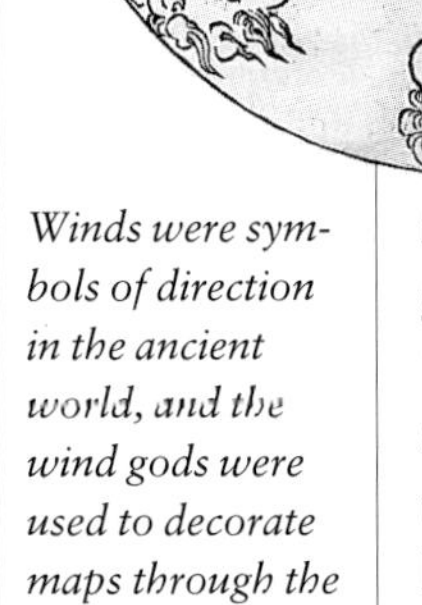

Winds were symbols of direction in the ancient world, and the wind gods were used to decorate maps through the 16th century.

When magnetic compasses came into wide use in the 13th and 14th centuries, the makers of sea charts had long been using wind roses to show sailing directions on their maps. Mariners began attaching their magnetic compass needles to cards on which these wind roses had been drawn, and compass roses were created.

SEE ALSO

Compass

Computers in mapmaking

MAPMAKING HAS been revolutionized many times over the centuries as the technology used to gather and display cartographic data has changed. Movable type, the marine chronometer, photogrammetry, flight—each new technology has changed the way maps are made. One instrument, however, has had a greater effect on mapmaking than any other. That instrument is the computer. Among the other changes it has brought to the field of cartography, the computer is blurring the age-old distinction between mapmakers and map users.

A computer is nothing more than a machine that weighs, checks, sifts, and blends facts and then displays them in whatever way it is told to do. Computers can perform more of these functions, and faster, than any cartographer ever born. Mapmakers could always copy existing maps fairly quickly, but maps based on new, more accurate land surveys took a long time to prepare. The Cassinis' map of France took four generations; the Great Trigonometrical Survey of India took nearly a century. Even using aerial photography, it can take several years to prepare and print a new map from new data. But once the database is established, a computer can create a new map in days or even hours.

The first published computer-generated map appeared in 1950. It was a black-and-white weather map, simple by today's standards, but it was the forerunner of the revolution. Computers are now part of every step in the mapmaking process. In surveying, computers are used to control electronic laser distance-measuring devices and to record data. Computers operate remote-sensing

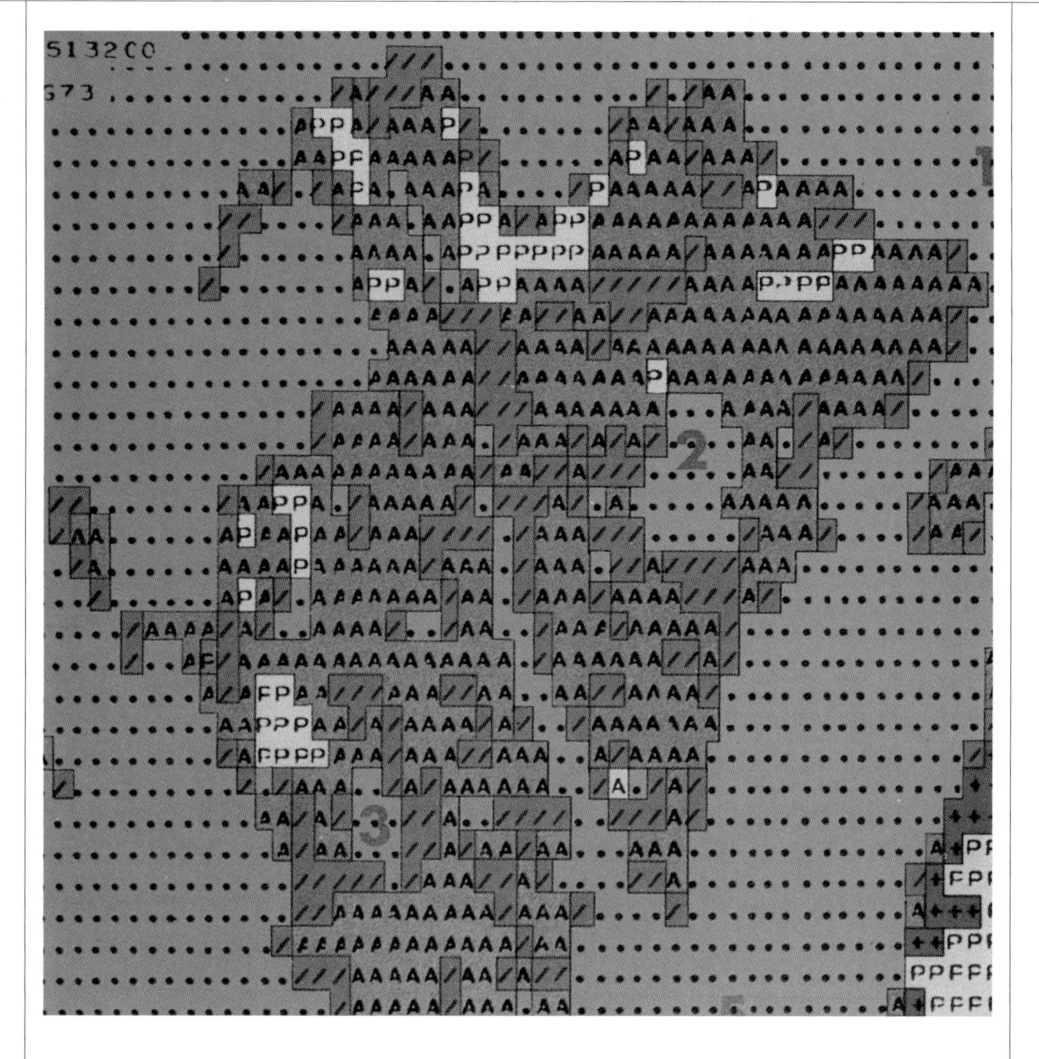

A computer-processed map of wetlands near Chincoteague Bay, Virginia. Vivid colors are used to show various types of terrain.

devices in orbiting satellites and process the data these devices yield into digital images that can be projected onto a screen or printed on a map. Computers are vital to the electronic navigational systems used by the pilots of aircraft and ships; they also guide "smart bombs" and other high-tech weapons, using map data developed by the U.S. Defense Mapping Agency. Computerized navigation systems are already available for automobiles, although they have certain limitations. In the future, cars will probably come with built-in computerized road maps that provide navigational guidance to the driver—or perhaps directly to the car's engine and wheels.

Computers are also changing the way maps are physically produced. Cartographers today are likely to spend as much time in front of a computer terminal as they spend at a drafting table. The U.S. Geological Survey (USGS) is just one of many agencies around the world that is converting its existing storehouse of paper maps into digital data—information in the computer's language, streams of numbers. The USGS hopes to finish the National Digital Cartographic Data Base by the year 2000. When the system is complete, cartographers at computer terminals will be able to arrange the most complete available data about population, resources, and a host of other factors onto topographically correct base maps and then print out those maps. Each map can be custom-made to suit a particular set of needs.

The way we look at maps is also changing, thanks to computers. By now, most people have seen picturelike "false color" images, taken from space using infrared cameras or other devices and then enhanced by the computer for ease of viewing. The computer operator can assign arbitrary colors to various data values to make different features stand out, so that a map might show vegetation in vivid red, cities and towns in pale pink, and water in yellow.

The computer has also brought about a three-way marriage of cartography, photography, and painting. Computer-generated landscape portraits combine topographic survey data with information about land cover (snow, trees, desert, and the like) from remote-sensing devices in space. Then, sitting at a terminal, the cartographer can choose a point of view, tilt the angle at which the landscape is viewed, distort any aspect of the topography to produce a special effect, and adjust the color values of the land cover so that they look attractive, resulting in a high-tech version of a landscape painting.

For centuries, the distinction between those who made the maps and those who bought or used them was quite clear. Today, however, anyone with a desktop computer can become a cartographer and produce maps whose quality and accuracy Mercator could not have imagined. Some computer applications of cartography are passive information sources, such as on-line database services, which a home user can tap into with a computer and a modem. Others are interactive, such as geographic video games and software programs that allow users to design maps. To create a map, the user makes choices from a menu of base maps and features. Once the map is assembled on-screen, the computer cartographer can print it out and put it in a report or tack it to the wall. Hundreds of such programs are available for educational and business use or simply for the computer owner who likes to play with maps. The most sophisticated and costly software programs are called geographic information systems (GIS). Designed for map publishers, land-use planners, and professional cartographers, these programs contain very large databases of topographic, political, land-cover, and census information that can be combined in an almost infinite number of ways.

The growing power of personal computers, combined with a surge of interest in geography, led to a proliferation of map-related software in the 1980s and 1990s. The USGS publishes a guide called *Sources for Software for Computer Mapping and Related Disciplines* that is available from the USGS Earth Sciences Information Center. The *Map Catalog* contains a briefer but more user-friendly review of cartographic and geographic software.

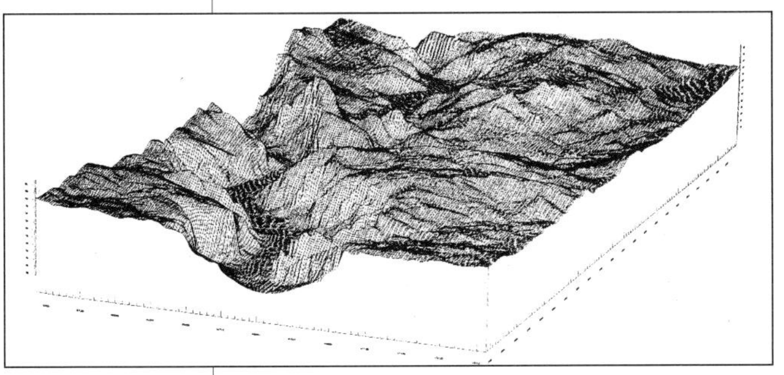

Computers allow mapmakers to create three-dimensional landscapes that can be turned, tilted, and viewed from every angle.

SEE ALSO

Satellites

FURTHER READING

Makower, Joel, ed. *The Map Catalog*. 3rd ed. New York: Vintage, 1992.

Monmonier, Mark. *Computer-Assisted Cartography: Principles and Prospects*. Englewood Cliffs, N.J.: Prentice-Hall, 1982.

———. *Technological Transition in Cartography*. Madison: University of Wisconsin Press, 1985.

Taylor, D. Fraser, ed. *The Computer in Contemporary Cartography*. Ann Arbor, Mich.: Books on Demand, n.d.

Conformal projection

SEE Projections

Conic projection

SEE Projections

Contarini, Giovanni Matteo

ITALIAN MAPMAKER

• *Active: late 15th century*

ALMOST NOTHING is known about Giovanni Matteo Contarini—not even whether he lived in Venice or Florence. Only one of his maps has survived; it was discovered in 1922 and is now kept in the British Library. It is a cone- or fan-shaped map of the world, printed from a copper engraving in 1506. Contarini's map occupies a special place in cartographic history, for it is the first known printed map to show the Americas.

Contarini's map shows Brazil—which had been known to Europeans for only a few years at the time the map was made—merged into a large landmass that reaches to the south pole. North America exists only as a vague outline of Labrador, grafted onto the eastern extremity of Asia. The ocean between the Americas and Asia is too small to be an accurate representation of the Pacific. The Caribbean islands explored by Columbus, however, are fairly well drawn, and Africa and the Indian Ocean are well

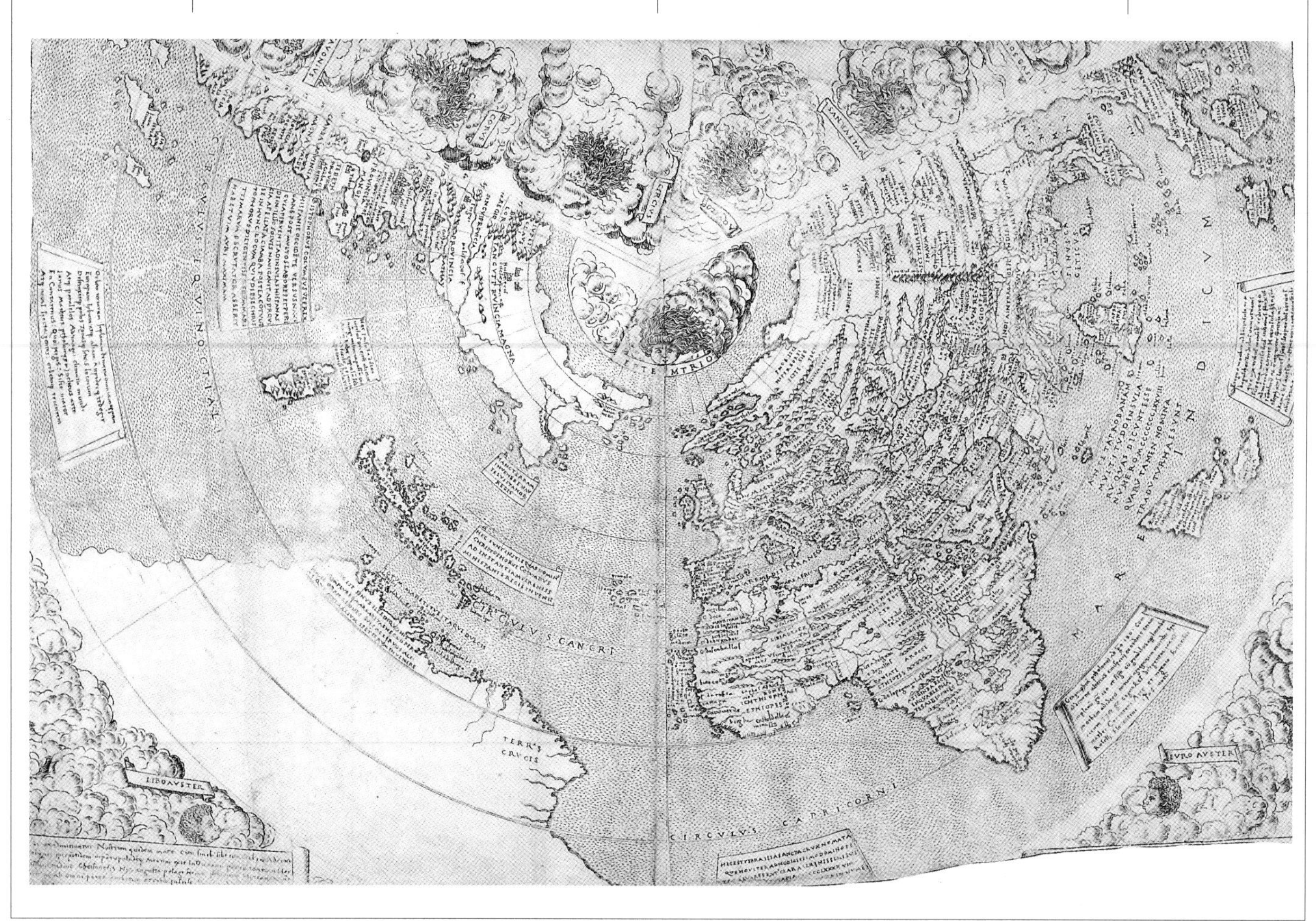

Contarini's fan-shaped world map of 1506 is the first known printed map to show the Americas.

outlined. In a note on the map, Contarini describes himself as a famous mapmaker, but no other maps by him are known today.

Continent

CONTINENTS ARE the large landmasses that make up the earth's land surface. There are seven continents: Africa, Antarctica, Asia, Australia, Europe, North America, and South America. The science of plate tectonics is the study of how these continents, which rest on large plates of the earth's crust, move slowly across the planet's surface. Geologists have determined the positions of the continents at various points in earth's history; for example, during some eras millions of years ago, the continents we know today were joined together into large supercontinents. Maps and globes of the earth as it appeared during these periods are a startling reminder that not even continents are permanent.

Contour lines

CONTOUR LINES are used by mapmakers to show relief on topographic maps. A contour line is a line created by mapmakers to connect points that have the same altitude—that is, points that are the same distance above (or below) sea level. A map with few contour lines, or contour lines that are very far apart, is a map of fairly flat terrain. Maps of hilly, cliffy, or mountainous terrain, on the other hand, show many contour lines, closely spaced. The closer together the lines are printed, the more steeply the ground rises. Small figures are printed on every fifth or tenth contour line to give its altitude.

SEE ALSO
Relief; Topographic maps

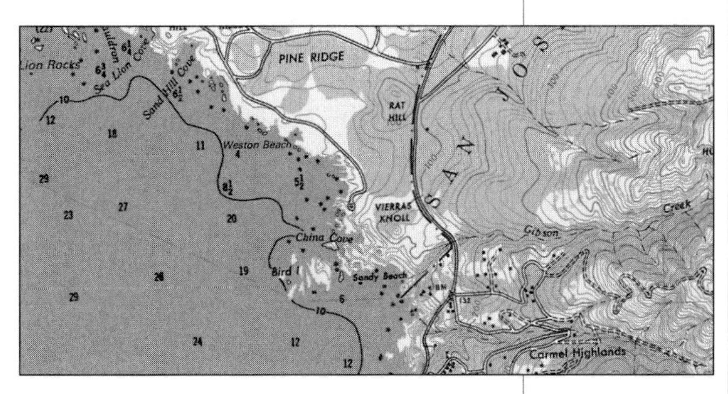

A stretch of California coastline, topographically mapped. Dense clusters of contour lines represent steep slopes and cliffs. With practice, a map user learns to "read" the terrain from contour lines.

Cook, Frederick

AMERICAN POLAR EXPLORER

- *Born: June 10, 1865, Hortonville, New York*
- *Died: Aug. 5, 1940, New Rochelle, New York*

FREDERICK COOK was a physician who turned his attention to exploration after serving on a Belgian expedition to Antarctica in 1897–99. Roald Amundsen, a Norwegian explorer who would later be the first person to reach the south pole, was also part of this expedition, and he and Cook became friends.

Cook later made several journeys into the Canadian Arctic. The most notorious of these journeys took place in 1907–8, when he spent 14 months traveling in the far northern Arctic with only two Native American companions. This journey added many details to the map of Ellesmere Island, Devon Island, and Axel Heiberg Island, a region of the far north that was almost unknown to the outside world. The trip, a remarkable feat of endurance, would have been hailed as heroic even without the claim

Frederick Cook, wearing a necklace of flowers, is hailed as a hero for conquering the north pole. Soon, though, he would be disgraced as a fraud.

that Cook added to it. He said that in 1908 he and his companions had ventured out onto the polar ice and reached the north pole—the goal of every Arctic explorer of Cook's time. Cook proudly declared triumph over "hostile, death-dealing Nature."

Cook had no proof to support his claim, which was vigorously challenged by Robert Peary, another American explorer who claimed to have been the first to reach the pole. And Cook's case grew weaker when it was proved that his earlier claim to have been the first person to climb Mount McKinley (also called Denali) in Alaska was a lie. Although some supporters continued to believe Cook's polar claim, most people rejected it within a year or two, and Peary was hailed as the discoverer of the pole. Cook later served a jail term for fraud in connection with some oil wells in Texas. He died a discredited and all-but-forgotten man.

SEE ALSO

Arctic, mapping of; Peary, Robert Edwin

FURTHER READING

Berton, Pierre. *The Arctic Grail: The Quest for the North West Passage and the North Pole, 1818–1909.* New York: Penguin, 1988.

Cook, James

ENGLISH EXPLORER OF THE PACIFIC

- *Born: Oct. 27, 1728, Marton-in-Cleveland, England*
- *Died: Feb. 14, 1779, Hawaii*

JAMES COOK, the son of a farm worker, was apprenticed as a boy to a shipowner. He studied navigation and mathematics, eventually joining Great Britain's Royal Navy. In 1757 he was made navigator of the *Pembroke,* which served in the Seven Years' War between France and England in Canada (1756–63). Cook was an excellent mapmaker and chart maker. His charts of the treacherous St. Lawrence River enabled the British forces to land safely at the French citadel of Montreal, thus contributing to the British victory over France.

Cook was the first European to map Hawaii, New Zealand, and many other Pacific islands.

Cook's next assignment was to chart the shoreline of Newfoundland, a large island in eastern Canada. This was a difficult task because Newfoundland's ragged coastline features thousands of large and small bays, headlands, and outlying rocky reefs and islets. Cook spent five years (1763–67) meticulously doing his job, producing maps so detailed and precise that they were used into the 20th century.

The British Admiralty selected Cook to lead an expedition into the Pacific Ocean in 1768. He had two missions: first, to carry a team of scientists to Tahiti to make astronomical observations that were supposed to determine the distance from the earth to the sun, and second, to probe the southern ocean for a sight of the Terra Australis, a large, unknown continent that geographers expected to find in the Southern Hemisphere.

Failing to find the Terra Australis, Cook sailed west and spent six months charting New Zealand, proving that it was a pair of islands rather than part of a continent. He then made the first recorded European landfall on Australia's east coast—in fact, he was shipwrecked there and barely managed to salvage his vessel *Endeavour* for the return trip.

Although Dutch explorers had begun to fill in the outline of western Australia, Cook reported that eastern Australia "was never seen or visited by any European before us." He claimed it for Britain and sailed home, bearing news of such marvels as the kangaroo. But the question of the Terra Australis remained unanswered, and in 1772–75 Cook made a second voyage, specifically to scour the southern seas for any sign of the hypothetical continent. He took with him the chronometer, or timekeeper, newly invented by John Harrison for the measurement of longitude at sea. On this voyage, Cook visited and determined the locations of many Pacific islands: Easter Island, Tonga, New Caledonia, and others. He also probed far south into polar waters. He was the first navigator to cross the Antarctic Circle, going farther south than anyone had yet ventured. He found no sign of the Terra Australis but speculated that land might lie beyond the wall of icebergs that blocked his progress.

The reports and maps that Cook made after his two voyages painted the clearest picture yet of the South Pacific Ocean, but geographers were also curious about the North Pacific. They still believed that a long-sought waterway called the Northwest Passage crossed North America from the Atlantic Ocean to the Pacific. Cook was sent out on a third voyage, this time to look for the western end of the passage in the North Pacific Ocean. He explored and mapped the northwest coast of North America from present-day Oregon to the Bering Strait, but he found no Northwest Passage. He did, however, make the first recorded European landfall in Hawaii, which he named the Sandwich Islands after the Earl of Sandwich. He was killed there in 1779 in a skirmish with the Hawaiians.

Cook was not only one of the greatest navigators history has known but also a figure of towering importance to geography. The Pacific Ocean is the largest geographic feature in the world, covering more area than all the earth's landmasses combined and stretching more than a third of the way around the planet. Before Cook, the geography of the Pacific was a jumble of guesswork, legend, and unreliable reports. Cook's three voyages clarified the main outline of the Pacific; later explorers had only to fill in the details. In his own day, Cook was regarded even more highly abroad than in his native England. His achievements placed him in the first rank of world explorers, and for many years cartographers of all nations traced his routes on the maps they made. His own maps and reports were widely reprinted. He was also admired for his humane treatment of his crews at a time when

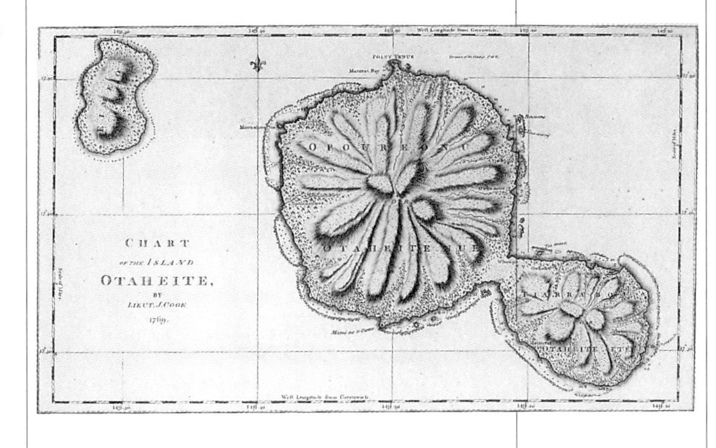

Cook's 1769 map of Tahiti.

naval discipline was often savage and shipboard conditions appalling.

SEE ALSO

Australia, mapping of; Harrison, John; Longitude; Northwest Passage; Pacific Ocean, mapping of; Terra Australis

FURTHER READING

Allen, Oliver E., and Time-Life editors. *The Pacific Navigators.* Alexandria, Va.: Time-Life Books, 1980.
Beaglehole, J. C. *The Life of Captain James Cook.* Stanford, Calif.: Stanford University Press, 1974.
Blackwood, Alan. *Captain Cook.* New York: Franklin Watts, 1987.
Cameron, Ian. *Lost Paradise: The Exploration of the Pacific.* Topsfield, Mass.: Salem House, 1987.
Dunmore, John. *Who's Who in Pacific Navigation.* Honolulu: University of Hawaii Press, 1991.
Haney, David. *Captain James Cook and the Explorers of the Pacific.* New York: Chelsea House, 1992.
Hoobler, Dorothy, and Thomas Hoobler. *The Voyages of Captain Cook.* New York: Putnam, 1983.
Smith, Bernard. *Imagining the Pacific.* New Haven, Conn.: Yale University Press, 1992.
Stefoff, Rebecca. *Scientific Explorers: Travels in Search of Knowledge.* New York: Oxford University Press, 1992.
Withey, Lynne. *Voyages of Discovery: Captain Cook and the Exploration of the Pacific.* New York: Morrow, 1987.

Cooley, William Desborough

BRITISH GEOGRAPHER

- *Born: 1795*
- *Died: 1883*

WILLIAM DESBOROUGH COOLEY'S career illustrates the dangers of rigid thinking. Cooley earned a high reputation as a historian of exploration. In 1846 he was one of the founders of the Hakluyt Society, named for Richard Hakluyt, a 16th-century publisher of travel narratives. Since its founding, the society has dedicated itself to publishing important works of travel, exploration, and geography from earlier periods.

At the peak of his career, in the 1830s and 1840s, Cooley was considered a leading authority on Africa. But he was a theoretical geographer whose vision of the world was shaped by abstract notions that did not always match reality. He made maps of Africa but failed to update them when new discoveries contradicted his theories. His downfall came when some European missionaries in Africa reported seeing snowcapped peaks near the equator, in the center of the world's hottest zone. Cooley firmly denied that such a thing was possible, but further exploration soon revealed that Mount Kenya and Mount Kilimanjaro, both located close to the equator, do indeed have snowy tops. (Their peaks remain cold because of their great height.) Cooley was ridiculed as an "armchair geographer," one who sits in his study and spins notions but is out of touch with reality. He died in poverty and obscurity.

Coordinates

COORDINATES ARE numbers or letters that locate a point in two dimensions. Map coordinates allow you to pinpoint the location of any place in the world.

Look up a city—your hometown, maybe, or somewhere you would like to visit—in the index of a map or atlas. You will find a set of coordinates that will allow you to find that place on the map. If you are using a road map that has its own grid, perhaps with letters on one side of the map and numbers on another, you may see coordinates like these: G-5. Find G and 5, then find the point on the

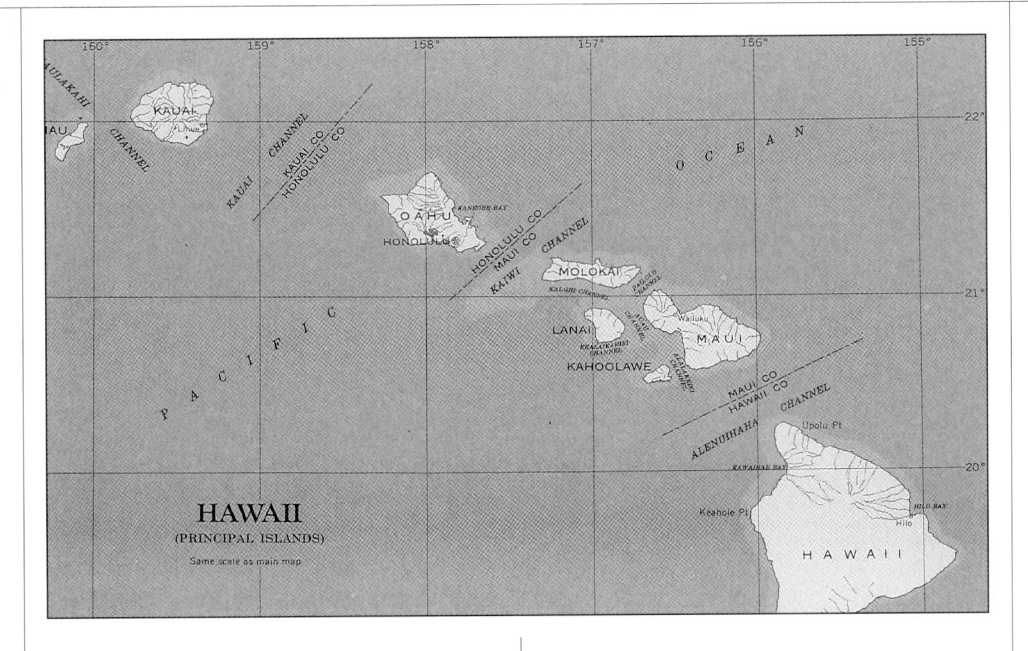

Honolulu, Hawaii's capital, is at coordinates 21° 19' N, 157° 52' W.

map where they intersect, and the city you are looking for should be nearby.

If you are using an atlas, the index will give you one number that tells you which map to use and another set of numbers that are coordinates of latitude and longitude, like these: 45.33 N 122.36 W (or perhaps 45° 33' N, 122° 36' W). The latitude is always given first, then the longitude. To use these coordinates, find a point on the margin of the map that is midway between the 45th and 46th degrees of latitude north, and find another point on a different margin that is midway between the 122nd and 123rd degrees of longitude west. Using a ruler or other straight edge as a guide, move inward from these points toward the center of the map along a horizontal and a vertical line. Your city will be at or near the place where the lines meet.

SEE ALSO
Latitude; Longitude

Cosmas Indicopleustes

SEE Ancient and medieval mapmakers

Cosmography

COSMOGRAPHY LITERALLY means the study of the universe. It usually refers to the study of the earth and of earth's place in the natural order. The beliefs of whole peoples or of individual philosophers and scientists are often called cosmographies. For example, a historian might speak of the cosmography of the ancient Babylonians—in which the earth is pictured as a giant, hollow mountain, surrounded by an ocean—or the cosmography of Claudius Ptolemy, who recognized that the earth is a sphere but believed that it was the center of the universe. Many early maps illustrate cosmographic theories. During the Middle Ages, for example, European thinking was dominated by Christianity; as a result, Jerusalem appeared at the center of most European maps because, as the birthplace of Christianity, it was thought to be the center of the world. Geographers sometimes used the term *cosmography* for their work. The *Universal Cosmography* (1544) of German scholar Sebastian Münster was a description of the known world, with maps.

Cresques, Abraham

CATALAN MAPMAKER

• *Active: 14th century, Spain*

ABRAHAM CRESQUES was a native of Palma, on the Mediterranean island of Majorca. As part of the Spanish province called Catalonia, Majorca contributed greatly to the Catalan school of navigation and chart making. The Catalan chart makers drew upon practical sailing experience and also upon the store of knowledge possessed by Islamic geographers in North Africa and the Middle East. Cresques, who was Jewish, may also have been familiar with traditional Jewish lore about Asia and North Africa. He is believed to have been the creator of the Catalan atlas, a magnificently detailed map of the known world that was produced in 1375.

SEE ALSO
Catalan atlas

Cross-staff

THE EARLIEST written reference to the cross-staff occurs in a manuscript written in 1342 by Levi ben Gerson, a Spanish scholar who drew on traditional Catalan and Jewish scholarship. Long before that time, however, astronomers had been employing this useful and simple instrument to measure the altitudes of stars above the horizon. The cross-staff could also be used for measuring the sun's altitude, from which a navigator could determine his latitude. In the 16th century, the cross-staff replaced the astrolabe as the preferred instrument for taking latitude sightings at sea.

In its simplest form, the cross-staff was a rod (the staff) with an eyepiece at one end and a smaller rod (the cross) made to slide along it. The cross was perpendicular to the staff and had a peephole in each end. To determine the angle of the sun above the horizon, the navigator would hold the staff at eye level and look through the eyepiece, sliding the cross back and forth until he could see the two peepholes at the same time, the upper one lined up with the sun and the lower one lined up with the horizon. He could then read off the angle from the position of the cross on the staff; the closer the staff was to the eyepiece, the greater the angle and the higher the altitude of the sun.

A navigator who used the cross-staff correctly could determine latitude to within about one degree of accuracy.

The cross-staff was a little easier than the astrolabe to use on the pitching deck of a ship. Its main drawback, however, was that the user had to squint directly into the sun's glare, which not only made it difficult to obtain accurate readings but could also damage the user's eyesight. The English explorer John Davis solved that problem in 1595 with the introduction of an improved cross-staff that he called the backstaff. To use the backstaff, the navigator stood with his back to the sun, lined up the staff with the horizon, and then moved the cross until it was lined up with the sun and cast a shadow onto a sighting plate at the far end of the staff. Later versions of the backstaff used lenses to capture the sun's image and project it onto the plate. Within a few years the backstaff had been refined into what was called the Davis quadrant or the English quadrant, with two movable sights on braces: one for the user to look through and one to cast the sun's image onto the sighting plate. This instrument remained in use into the 18th century, when it was replaced by the sextant.

Cusanus, Nicolaus

GERMAN SCHOLAR

- *Born: 1401, Cues, Germany*
- *Died: 1464, Todi, Italy*

Nicolaus Cusanus was born Nicolaus Khrypffs but was called Cusanus in honor of his birthplace, the German town of Cues (now called Bernkastel-Kues). He was a well-traveled and learned man, educated in Germany and Italy and familiar with Rome, Paris, Hungary, and Constantinople. He was knowledgeable in both mathematics and astronomy and, like many other learned people of the time, he made his own maps.

Cusanus was made a cardinal of the Roman Catholic church in 1448; two years later he was appointed bishop of Brixen, Austria (now Bressanone, Italy). He died in 1464, and although he was buried in Rome, as befitted an official of the church, he left his map manuscripts to the town of Cues. One of these maps, published in 1491, shows central Europe in considerable detail. On the map's northern edge are the coasts of Norway and Sweden; the map's southern edge shows part of Italy. Focusing on the lands between, Cusanus's map is considered the first modern map of Germany. Some of his maps of eastern Europe reflected geographic knowledge gained during the Crusades.

FURTHER READING

Campbell, Tony. *The Earliest Printed Maps, 1472–1500*. Berkeley and Los Angeles: University of California Press, 1987.

Cylindrical projection

SEE Projections

Dalrymple, Alexander

SCOTTISH GEOGRAPHER AND HYDROGRAPHER

- *Born: July 24, 1737, Edinburgh, Scotland*
- *Died: June 19, 1808*

ALEXANDER DALRYMPLE went to work for the British East India Company in Madras, India, in 1752. By 1779 he had advanced to the position of hydrographer to the company; he was its expert on bodies of water and its adviser on geographic questions. From 1795 until just before his death, he held a similar post at the British Admiralty.

Alexander Dalrymple was a good geographer, but he is remembered for a colossal error: his belief in a huge southern continent that does not exist.

Dalrymple's biggest contribution to geography and mapmaking came about because he believed in something that did not exist. He argued that there must be a very large continent somewhere in the Southern Hemisphere to balance the weight of Europe and Asia in the Northern Hemisphere—the same argument that Ptolemy had advanced in the 2nd century to support his theory of the Terra Australis, or Southern Land.

So convincing was Dalrymple's case for the existence of the Terra Australis, and so alluring was his picture of 50 million British colonists settling it, that the British government decided to send someone south to make a determined search for the undiscovered country. Thus, Captain James Cook set out on his three Pacific voyages, which disproved forever the theory of the Terra Australis but resulted in the first complete maps of the world's largest ocean.

SEE ALSO

Cook, James; Ptolemy; Terra Australis

Dampier, William

ENGLISH PIRATE AND EXPLORER

- *Born: about 1652, East Coker, England*
- *Died: March 1715, London, England*

WILLIAM DAMPIER began his career as a privateer in the Caribbean Sea and along the western American coast before crossing the Pacific Ocean to the west coast of New Holland, as Australia was then known, in 1688. He is believed to be the first Englishman to have visited Australia. After returning to England in 1691, he wrote *A New Voyage Round the World,* which was published in 1697. A map included in the book shows that Dampier had a good grasp of the geography of Southeast Asia and Indonesia, although he believed that New Holland was part of the unknown southern continent called Terra Australis. The book won Dampier a government commission to lead a naval expedition in 1699; he explored part of Australia's northern coast and added the island of New Britain to the map. He made a third Pacific voyage in 1703, after returning to privateering.

Pirate, explorer, and author William Dampier was the first to map the winds of the Pacific Ocean.

SEE ALSO

Australia, mapping of

FURTHER READING

Botting, Douglas, and Time-Life editors. *The Pirates.* Alexandria, Va.: Time-Life Books, 1978.

Lloyd, Christopher. *William Dampier.* Seattle: Shoestring Publications, 1966.

Davis, John

ENGLISH NAVIGATOR

- *Born: about 1550, Sandridge, England*
- *Died: Dec. 1605, at sea near Sumatra*

JOHN DAVIS was one of the many navigators who looked for the Northwest Passage, the navigable waterway that was believed to cut across North America, linking the Atlantic and Pacific oceans. In three expeditions to the northwestern Atlantic, beginning in 1540, he explored the coast of Greenland and the nearby waters, but he found no passage west. Davis Strait, between Greenland and Baffin Island, was named for him. In 1588 Davis was in English waters, battling the Spanish Armada. In 1591 he entered the Pacific, hoping to find the western end of the Northwest Passage. He made three voyages to the islands of present-day Indonesia, where he was killed by Japanese pirates.

Davis's legacy to geography includes the invention of the backstaff, a simplified version of the cross-staff, a sighting instrument that had been used by navigators since ancient times. He also wrote two books: *The Seaman's Secrets* (1584), about navigation, and *The World's Hydrographical Description* (1595), about the Northwest Passage.

SEE ALSO

Cross-staff; Northwest Passage

FURTHER READING

Brown, Warren. *The Search for the Northwest Passage.* New York: Chelsea House, 1991.

Lehane, Brendan, and Time-Life editors. *The Northwest Passage.* Alexandria, Va.: Time-Life Books, 1981.

Scott, J. M. *Icebound: Journey to the Northwest Sea.* London: Gordon & Cremonesi, 1977.

Dead reckoning

NAVIGATING BY estimate and judgment rather than by precise measurement is called dead reckoning. The term *dead reckoning* is thought to be related to the nautical one *dead wind,* which refers to a wind blowing in the exact opposite direction to a ship's course; *dead* thus came to mean "direct" or "straight," and dead reckoning is the practice of keeping track of a ship's position by tracing each segment of the course as a straight line on a sailing chart. The name is grimly appropriate in another way, however, for many mariners have died in shipwrecks or other disasters because they lacked accurate knowledge of their positions. Yet dead reckoning can be surprisingly accurate when performed by a skilled pilot. Pacific Islanders in Polynesia and Arab traders in the Indian Ocean, among others, routinely made long voyages over turbulent seas with pinpoint accuracy, using a highly refined sense of direction and a thorough knowledge of wind and wave patterns.

The hourglass allowed mariners to measure time, one of the three factors needed for dead reckoning; the other two factors were speed and direction.

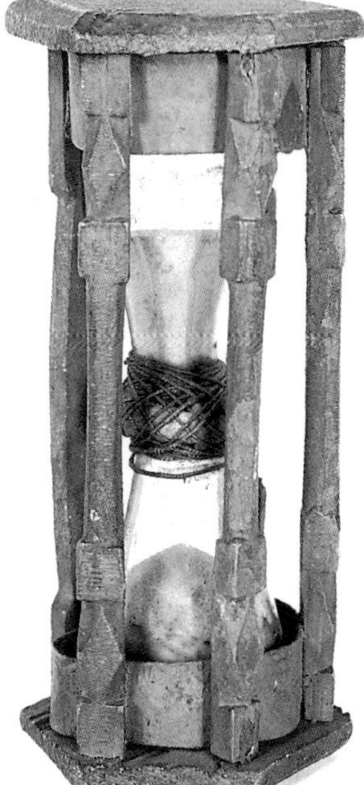

Since ancient times, mariners have been able to judge latitude fairly accurately, using astrolabes and cross-staffs, which have been around for thousands of years. But before modern navigational instruments were invented, longitude was much more difficult to measure. Pilots made do with a simple system of estimating longitude, using the compass to tell direction, the hourglass to mark the passage of time, and the log line—a rope with a piece of wood tied to it—to measure the ship's speed. (By throwing the log overboard and then seeing how long it took for the log to disappear, a seaman could guess the speed of travel; log-line observations were written down, and eventually all the written notes of a voyage came to be called the "log"). A pilot who knew the direction, duration, and speed of each day's travel could make a reasonably good estimate of the ship's location at the end of the day. The next day he would do the same thing again, using the previous day's position as the starting point for the new calculation, and so on.

Dead reckoning offered many opportunities for error. Something as simple as a sailor bored with night watch and eager to catch up on his sleep could throw off the daily estimate. The tired seaman might secretly hold the hourglass under his shirt to warm it. This would make the glass expand, so the sand would flow through faster and his turn on duty would end sooner. That day's estimate of travel time—and therefore the estimate of longitude—would be a little bit off. But each day the error would grow larger, because each day's estimated position was based on that of the day before.

Mariners using dead reckoning did not expect to reach the desired port every time. Often they were glad simply to sight a familiar shore landmark or coastline, along which they could sail toward their destination.

Declination

THE TERM *declination* has two meanings related to maps and navigation. First, celestial declination refers to the angle of the sun, the moon, or a star

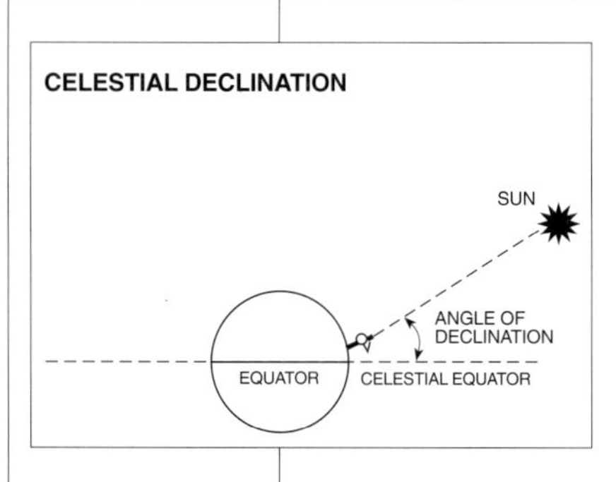

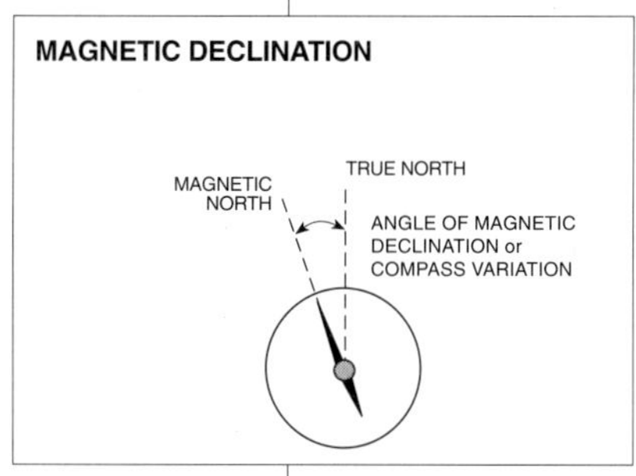

above the equator; for example, someone taking a sighting of the sun to use in calculating latitude might say that the sun had a declination of 68 degrees. Declination also refers to a second angle, this one connected with magnetic compasses. Compass declination is the angle between magnetic north and true north at a given point. It is often called compass variation to distinguish it from celestial declination.

SEE ALSO
Compass

Decoration

MAPS OFTEN contain elements that are not cartographic at all—they are there to make the map look prettier or more interesting. But some features that at first glance seem purely decorative by modern standards also convey information that supplements the map itself.

Many medieval maps are studded with pictures. The illustrations on the Catalan atlas, for example, are quite beautiful, clearly intended to make the map a fit gift for a king. Yet some map decorations were also meant to communicate information; they were a visual shorthand in an era when many people could not read. A drawing of the pope meant Rome; a drawing of a biblical scene marked the Holy Land; and a drawing of fierce animals or of the strange beings described by Solinus meant "unknown and probably dangerous lands."

It is impossible to say whether people interpreted these drawings as literal truth or as symbols. When 16th-century Swedish geographer Olaus Magnus ornamented one of his woodcut maps with a picture of a huge sea serpent twining around a ship and crushing it, was he saying that such things had really happened in the North Atlantic Ocean? Or merely that the seas were full of perils unknown to land dwellers? Decorative illustrations straddled the borderline between fact and fancy, and many of them crossed it into out-and-out fiction, as when mapmakers took the liberty of filling the unknown interior of Africa with rivers, lakes, and mountains.

Decorative mapmaking reached its peak in the 17th century. A plain, functional map would have seemed ridiculously barren to cartographers who filled their maps with elaborate ornamentation. Several decorative elements were especially important. The cartouche, or titlepiece, was often very elaborate, with flourishes, scrolls, figures from classical mythology, and pictures of plants and animals in addition to information about the map itself. Borders became ornamental, designed to look like ribbons,

The seas of Olaus Magnus's maps teemed with lively sea creatures. Such images were decorative, but they were also symbols of mystery and peril.

wreaths of flowers, picture frames, or columns. The sea continued to be a fertile area for decoration: spouting whales, flying fishes, ships in full sail, and mermaids dotted its surface, and some mapmakers filled in the sea with row after row of tiny waves. As geographic knowledge expanded and maps grew more crowded with place names, the small scenes called vignettes moved off the map surface itself and into insets or panels along the edges of the map.

The 17th-century Dutch cartographers elevated map decoration to a new high with a special kind of map called the *carte à figures* (French for "map with figures"—at that time, French was commonly used for trade and communication among people of various European nationalities). The *cartes à figures* were bordered with vignettes. Square or rectangular panels on each side contained pictures of the region's typical inhabitants in their national costumes (Native Americans were especially popular). Oval panels along the top of the map, and sometimes the bottom as well, contained pictures of cities or notable landscape views from the region. The illustrations on the *cartes à figures,* like other map vignettes, combined information with entertainment. This style of mapmaking was copied by cartographers in other countries.

In the 18th century, French cartographers introduced a plainer, more scientific style of mapmaking, which gradually replaced the ornate cartography of earlier times. But cartouches remained elaborate on many maps until well into the 19th century, and as late as 1851 the British mapmaker John Tallis published an atlas of maps illustrated in the old decorative style. This was the last ornamental atlas. Most modern maps, unless they are designed as novelties or in imitation of earlier mapmaking styles, are straightforward and unadorned, with the emphasis on content and not on decoration. Yet many people find elegance and beauty in a simple, well-executed modern map, even if it lacks sea monsters and beribboned borders.

SEE ALSO

Cartouche; Solinus, Gaius Julius

FURTHER READING

Robinson, Arthur H. *The Look of Maps: An Examination of Cartographic Design.* Madison: University of Wisconsin Press, 1986.

Skelton, R. A. *Decorative Printed Maps of the 15th to 18th Centuries.* London: Spring Books, 1965.

Dias, Bartolomeu

PORTUGUESE NAVIGATOR

- *Born: mid-15th century, Portugal*
- *Died: May 29, 1500, at sea off southern Africa*

BARTOLOMEU DIAS was part of the century-long Portuguese effort to find a sea route around Africa to India that began in the early 14th century under Prince Henry. Few details of Dias's life are known, but by 1482 he was in command of a Portuguese vessel on a voyage to West Africa. In 1487, Dias was sent south on a special mission by King João II of Portugal, who was convinced that the passage around the southern tip of Africa was not only possible but within his grasp. The king gave Dias three ships and instructed him to find the passage.

Dias sailed south, battling currents and winds, and then—either by design or by accident—he ordered his ships to stand away from the African continent, in a wide arc out into the South Atlantic Ocean. There they picked up winds blowing east that carried them neatly around the southern tip of Africa. Dias sailed north for a little distance, enough to be certain that the coastline continued unbroken in that direction, and then he returned to Portugal in 1488 with the glad news that he had rounded Africa. He had added 1,260 miles (2,027 kilometers) of new coastline to the map. Later, Dias sailed with the Portuguese fleet of Pedro Álvars Cabral, bound for India by way of the Cape of Good Hope. Dias was lost at sea in a storm.

Dias's 1487–88 voyage was the first verified European voyage from the Atlantic to the Indian Ocean. (Although Herodotus and other ancient writers told of similar voyages by Egyptian or Phoenician mariners, details are vague, and these journeys may not really have taken place.) By completing his mission and returning to tell about it, Dias disproved the theory advanced by the geographer Ptolemy centuries earlier that the Indian Ocean was really a huge landlocked sea. Dias named the southern tip of Africa Cabo Tormentoso, the Cape of Storms, because of the wild weather he encountered there; King João changed its name to the Cape of Good Hope to honor the prospects of trade with India. A decade later Vasco da Gama pioneered the route to India, but it was Dias who paved the way by rounding the cape.

Dias's discovery of the sea passage around Africa from the Atlantic to the Indian Ocean was soon reflected in new maps made by the cartographers of Europe. The discovery also appears on the world's oldest surviving globe, made in 1492 by Martin Behaim.

SEE ALSO

Behaim, Martin; Gama, Vasco da; Henry, Prince of Portugal; Ptolemy

FURTHER READING

Divine, David. *The Opening of the World: The Great Age of Maritime Exploration.* New York: Putnam, 1973.

Stefoff, Rebecca. *Vasco da Gama and the Portuguese Explorers.* New York: Chelsea House, 1993.

Direction

NO ONE knows where or when the concept of "direction" first came into use, but it is likely that all ancient peoples developed some concept of direction, a way to describe how various places were located relative to one another, or to guide travelers. The earliest uses of directions were probably as simple as "My village lies in the direction of the setting sun" or "Walk for three days toward the rising sun."

The most basic set of directions grew out of people's familiarity with the daily cycle of the sun, which rises in the east and sets in the west. Two other directions, north and south, probably came from the sun's annual cycle. In the Northern Hemisphere, the sun's path moves from the northern to the southern part of the sky as summer gives way to winter; this pattern is reversed in the Southern Hemisphere. Today, the four principal directions—often called the cardinal points of the compass—are still north, south, east, and west.

Nearly all modern maps are positioned so that north is at the top of the map. This arrangement, called by cartographers the orientation of the map, is taken for granted by map users, who automatically expect each new map they see to be oriented to the north. Even the words we use when speaking about maps and directions reflect how accustomed we are to picturing the world with north at the top of the picture. For example, someone in Sydney, Australia, might speak of going "up the coast" to Brisbane, rather than "north," just as a New Yorker might talk of driving "down" to Florida, instead of "south."

But maps have not always been oriented to the north. Claudius Ptolemy, a Greek scholar in the 2nd century A.D., placed north at the top of his maps, as did most mapmakers in ancient China, but the ancient Romans generally oriented their maps to the east—that is, east was at the top of the maps. Many Islamic maps had south at the top. Japanese maps varied; some were oriented to the south, somc to thc cast. Native American maps, too, were oriented in different ways. The direction of travel was generally placed at the top of the map, so that a given map could be oriented northwest, southeast, or any direction. Nor were medieval European maps consistent. Many were oriented to the north, but there were exceptions—for example, Fra Mauro's world map of 1459, one of the last great expressions of the medieval mapmaking tradition, has south at the top. The northern orientation began to dominate after European mariners started using magnetic compasses—which point northward—in the 13th and 14th centuries. The widespread popularity of Ptolemy's geographic theories in Europe after the 15th century also reinforced the northern orientation, for Ptolemy's maps had north at the top. By the end of that century, the printing press was being used to produce many copies of Ptolemy's *Geography,* and the northern orientation of maps was becoming standard.

SEE ALSO
Compass rose; Mauro, Fra; Ptolemy

Drake, Francis

ENGLISH NAVIGATOR AND PRIVATEER

- *Born: about 1543, Tavistock, England*
- *Died: Jan. 27, 1596, at sea in the Caribbean*

A LIFELONG seaman, Francis Drake received his first command in 1567, when he was made captain of a slave-trading ship. Later, he turned privateer, attacking Spanish ships and colonies with the permission of the English crown during an era when England and Spain were fierce enemies. In 1577 he set out for the Pacific Ocean with five ships, intending to raid Spanish settlements on the Pacific coast of South America.

On this voyage Drake's ship, the *Golden Hind,* made the first English passage through the Strait of Magellan at the southern tip of South America. Drake also explored a region that appears on modern maps as the Drake Passage: the waters south of Tierra del Fuego, the island that lies just off continental South America. According to an account published in London soon after his return, Drake determined that there was "no Continent" in those waters, only "broken Islands," although many geographers continued to believe that Tierra del Fuego was part of a large southern landmass.

Drake sailed north along the west coast of the Americas, looting Spanish outposts as he went, to a point north of California. All the way, he looked for a passage eastward back into the Atlantic, but no such passage existed until the Panama Canal was built centuries later. Drake claimed northern California, which he called New Albion, for England. Then he crossed the Pacific to the Moluccas, a group of islands in what is now Indonesia. At that time, the Moluccas were called the Spice Islands because they were the source of cloves and other prized spices. From the Spice Islands, Drake made his way back to England, which he reached in 1580. He was the first English mariner to circumnavigate, or sail around, the globe, and Queen Elizabeth I knighted him on the deck of the *Golden Hind.*

Drake's route through the Pacific and Indian oceans is traced on this portion of Hondius's 1590 map. "Nova Albion" is New Albion, the stretch of California coast that Drake claimed for Queen Elizabeth I of England.

An important world map published around 1590 by the Dutch cartographer Jodocus Hondius shows Drake's route, complete with small pictures of the voyage. One vignette, for example, shows the *Golden Hind* being towed into harbor by the war canoes of the sultan of Ternate, in the Moluccas, while the sultan listens in amazement to Drake's trumpeters. The map also shows New Albion, which began appearing on many maps of North America, although the English never colonized California the way the French settled New France and the Spanish settled New Spain.

Drake made no further voyages of exploration, although he continued to battle the Spanish at sea in European and American waters. He died at sea after a failed piratical raid. For years after his death, mapmakers honored him by showing the route of his circumnavigation on their maps. A tiny portrait of Drake himself is one of the decorations on a world map made by the English cartographer John Speed in 1627.

SEE ALSO
Americas, mapping of

FURTHER READING

Gerrard, Roy. *Sir Francis Drake.* New York: Farrar, Straus & Giroux, 1989.
Smith, Alice. *Sir Francis Drake and the Struggle for an Ocean Empire.* New York: Chelsea House, 1993.
Sugden, John. *Sir Francis Drake.* New York: Simon & Schuster, 1992.

Dudley, Robert

BRITISH ENGINEER, GEOGRAPHER, AND CHART MAKER

- *Born: 1574*
- *Died: 1649, Florence, Italy*

ROBERT DUDLEY, duke of Northumberland and earl of Warwick, was the illegitimate son of the earl of Leicester, a favorite of Queen Elizabeth I. Dudley grew up around the navigators and explorers of the Elizabethan court; by the age of 21 he had become a mariner and privateer, sailing to the Caribbean Sea to prey upon Spanish shipping. Later, he fell out of favor at court and left England. He settled in Italy, where he became a Roman Catholic and devoted himself to compiling an atlas of sea charts. *Dell' Arcano del Mare (The Mysteries of the Sea)*, containing 146 charts, was published in several installments in Florence in 1646–47. The maps were engraved by Antonio Francesco Lucini, who reported that the copper plates from which they were printed weighed two and a half tons.

Although it was initially published in Italian, Dudley's was the first sea atlas to be compiled by an Englishman. It was also the first sea atlas to cover the whole world and the first to use the new projection invented by Gerardus Mercator.

SEE ALSO
Projections

Dulcert, Angelino

CATALAN OR MAJORCAN CHART MAKER

- *Active: about 1339, Spain*

ANGELINO DULCERT is known only from one map, the earliest known portolan in the style of Catalan mapmakers such as Abraham Cresques. It is dated 1339 and reflects Portuguese discoveries of the time, such as the Canary Islands off the African coast.

Dutch mapmakers

MAP COLLECTORS and historians have called the period from 1570 to 1670 the golden age of cartography. The Dutch were the leading mapmakers of this era, producing many maps and atlases that were widely circulated and imitated. Gerardus Mercator, probably the best-known cartographer of all time, was active, but so were many others who are less familiar but equally important in the history of cartography.

Before the modern era, the vigor of a nation's cartography tended to keep pace with its status as a world power, because countries active in trade, exploration, or warfare not only needed maps for use in these enterprises but also gathered new information to make new maps. By the middle of the 16th century, the Low Countries—the region that now includes the Netherlands, Belgium, and Luxembourg—had become a major sea power. Dutch ships plied the waters of the East Indies and probed the coast of North America. As Dutch fortunes rose, so did the Dutch mapmaking industry.

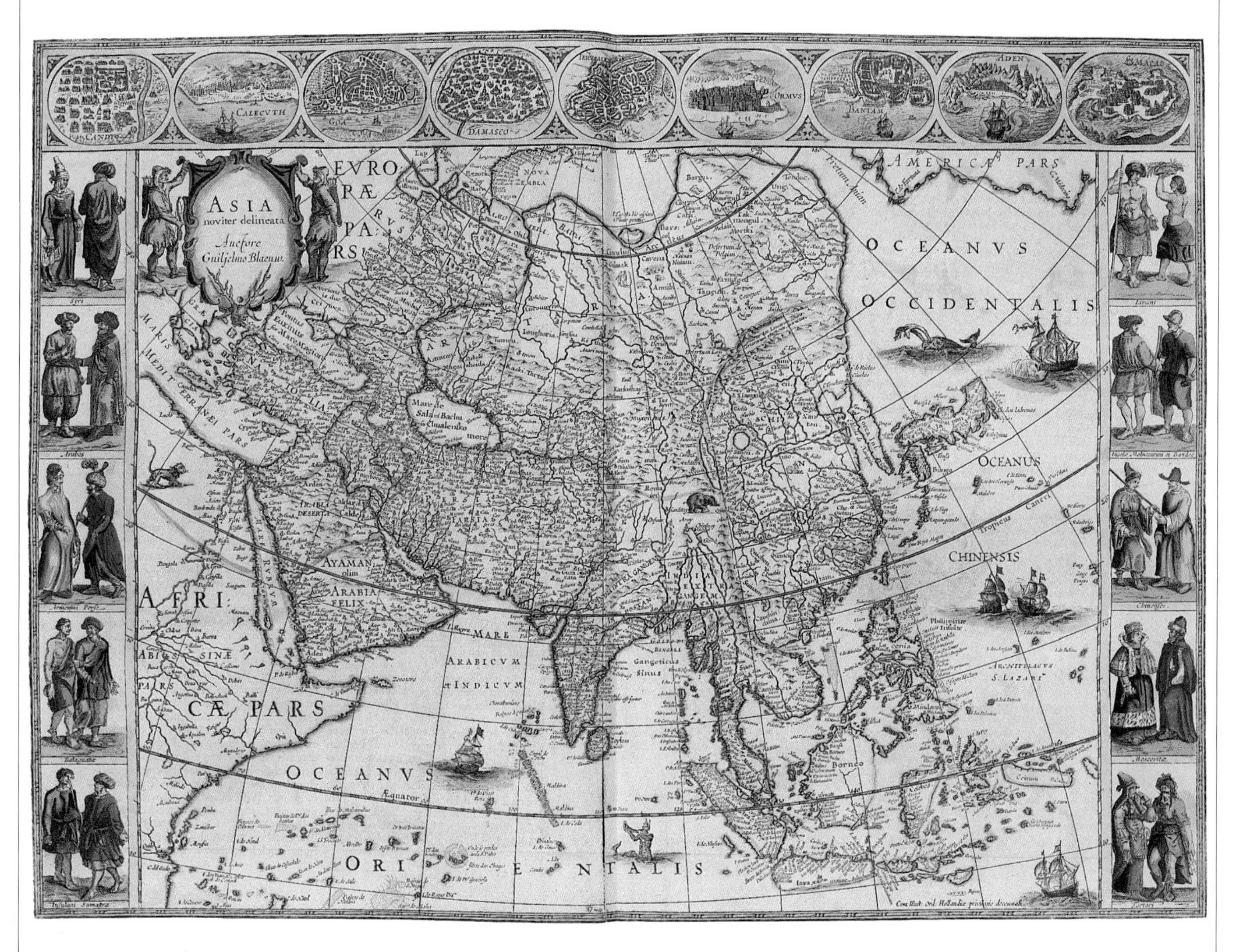

A map of Asia from the Blaeu Grand Atlas, *decorated with pictures of Asian people and cities such as Jerusalem and Goa. North America, called "Americae Pars," appears in the upper left corner, almost touching the Asian coast.*

Mercator, his friend Abraham Ortelius, and other cartographers were at work in the Low Countries by mid-century. Mercator introduced the revolutionary map projection that bears his name in 1569.

Although all of the Dutch cartographers produced individual maps and many of them made globes, the centerpiece of the Dutch mapmaking trade was the publishing of atlases. Map publishing concerns were family businesses, and the set of plates that was used to print an atlas was often reused by several generations of the family. In 1570, Ortelius issued the first modern atlas, the *Theatrum Orbis Terrarum (Theater of the World).* It became enormously popular, as did Mercator's own atlas, which appeared in three parts from 1585 to 1595. (Mercator was the first cartographer to use the word *atlas* for his collection of maps.) These atlases, each of which was reprinted many times, overshadowed the atlases produced by rival Dutch cartographers Gerard de Jode and his son Cornelis de Jode.

The next generation of atlas publishers included the Hondius family, which took over Mercator's atlas in the early 17th century; Jan Jansson; the Visscher family, who published atlases into the 18th century; the Allard family; and the Blaeu family, whose *Atlas Major (Grand*

Atlas) was the most lavish map collection of its time. The works of these mapmakers are notable for their splendid decorations, especially for the maps called *cartes à figures,* which were ornamented with panels containing small scenes of people and landscapes.

The Dutch also excelled at making sea charts. Dutch presses produced many fine marine atlases. The first of these was issued by Lucas Janszoon Waghenaer in 1584; it was republished often and translated into Latin, English, French, and German. A century later, between 1681 and 1696, Gerard van Keulen issued a marine atlas in five volumes. Although van Keulen's charts contained some decorative elements—polar bears in Greenland, for example, or cartouches nestled among figures and plants—they were intended for practical use at sea. They contained the best navigational information of their time.

In the early 18th century, French cartographers began to eclipse the Dutch. One of the last great ventures in Dutch atlas publishing was that of Pieter van der Aa, who issued 66 volumes, containing a total of more than 3,000 maps and illustrations, in 1729. Most of his maps were reprints of the *cartes à figures* and other maps from the great 17th-century atlases. A century later, in 1827, Philippe Vandermaelen published a six-volume atlas in Brussels, Belgium. Vandermaelen's atlas is a curiosity: It was the first atlas ever published in which every map was drawn on the same projection and to the same scale, so that the individual sheets could be taken out of the atlas and fitted together. As far as is known, only one of Vandermaelen's atlases was ever taken apart and reassembled in this way; the result was a globe more than 20 feet (6.1 meters) across.

SEE ALSO

Allard family; Atlas; Blaeu family; Decoration; Hondius family; Jansson, Jan; Mercator family; Ortelius, Abraham

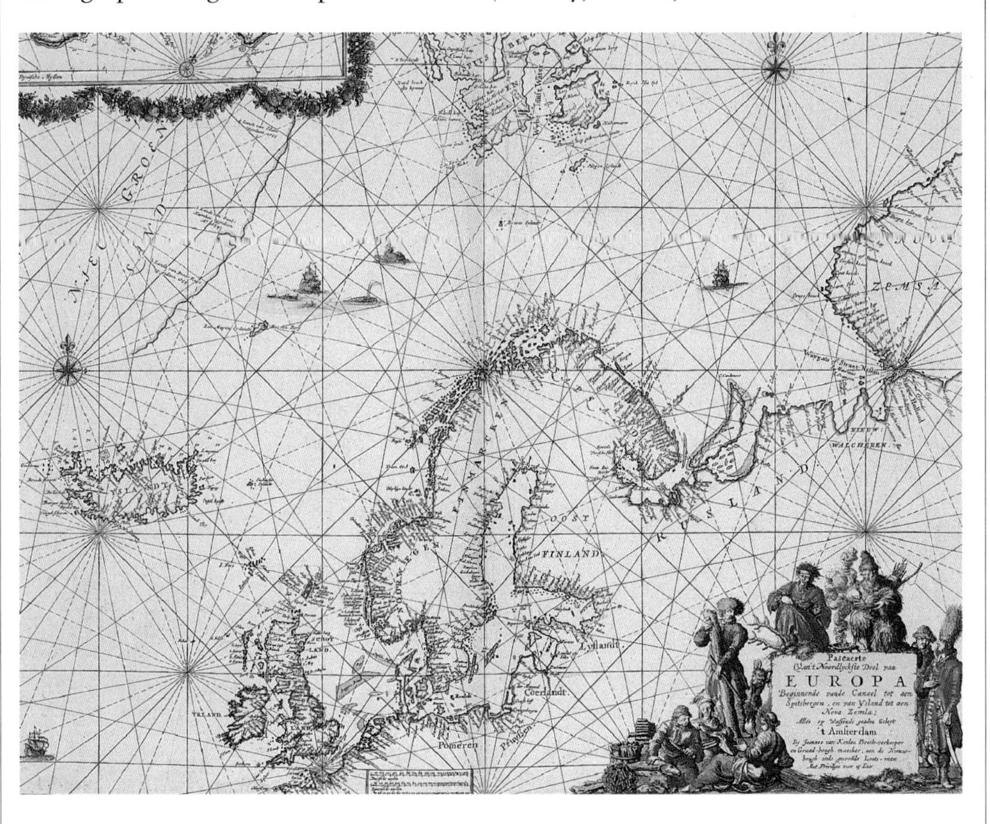

Figures representing the people and resources of the northlands surround the cartouche on this chart of the North Sea and the northern European coasts by Gerard van Keulen. Iceland ("Yslandt") is well mapped, but Greenland ("Nieu Groen Land") is only a sketchy fragment.

The Ebstorf map, a medieval mappa mundi, *was full of Christian symbolism. With Jerusalem at its center, the map was oriented to the east.*

Ebstorf map

THE EBSTORF MAP was discovered in 1830 in a monastery in Ebstorf, Germany, where it had been used as an altarpiece. Made in 1284, it was a circular world map nearly 12 feet (3.65 meters) across, covered with drawings and descriptive labels. The Ebstorf map was destroyed during World War II (1939–45) and is now known only from photographs. In its overall design and level of geographic accuracy, it was similar to another, smaller medieval map called the Hereford map.

SEE ALSO

Hereford map

Elevation

ELEVATION REFERS to the height of a given point above sea level. For example, the summit of Mount Everest, the world's tallest peak (called Sagarmartha in Nepal and Chomolungma in Tibet), has an elevation of 29,028 feet (8,848 meters).

Elliptical projection

SEE Projections

Engraving

ENGRAVING IS the technique of cutting an image into a metal plate with acid or with a steel tool called a burin. A printmaker can then reproduce the image by covering the plate with ink so that the engraved lines are filled with ink. The plate is then wiped and pressed to a sheet of paper in a press. The ink remaining in the engraved lines forms a picture on the paper. A skilled engraver can produce a remarkable variety of lines that allow complex shading and texture in the finished prints. During the heyday of line engraving, from the 17th to the mid-19th century, many map engravers were renowned for their distinctive styles.

Printing from engraved copper plates began to replace woodblock printing as the preferred way of reproducing maps around 1550. At that time the Netherlands had many excellent engravers, and it was these engravers, as much as the cartographers, who made Dutch mapmaking supreme in the late 16th and 17th centuries. The role of the engraver is often overlooked by students of mapmaking, who focus on the geographic content of maps over the years, but students of art history know that the engraver's contribution was crucial to the success of a map. Some mapmakers were both cartographers and engravers, but many maps were collaborations between the cartographer, who designed the map and determined its content, and the engraver, who produced the plate from which it was printed. Engraved plates were highly prized and were often bought and sold by mapmakers,

An 18th-century engraver's workshop. The standing man is heating the varnish that coats a copper printing plate; at the far left, an engraver cuts a design with a burin.

although they could not be used forever—the copper became worn, and eventually the prints made from the plates began to be blurry or incomplete. Plates were often recut or touched up, either to prolong their usefulness or to make changes in the maps.

Engraving is no longer used in cartography. Its place was taken first by lithography, in which grease or wax is used to hold the ink on the printing surface; then by photographic reproduction; and most recently by computerized design and printing programs.

SEE ALSO
Lithography; Printing; Woodcut

Equal-area projection

SEE Projections

Equator

THE EQUATOR is an imaginary circle around the earth, midway between the north and south poles, with a latitude of zero degrees. The Northern and Southern hemispheres are separated by the equator.

0°
EQUATOR

Equivalent projection

SEE Projections

Eratosthenes

SEE Ancient and medieval mapmakers

Eriksson, Leif

SEE Viking explorers

Erik the Red

SEE Viking explorers

Etzlaub, Erhard

GERMAN MAPMAKER

- *Born: about 1460*
- *Died: 1532*

ERHARD ETZLAUB was a compass maker and cartographer in Nürnberg, Germany. Sometime around 1492 he published *Roma Weg* (*The Way to Rome*), a guide for pilgrims on their way to Rome, the center of the Roman Catholic church. *Roma Weg* was an itinerary, combining a written description of roads, inns, and the like with an early form of road map; it was concerned only with the pilgrimage routes, not with general geography. Many similar itineraries were published in the Middle Ages. They were followed a few centuries later by the strip maps of John Ogilby and others, ancestors of today's utilitarian travel maps.

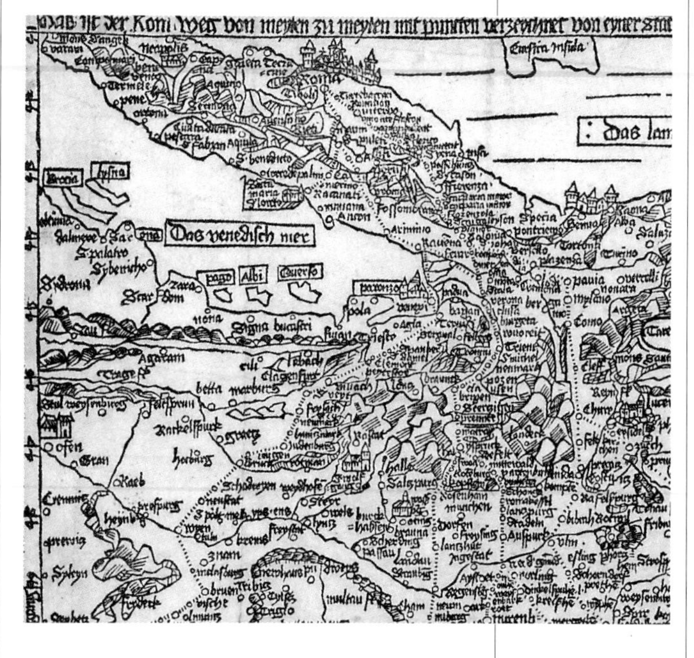

A section of Etzlaub's itinerary. Oriented to the south, it shows part of central Europe, with Italy at the top. Travel routes are shown by dotted lines.

SEE ALSO
Ogilby, John

FURTHER READING

Campbell, Tony. *The Earliest Printed Maps, 1472–1500.* Berkeley and Los Angeles: University of California Press, 1987.

Evans, Lewis

WELSH-BORN MAPMAKER IN AMERICA

- *Active: mid-18th century, Pennsylvania*

CARTOGRAPHER LEWIS EVANS came from Wales to the British colonies in America, where he worked for Benjamin Franklin. In 1749 he published a map of the Pennsylvania, New York, New Jersey, and Delaware colonies; it was revised in 1752. Evans's map was used by many colonists who came to settle in America because it accurately showed the position of roads and settlements. In 1755, Evans produced a map called "The Middle British Colonies." The text that accompanied the map explained that the information on which it was based had been gathered largely from settlers' and traders' reports and journals. Evans explained that scientific survey methods could not yet be used with much success in the American colonies, "the Woodes being yet so thicke." This map was printed in the Philadelphia print shop of Benjamin Franklin. Despite its patriotic background, however, the map was quite useful to the British redcoats during the American Revolution.

Everest, George

SEE Great Trigonometrical Survey of India

Filchner, Wilhelm

GERMAN EXPLORER

- *Born: 1877, Munich, Germany*
- *Died: 1957, Zurich, Switzerland*

WILHELM FILCHNER explored in northeastern Tibet in 1903; in 1911–12 he led a German expedition to Antarctica. He then returned to Tibet and the Himalaya Mountains, leading three expeditions to that region between 1926 and 1939. In 1957 he published *Route-Mapping and Position-Locating in Unexplored Regions,* sharing his experience in making maps and magnetic measurements with other surveyors. His studies influenced the field-workers who carried out surveys in Asia and South America in the 1960s.

Fine, Oronce

FRENCH ASTRONOMER AND MAPMAKER

- *Born: 1494, Briançon, France*
- *Died: 1555, Paris, France*

ORONCE FINE was a professor of mathematics in Paris. In 1525 he made a woodcut map of France, but he is better remembered for a world map he made for Peter Apian's 1530 atlas. This map used a projection called the cordiform (*cordiform* means "heart-shaped"). The cordiform projection was invented about 1515 by a German mathematician named Johann Werner, but Fine created the best-known example of it. Fine's map showed the Northern and Southern hemispheres in a curved projection centered around the north pole. Gerardus Mercator copied this projection for a

Peter Apian published Fine's cordiform map of the world in 1530. Drawings in the upper corners show how the projection relates to the globe. The map is centered on the Indian Ocean.

world map published in 1538. Seldom used since the 16th century, the cordiform projection creates a striking map that resembles a conventional heart shape, or two shells with their bases touching. In 1551, Fine published a book about the cordiform projection that also contains guidelines for topographic surveyors.

SEE ALSO

Apian, Peter

Flinders, Matthew

ENGLISH EXPLORER OF AUSTRALIA

- *Born: Mar. 16, 1774, Donington, England*
- *Died: July 19, 1814, England*

AS A SEAMAN in the British navy, Matthew Flinders made several Pacific voy-

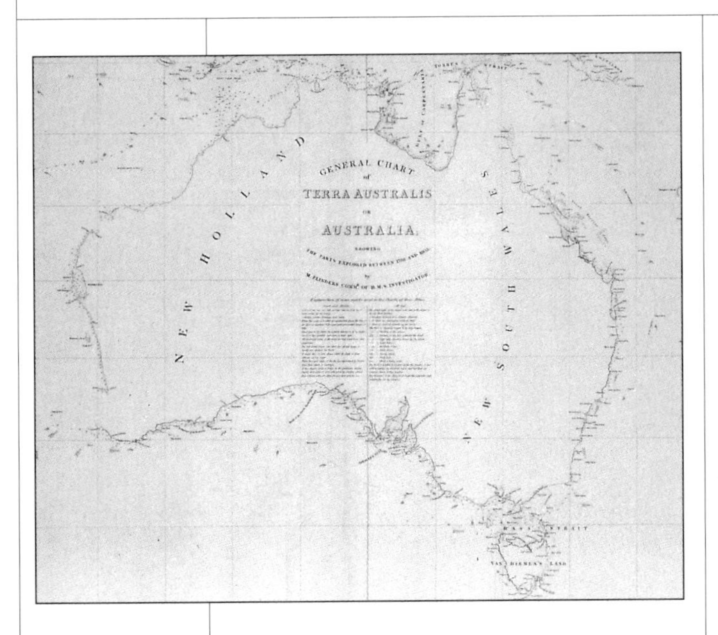

Flinders gave the island continent the name that had long been used for a mythical southern land—the Terra Australis—and shortened it to Australia.

ages in the 1790s. In 1798 he and another navigator sailed around Tasmania, charting the shoreline of this island south of Australia and proving that Tasmania and Australia were separated by a strait, or channel.

In 1801 the British Admiralty assigned Flinders to survey the coast of Australia, which was still imperfectly known. He performed this task diligently and in three years had mapped most of the continent's coastline. On his way back to England, he was held as a prisoner of war by the French for more than six years. He returned home in poor health and devoted his energies to a book called *Voyage to Terra Australis,* which was published, together with accompanying maps, in 1814; the unfortunate Flinders died, however, on the very day his book appeared in print.

Flinders's map of Australia was significant for two reasons. First, by charting the outline of Australia in detail, he revealed the vastness of the blank, unknown interior; this awakened people to the size of the continent and made them eager to explore and settle it. Second, he labeled the continent "Terra Australis, or Australia." Flinders was the first to use the name Australia, and his maps were so widely circulated that the name passed into general use within a few years. The continent is still called Australia, and Flinders's own name has been given to a small island off the coast of Tasmania and to a section of Australia's Great Barrier Reef.

SEE ALSO
Australia, mapping of; Terra Australis

Foxe, Luke

ENGLISH EXPLORER OF THE ARCTIC

- *Born: 1586, Hull, England*
- *Died: 1635*

A SAILOR since boyhood, Luke Foxe (sometimes spelled Fox) joined the race to find the Northwest Passage, the waterway that was believed to lead from the Atlantic Ocean into the Pacific. In 1631, with backing from a group of London merchants known as the North-West Company, he launched an Arctic expedition to search for the Northwest Passage. Carrying a letter from the king of England to the emperor of Japan, Foxe entered Hudson Bay and surveyed its southern and western shores, gathering data about water depth, currents, and harbors. When the sought-after passage failed to appear and his crew began suffering from scurvy, Foxe returned to England—without delivering the king's letter to the emperor, of course, because he had come nowhere near Japan. In 1635, the year of his death, Foxe published an account of his voyage that included the first history of Arctic exploration. The information in Foxe's book was useful some years later, when the Hudson's Bay Company began establishing outposts around the bay. "Northwest" Foxe, as the explorer called himself, also produced a map of his discoveries. He is commemorated on modern maps by two features in the Canadian Arctic: Foxe Basin and Foxe Peninsula.

SEE ALSO

Arctic, mapping of; Northwest Passage

FURTHER READING

Brown, Warren. *The Search for the Northwest Passage.* New York: Chelsea House, 1991.

Lehane, Brendan, and Time-Life editors. *The Northwest Passage.* Alexandria, Va.: Time-Life Books, 1981.

Scott, J. M. *Icebound: Journey to the Northwest Sea.* London: Gordon & Cremonesi, 1977.

Franklin, John

BRITISH NAVAL OFFICER AND ARCTIC EXPLORER

- *Born: Apr. 16, 1786, Spilsbury, England*
- *Died: June 11, 1847, in the Arctic*

JOHN FRANKLIN got a taste of exploration and mapmaking early in his career in the Royal Navy: He served on the ship in which Matthew Flinders surveyed the Australian coast. He then fought in the Napoleonic wars before making his first trip to the Arctic in 1818 as part of a naval expedition to Spitsbergen.

The Royal Navy was eager to settle the long-standing question of the Northwest Passage once and for all. Did a navigable waterway connect the Atlantic and Pacific oceans somewhere in Canada or not? Between 1819 and 1827 the navy sent Franklin on several overland journeys in northern Canada, hoping to find the answer. Franklin charted much of the Arctic Ocean shoreline, but at a high cost—on his first expedition, starvation killed 11 men.

Franklin was next sent to Tasmania to serve as governor of the British prison colony there, but when the Royal Navy planned another assault on the Northwest Passage in 1845, he was selected as the expedition's leader. He took two ships, the *Erebus* and the *Terror,* into the high latitudes and attempted to trace a passage through the maze of islands and channels that is the Canadian Arctic. The ships became frozen in the ice in a channel called Victoria Strait. Months passed, and the men began to die; Franklin perished in June 1847. After a third winter in the ships, the survivors abandoned the derelict vessels and fled south on foot across the ice, hoping to reach some settlement on the mainland. All died, although the expedition's fate was not known until the late 1850s, when searchers began finding bodies and other relics of the lost expedition in a remote part of the Canadian Arctic; letters found with the bodies told the sad tale of the expedition's end.

A search team seeks traces of the missing Franklin expedition in 1853. In the end, the searchers did what Franklin had been unable to do: they found the Northwest Passage.

Meanwhile, the many expeditions that searched for Franklin, urged on by his wife, did what Franklin had failed to do. They completed the map of the Canadian Arctic—including the Northwest Passage.

SEE ALSO

Arctic, mapping of; Northwest Passage

FURTHER READING

Berton, Pierre. *The Arctic Grail: The Quest for the North West Passage and the North Pole, 1818–1909.* New York: Penguin, 1988.

Brown, Warren. *The Search for the Northwest Passage.* New York: Chelsea House, 1991.
Lehane, Brendan, and Time-Life editors. *The Northwest Passage.* Alexandria, Va.: Time-Life Books, 1981.
Maxtone-Graham, John. *Safe Return Doubtful: The Heroic Age of Polar Exploration.* New York: Scribners, 1988.
Scott, J. M. *Icebound: Journey to the Northwest Sea.* London: Gordon & Cremonesi, 1977.

Frémont, John Charles

AMERICAN SOLDIER AND EXPLORER

- *Born: Jan. 21, 1813, Savannah, Georgia*
- *Died: July 13, 1890, New York, New York*

JOHN CHARLES FRÉMONT taught mathematics aboard a naval ship before 1838, when he entered the U.S. Bureau of Topographical Engineers, which was engaged in the exploration of the American West. In 1842 Frémont was assigned to map the Oregon Trail through the Rocky Mountains. In two expeditions he surveyed the trail to Oregon and also surveyed a route south from Oregon into California. His report of these journeys was accompanied by a map by the German cartographer Karl Preuss, whom Frémont had met and befriended in Washington, D.C., and who accompanied Frémont on his expeditions. The report was presented to Congress in 1845 and widely circulated; Preuss's map appeared in 1846. The report cleared up many misconceptions about the Northwest, and the map was a powerful aid to American settlement and expansion in the region.

Frémont became a popular hero. His later career included three more expeditions (none of which led to important discoveries or publications), a court-martial for mutiny, service in the Union army during the Civil War, and an unsuccessful candidacy for the U.S. Presidency.

SEE ALSO

Americas, mapping of

FURTHER READING

Goetzmann, William H. *Exploration and Empire: The Explorer and the Scientist in the Winning of the American West.* New York: Norton, 1978.
Harris, Edward D. *John Charles Frémont and the Great Western Reconnaissance.* New York: Chelsea House, 1990.
Viola, Herman. *Exploring the West.* Washington, D.C.: Smithsonian Books, 1987.

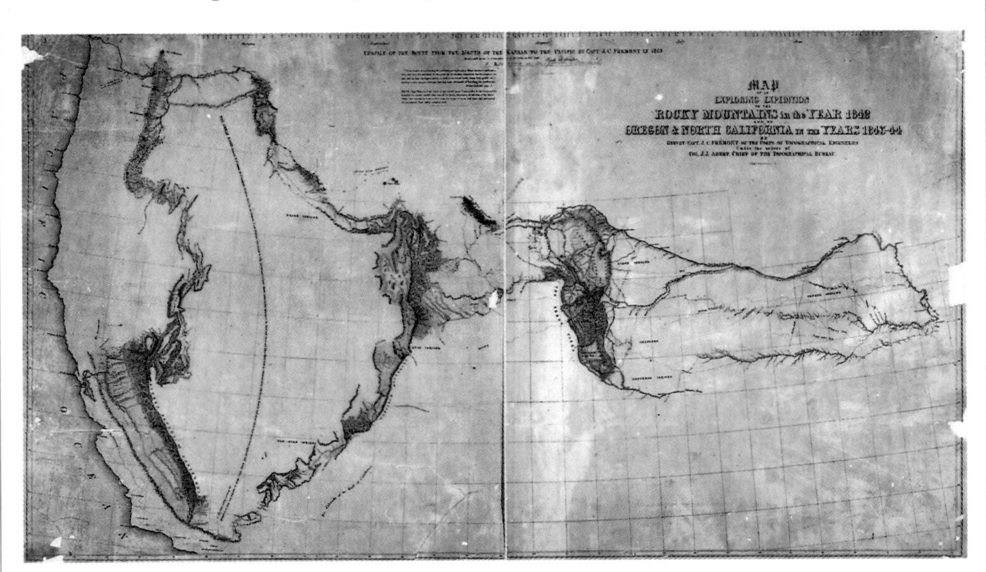

Frémont's map of 1845, one of several maps of the American West published after his first two expeditions. The map shows only territory that Frémont himself had surveyed—nothing is based on rumors or guesswork.

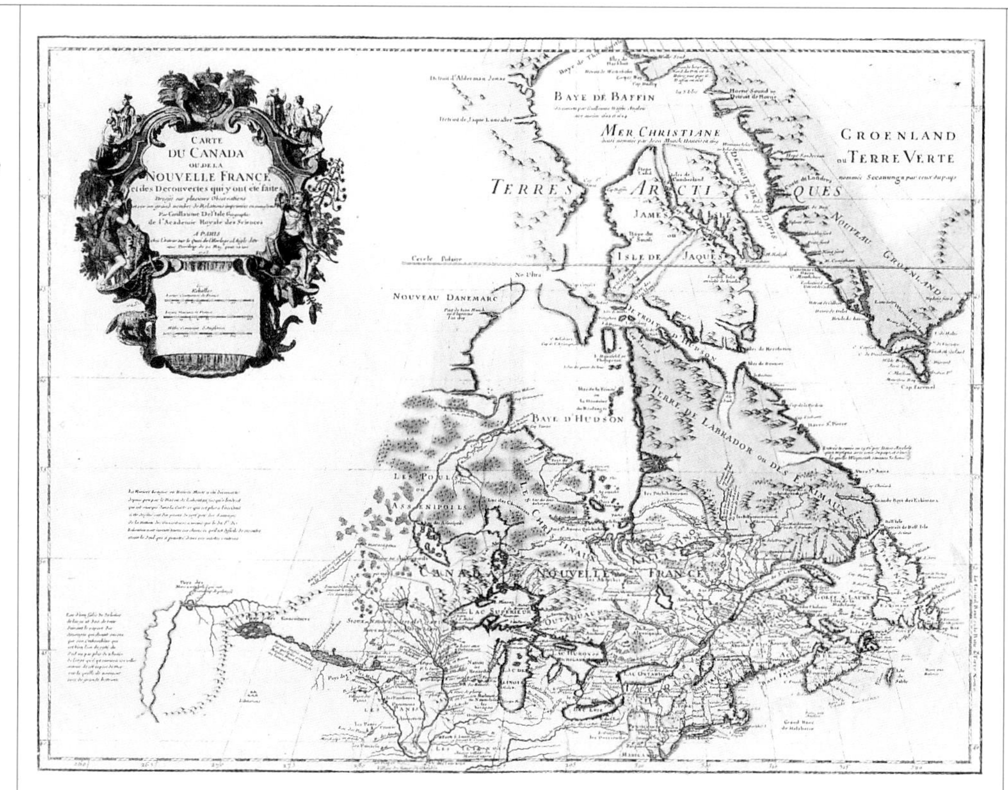

The French colony in Canada, as mapped in 1703 by Guillaume De L'Isle. Although the borders of many French maps were magnificently decorated, the maps themselves tended to be clear and practical rather than ornamental.

French mapmakers

FRANCE'S FIRST major contribution to cartography came in the 16th century, when a school of French portolan makers produced worthy sea charts and some world maps. Aside from the portolan makers, the leading French mapmaker of the 16th century was Oronce Fine, who produced world maps in distinctive heart-shaped projections and also made several maps of France; his 1525 woodcut map of France on four sheets was often copied. Local and regional maps were a specialty of many 16th-century French mapmakers. Gilles Boileau de Bouillon made maps of Germany, Peru, the region around Rome, and Saxony (part of Germany). Other cartographers mapped French territory, producing detailed maps of individual provinces and estates. Abraham Ortelius and other Dutch map publishers included these provincial maps in the great Dutch atlases of the late 16th and 17th centuries.

At this time France was busy exploring and colonizing New France, now known as Canada. Some explorers, such as Samuel de Champlain, published their own maps. In other cases, cartographers studied manuscript maps, drawings, and reports from the colonies and incorporated the latest information into their speculative maps of North America. These maps tended to feature a lot of detail in the French territory around the St. Lawrence River and the Great Lakes, but their portrayal of other parts of the Americas was often sketchy.

Seventeenth-century French map publishing was dominated by the Sanson family, who produced a series of atlases beginning with a 1654 atlas that had 100 maps. Sanson maps have a distinctive style: they are less elaborately decorated than the Dutch maps of the period, with the emphasis on clarity rather than on a profusion of eye-catching pictures.

Alexis Hubert Jaillot (1632–1712) founded the second dynasty of French mapmakers. He married the daughter of a map artist and collector; he also acquired the Sanson family's copper printing plates. Combining maps from these and other sources, he began to issue atlases in 1681. His maps were ornamented with hand-painted pictures; for special customers, he used gold paint. Jaillot's descendants continued the family map-publishing business until 1780.

The Cassini family emerged into importance at the end of the 18th century. Although they made maps, Jean Dominique Cassini and his descendants are more often remembered for their advances in scientific surveying techniques. They carried out the first triangulation survey of a whole country, spending four generations in the mapping of France by scientific principles.

Guillaume De L'Isle, whose first atlas was published when he was 25 years old, was the dominant figure in French mapmaking from 1700 to his death in 1726. L'Isle's maps are notable for their absence of guesswork; he tried to check the reliability of his data. L'Isle adopted a clean, precise style that was carried on by Jean Baptiste d'Anville. These mapmakers banished fanciful fripperies from their maps in the interest of stern science, foreshadowing the major trend in cartography in the 19th century. L'Isle's son-in-law Philippe Buache, however, was a believer in what came to be called "theoretical" cartography, in which mapmakers filled in blank spaces on the maps with what they thought was the most logical geography—for example, by extending the coastline of Australia to include New Zealand and New Guinea.

French cartographers continued to produce maps and atlases throughout the 19th century, but they were no longer the dominant force in the field of mapmaking. Great Britain, Europe's foremost naval and imperial power during the 19th century, took the lead in cartography away from France.

SEE ALSO

Anville, Jean Baptiste Bourguignon d'; Cassini family; Fine, Oronce; L'Isle family; Sanson family

FURTHER READING

Konvitz, Josef W. *Cartography in France, 1660 to 1848: Science, Engineering, and Statecraft.* Chicago: University of Chicago Press, 1987.

Tooley, Ronald V. *Maps and Map-Makers.* 1949. Reprint. New York: Dorset Press, 1990.

Frézier, Amédée-François

FRENCH MAPMAKER

- *Born: 1682, Chambery, France*
- *Died: 1773*

AMÉDÉE-FRANÇOIS FRÉZIER was supposed to become a priest, but he took up military engineering instead. His superior officers sent him to the Americas in 1712 to spy out the forts and other defenses of the Spanish colonies. On his voyage, which took him around Cape Horn at the tip of South America, Frézier passed the Falkland Islands and made the first accurate chart of this South Atlantic island group.

Frobisher, Martin

ENGLISH NAVIGATOR

- *Born: about 1539, Normanton, England*
- *Died: Nov. 22, 1594, Plymouth, England*

MARTIN FROBISHER'S seafaring career began in 1553 with a voyage to

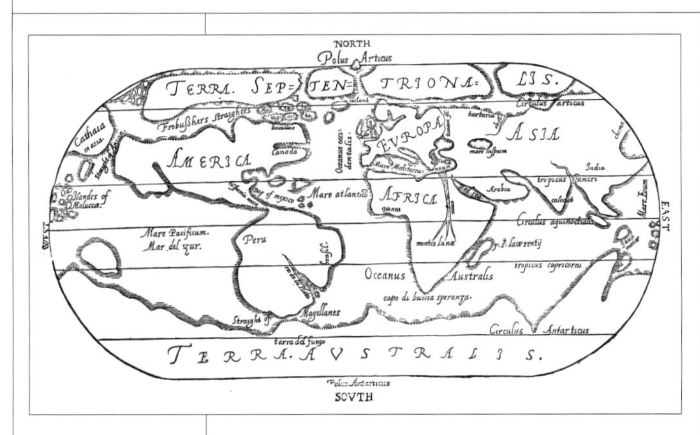

Frobisher's shipmate George Best drew a map on which the nonexistent "Frobishers straightes" seem to offer clear sailing to Cathaia (China) and the Molucca or Spice Islands.

West Africa; over the next decade he made annual trips to North Africa and the Middle East. Then his thoughts turned to the north. He believed that the discoverer of the Northwest Passage, the hoped-for sea route from England to China, would win both glory and fortune.

In 1576 Frobisher made the first of three voyages to the waters west of Greenland. He entered a large bay that he mistook for the entrance to the passage and named it Frobisher's Strait; today it is called Frobisher Bay. There he found rocks that he thought contained gold. Abandoning the search for the passage, he made several trips to gather these rocks, which he hauled back to England. To his dismay (and the fury of sponsors who had backed his venture), the rocks turned out to be worthless.

Frobisher never returned to the Arctic but spent the rest of his career fighting Spain, in which cause he was killed. His ideas about the geography of the Arctic and the Northwest Passage were communicated to the public by George Best, who sailed with him on his three voyages to Frobisher Bay. In 1578 Best published a book on Frobisher's discoveries. It included a world map that shows "Frobishers straightes" as a broad waterway leading from eastern Canada to China. No such waterway exists, but explorers continued to search for it—and cartographers continued to include it on their maps—for two centuries after Frobisher's death.

SEE ALSO

Arctic, mapping of; Northwest Passage

FURTHER READING

Brown, Warren. *The Search for the Northwest Passage.* New York: Chelsea House, 1991.

Lehane, Brendan, and Time-Life editors. *The Northwest Passage.* Alexandria, Va.: Time-Life Books, 1981.

Morison, Samuel Eliot. *The Great Explorers: The European Discovery of America.* New York: Oxford University Press, 1978.

Scott, J. M. *Icebound: Journey to the Northwest Sea.* London: Gordon & Cremonesi, 1977.

Gama, Vasco da

PORTUGUESE NAVIGATOR

- *Born: about 1460, Sines, Portugal*
- *Died: Dec. 24, 1524, Cochin, India*

VASCO DA GAMA was trained in navigation before becoming an officer of the Portuguese navy in 1492. For most of the 15th century, Portugal had been seeking an eastern sea route to India, which was a center of trade in spices, silks, and gems. Portuguese navigators had inched south along the coast of Africa until, in 1488, Bartolomeu Dias rounded the Cape of Good Hope, proving that ships could reach the Indian Ocean from the Atlantic Ocean. In 1497 King Manoel I chose Gama to lead the first trading expedition to India along the route that Dias had pioneered.

As the helmet and the rolled map suggest, Gama was both an explorer and a conqueror.

Gama's fleet consisted of four ships and about 170 men. Gama knew from the reports of earlier mariners that the winds and currents along the

African coast ran steadily northward. Dias had advised him not to battle these currents but instead to swing out into the Atlantic Ocean in a wide arc before heading east. This is what Dias had done, and he had discovered mid-ocean winds and currents that carried him in the direction he wanted to go. Gama followed the same plan, but more boldly. He stood sharply out to sea and sailed west to within 600 miles (965 kilometers) of Brazil before allowing the currents to sweep him south and east toward the tip of Africa. By the time he made landfall in South Africa, he had been out of sight of land for 93 days—the longest open-sea voyage on record by European sailors up to that time. He rounded Africa and made his way north along the coast of East Africa and across the Indian Ocean to India, where he laid the foundation for Portuguese domination and trade. He returned to Portugal in 1502.

Gama made several later trips to India and briefly served as the governor of the Portuguese colony there before his death, but his major contribution to geography was his first voyage, which connected Europe and Asia in a new way. That voyage forever overturned the worldview of the Greek geographer Ptolemy, which had dominated European mapmaking since the 2nd century A.D. Gama ushered in a new age not only of exploration and colonization but also of cartography. Although the Portuguese government tried to keep the details of Gama's voyage a secret in order to protect the newly discovered trade route, word leaked out. A map made in 1508 by Dutch cartographer Johan Ruysch gives a very accurate picture of southern Africa and the Indian Ocean, showing that Gama's discoveries were no longer a well-kept secret.

SEE ALSO

Dias, Bartolomeu; Ptolemy

FURTHER READING

Correa, Gaspar. *Three Voyages of Vasco da Gama.* New York: Franklin, Burt, 1964.
Hart, Henry T. *Sea Road to the Indies: An Account of the Voyages & Exploits of the Portuguese Navigators, Together with the Life and Times of Dom Vasco da Gama, Capitão-Mor, Viceroy of India & Count of Vidigueira.* Westport, Conn.: Greenwood Press, 1950.
Jones, Vincent. *Sail the Indian Sea.* London: Gordon & Cremonesi, 1978.
Ravenstein, E. C., ed. and trans. *A Journal of the First Voyage of Vasco da Gama, 1497–1499.* London: Hakluyt Society, 1898.
Stefoff, Rebecca. *Vasco da Gama and the Portuguese Explorers.* New York: Chelsea House, 1993.

Games and novelties

MAPS HAVE appeared on games at least since 1590, when William Bowes of England produced sets of playing cards with maps of the English counties on them. By the middle of the 17th century, French publishers were issuing simple board games based on maps: the players threw dice and moved markers around a board on which each position represented a country.

In the years that followed, such games became considerably more sophisticated. They were played on boards resembling real maps, and sometimes the players' pieces encountered travel hazards, such as shipwrecks in the Bahamas. The point of the game was to race the other players around the board, and the rules were sometimes highly complicated. Such games were popular in France and especially in England, where they were promoted as having educational value for children. For example, the Crystal Palace Game, produced in England in 1851, was subtitled "A Voyage Round the World: Geography Is Made Easy." These travel and geography games have several kinds of modern

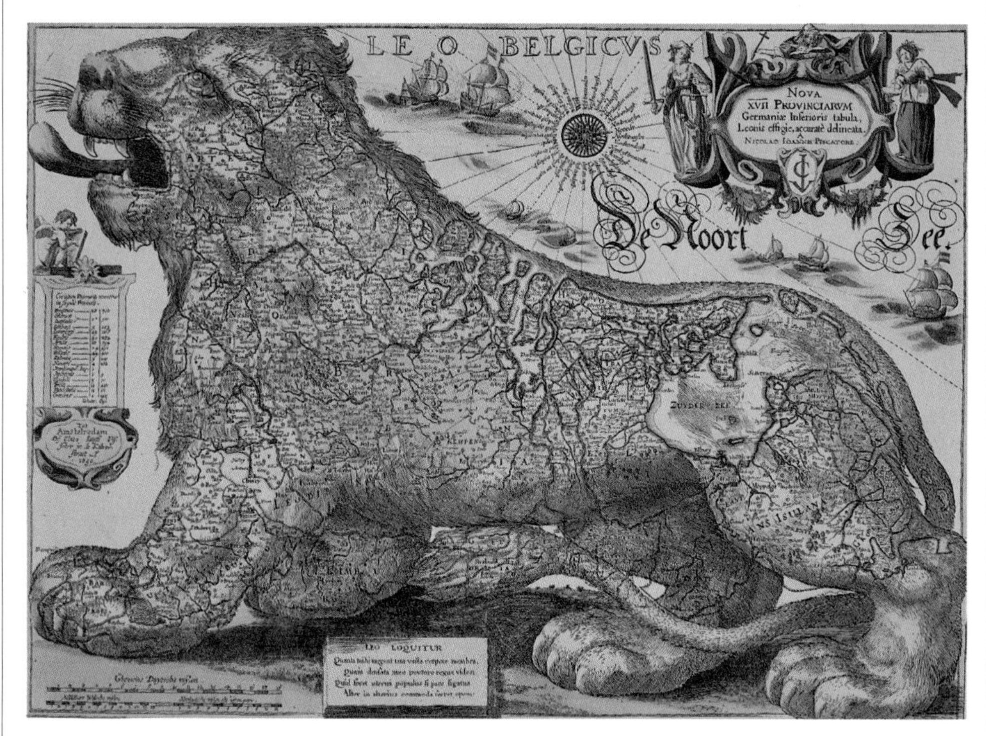

A Leo Belgicus map by Claes Janszoon Visscher, 1656. The lion is facing southwest; the North Sea is above its back. The large Dutch lake called the Zuyder Zee appears on its hindquarters.

descendants—board games in which players must "travel" across a route, such as the French game "Mille Bornes" (A Thousand Milestones), and computerized geography-based games, such as "Where in the World Is Carmen Sandiego?"

Maps have also been used to make jigsaw puzzles. These began appearing in the 18th century, and puzzle maps of the world and of various countries and states are still made today. Some of the educational puzzles for children are cut along political borders, such as county lines or international boundaries. Others, far more difficult to complete, are made of topographic maps, with muted tones and thousands of faint contour lines and tiny symbols.

Some cartographic novelties use the concept of mapping for purposes of symbolism, amusement, or satire. Starting in the 16th century, the continents were often portrayed as people or animals, with the outline of a figure superimposed onto the map of countries, rivers, and other features. Europe was sometimes drawn as a woman, with Spain as her head and Italy and Denmark as her arms. The Netherlands and Belgium were often mapped on the figure of a lion; this design was called the *Leo Belgicus* (*Lion of Belgium*). In the 19th century, maps were used as the basis for political cartoons. A satiric map of 1877 shows Russia as a large octopus grabbing up the European nations, each of which is drawn as a figure that represents that country's political situation at the time.

Another novelty map that was popular in the 19th century was called the "love map." It appeared in many versions, each of which used a fictional landscape to portray the course of love and marriage, with such features as the River of Tears, the Rocks of Jealousy, and St. Brides Bay.

A puzzle depicts the Oregon Trail, which runs from Independence, Missouri, to Portland, Oregon. Many settlers made the journey with less information about the route than appears on this novelty map.

Not all map novelties are intended to educate or entertain. Many people find the appearance of maps attractive, even without regard to their content, and for this reason maps have been used as designs on many things, from scarves issued in honor of Queen Victoria's 50th year on the throne to bed sheets marketed in the 1980s. The map lover today can find cartographic designs on hats, shirts, jackets, clocks, dishes, umbrellas, beach balls, stationery, and many other objects.

FURTHER READING

Hill, Gillian. *Cartographical Curiosities.* London: British Library, 1978.

Makower, Joel, ed. *The Map Catalog.* 3rd edition. New York: Vintage, 1992.

Gastaldi, Giacomo

ITALIAN MAPMAKER

- *Born: about 1500, Villafranca, Italy*
- *Died: about 1565*

GIACOMO GASTALDI, who worked in Venice, helped make Italy the leader in map production in the mid-16th century. Gastaldi was named the official cosmographer to the republic of Venice. He made numerous maps that were widely reprinted and copied, thus influencing other cartographers, including Abraham Ortelius and Gerardus Mercator. Gastaldi is thought to have made more than 200 maps between 1544 and 1565. Among these were maps of regions, countries, continents, and the world. In 1548 he produced the first miniature, or pocket, atlas. He also published an edition of Ptolemy's *Geography* with maps that remained the standard Ptolemaic ones until the end of the 16th century.

SEE ALSO

Italian mapmakers; Ptolemy

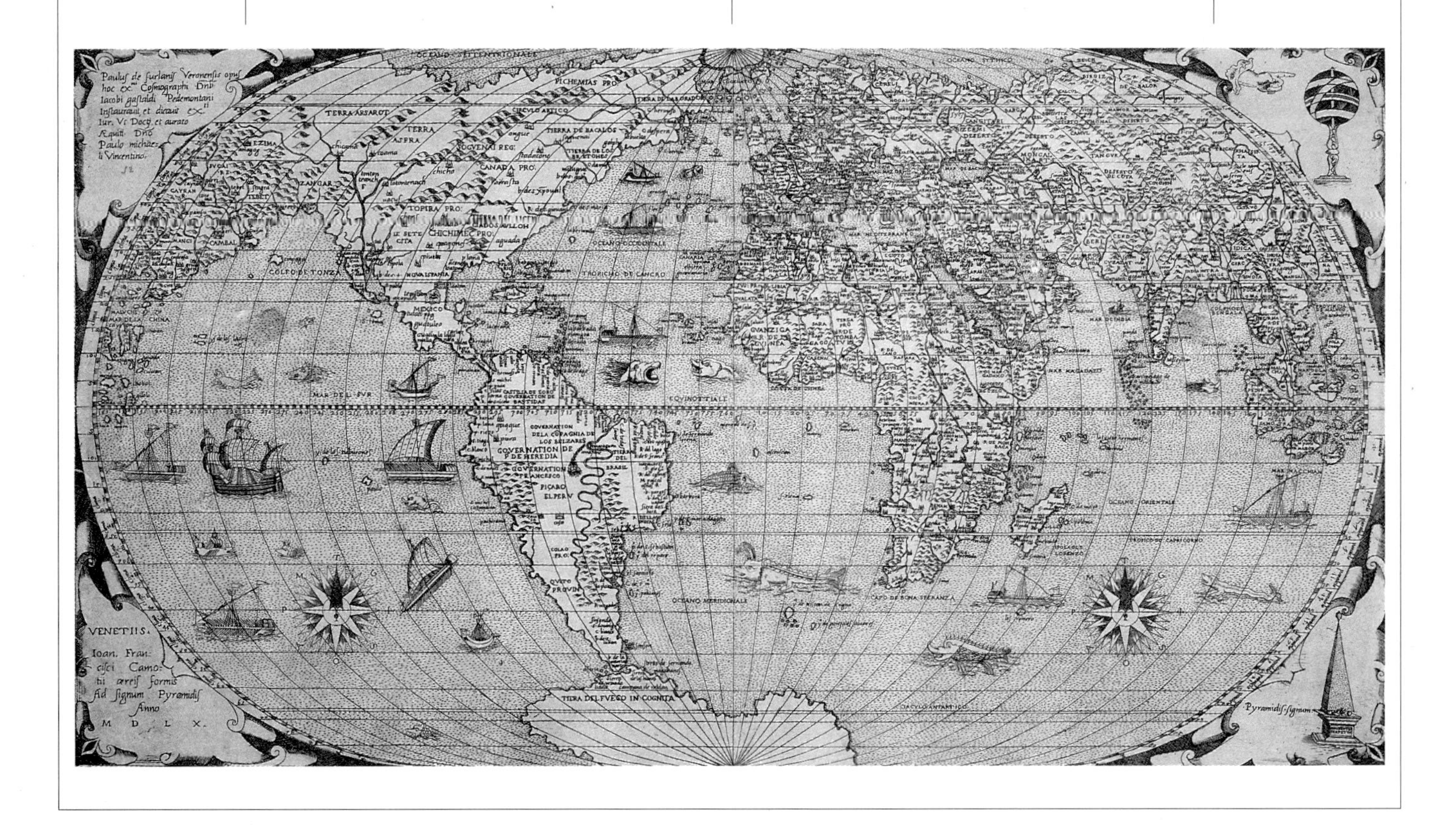

A 1560 world map by Gastaldi. Here, western North America merges into Asia, but on other maps Gastaldi separated the continents. The region's geography was a mystery to Europeans.

Gazetteer

A GAZETTEER is a dictionary of places. The names of countries, cities, towns, and topographic features are listed alphabetically, so that any place can easily be looked up. Each entry is followed by a brief description of the place or feature named. Some gazetteers also include maps. Gazetteers are often published under the name "geographical dictionaries."

Gemma Frisius, Regnier

DUTCH ASTRONOMER AND GEOGRAPHER

- *Born: 1508, Dokkum, Netherlands*
- *Died: 1555, Louvain, Belgium*

REGNIER GEMMA FRISIUS, born in the Netherlands, became a professor of astronomy in France and royal mathematician to King Charles V; he also taught Gerardus Mercator. Gemma Frisius made several maps and globes, but his biggest contribution to cartography was a treatise he published in 1533. Called *A Little Book on a Method for Delineating Places,* it was an early handbook on triangulation, the use of trigonometry to measure distances on land. Gemma Frisius's guidelines became the basis of surveying; they still form the foundation of modern triangulation.

SEE ALSO
Surveying; Triangulation

Geodesy

GEODESY IS the science of measuring the earth: its exact shape (it is not a perfect sphere but is slightly flattened at the poles), size, weight, magnetic properties, and density. The term is also used to describe large-scale surveying activities, when the exact location and elevation of points are determined over such a big area that irregularities in the earth's shape must be taken into account. People who practice geodesy are called geodesists.

Geographic information system (GIS)

SEE Computers in mapmaking

Geography

GEOGRAPHY IS the systematic study of the earth's surface and its features. Those who practice geography are called geographers.

For many centuries, geographers concentrated on physical geography, the study of the physical features of land and oceans, including not just the locations of places but also their characteristic climate, vegetation, and wildlife. Political geography also has a long history. It focuses on political divisions, such as national borders and territorial claims, and also on place names.

In the 19th and 20th centuries, new approaches to geography emerged. Economic or commercial geography is concerned with such matters as resources, trade, and industry and how they are related to the earth. Human geography studies how and where human societies live, with emphasis on the connections between human life and physical geography. Cultural geography deals with the distribution of human cultural characteristics, such as languages and religions, across the earth's surface. And environmental geography focuses on ecology, or the relationships that exist among plants, animals, people, and the physical landscape. In all types of geography, however, maps are a vitally important tool.

FURTHER READING

Bell, Neill. *The Book of Where: Or How to Be Naturally Geographic.* Boston: Little, Brown, 1982.
Davis, Kenneth. *Don't Know Much About Geography: Everything in the World You Need to Know about Where in the World You Are.* New York: Morrow, 1992.
DeBlij, Harm J. *The Earth: A Physical and Human Geography.* New York: Wiley, 1987.
Demko, George. *Why in the World: Adventures in Geography.* New York: Anchor, 1992.
Gould, Peter. *The Geographer at Work.* New York: Routledge & Kegan Paul, 1985.
Haggett, Peter. *Geography: A Modern Synthesis.* New York: Harper & Row, 1983.

Geologic maps

GEOLOGIC MAPS are special-purpose maps that show the geologic features of an area. Unlike topographic maps, which show the surface of the terrain in all its contours and configurations, geologic maps show not just the surface but also what is beneath it. Contour lines, shading, and symbols are used to communicate geologic information, including the type of rock masses and mineral deposits, how these masses and deposits are distributed, the age of the rock, and the

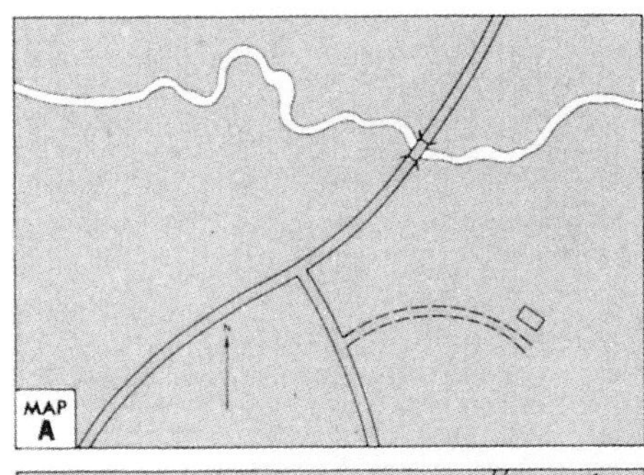

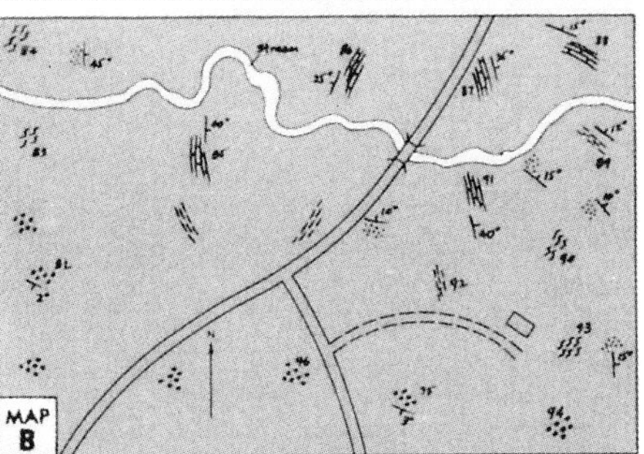

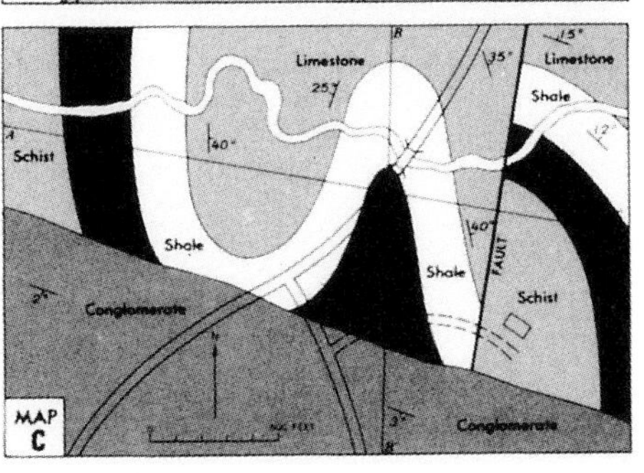

Stages in the making of a geologic map: a base map showing roads and a stream (top); the base map with field notes added to show the type of rock found at many locations (center); and the finished map, showing the rock formations beneath the surface features (bottom).

evidence of glacial flows, floods, and the like. Many geologic maps are in two parts: a landform map that shows the geologic surface of the terrain, and a cross-section of the rock formations under the surface.

The first geologic maps were made by the British geologist William Smith (1769–1839), who studied fossils and rocks and noticed that rock and soil occur in layers, called strata. Stratigraphy, or the measuring of these strata, has been an important element of geology ever since; it also plays an important part in determining the age of fossils. Smith published a geologic map of England and Wales in 1815 and an atlas of geologic maps in 1819. Their modern successors include hundreds of geologic atlases covering all regions of the world, and new geologic mapping projects are constantly being undertaken. Since 1990, for example, the U.S. Geological Survey (USGS) and the Circum-Pacific Council for Energy and Mineral Resources, an international organization formed by the countries that border the Pacific Ocean, have been working on a series of new geologic maps of the entire Pacific basin, from the Arctic to the Antarctic. This project will result in a geologic atlas of the Pacific Ocean and the land around it, showing landforms, the seafloor, and mineral and energy resources.

Today, many nations have geologic mapping programs. In the United States, the USGS produces nearly a dozen different series of geologic maps for a variety of purposes. The Department of Mines and Technical Surveys performs a similar function in Canada and distributes its maps through the Canada Map Office; in the United Kingdom, the Institute of Geological Sciences, based in London, produces geologic maps in cooperation with the Ordnance Survey, the nation's topographic mapping agency. Oil and mining companies use these maps and also commission their own maps from scientists, for they have learned that geologic maps—often based on satellite data—can help them pinpoint new deposits of oil, gas, coal, and other resources.

German Atlantic Expedition

THE GERMAN Atlantic Expedition was the first oceanographic expedition to make a detailed survey of an entire ocean—the South Atlantic. From 1925 to 1927, a team of scientists on the ship *Meteor,* using an early form of sonar, crossed the Atlantic Ocean 14 times between the equator and the Antarctic Circle, taking thousands of depth soundings and making a complete chart of the currents. The Germans made one of the first maps of the Mid-Atlantic Ridge, an undersea mountain range that runs the entire length of the ocean.

SEE ALSO
Undersea mapping

German mapmakers

GERMANY HAS never dominated mapmaking for an entire era, as Italy, the Netherlands, England, and France have all done at various times. But German geographers and mapmakers made vital contributions to cartography, although many of them left their native land to work in courts, universities, and cities throughout Europe.

The first landmarks in German mapmaking came in the 15th century. Nicolaus Cusanus made a number of maps that were published after his death in 1464; one of them is the first known modern map of Germany. An even bigger milestone occurred in 1482, when Dominus Nicolaus Germanus published a new edition of Ptolemy's *Geography* with woodcut maps. It is called the Ulm edition of Ptolemy after the city of Ulm, in which Germanus worked. The Ulm edition is especially important in the history of mapmaking because it contains not only the maps that were traditionally associated with Ptolemy but also five new maps of the modern world, including the first printed map of Scandinavia (earlier maps of the area were hand-drawn).

The 15th century ended with two important German cartographic productions: a world map by Henricus

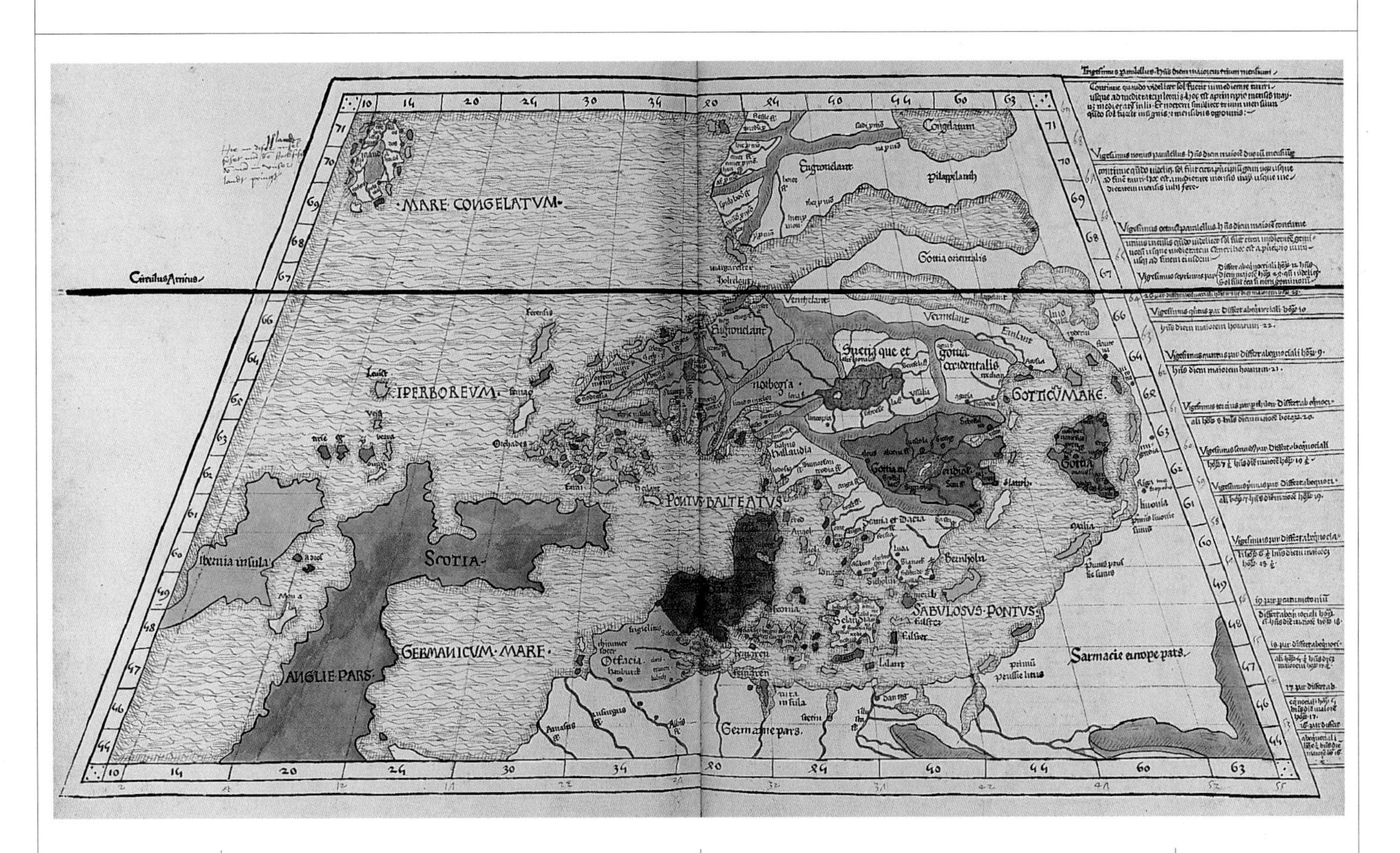

The first printed map of northern Europe, from the 1482 Ulm edition of Ptolemy's Geography. *Southern Sweden and Norway are at the upper right, Iceland at the upper left.*

Martellus and a globe by Martin Behaim. The Behaim globe was completed in 1492 and presents a remarkably complete picture of the world as it was known before Columbus returned from his first voyage to the Americas.

In the 16th century, advances in printing technology, along with the spread of learning that was part of the cultural movement known as the Renaissance, brought new life to cartography. Germany, the birthplace of the printing press in the West, became an important center of map production and soon surpassed Italy in mapmaking. Part of the success of the German mapmakers was due to the work of Peter Apian and others who created new, more accurate tools for surveying. German mapmakers were noted for making useful, practical maps, especially maps of Germany and Europe. These were highly prized by the merchants of German trading cities such as Hamburg and Bremen, where geographic knowledge was a business asset.

As mapmaking flourished in 16th-century Germany, artisans in such cities as Nürnberg and Strassburg gained renown for their globes, compasses, and other geographic tools. Peter Apian and other German cartographers made many important maps. Erhard Etzlaub pioneered the route map, or road map, and Johann Schoner (1477–1547), who wrote dozens of works on medicine, astronomy, and other subjects, produced a series of four globes between 1515 and 1533. Albrecht Dürer of Nürnberg, one of the world's foremost masters of the arts of engraving and printmaking, made a map of the Eastern Hemisphere in 1515. In Cologne, another center of geographic study, Kaspar Vogel (1511–1564) printed a set of globe gores, tapered map segments that could be glued to a sphere to form a globe, in 1542; he produced a map of the Rhine River and a large-scale map of Europe in 1555.

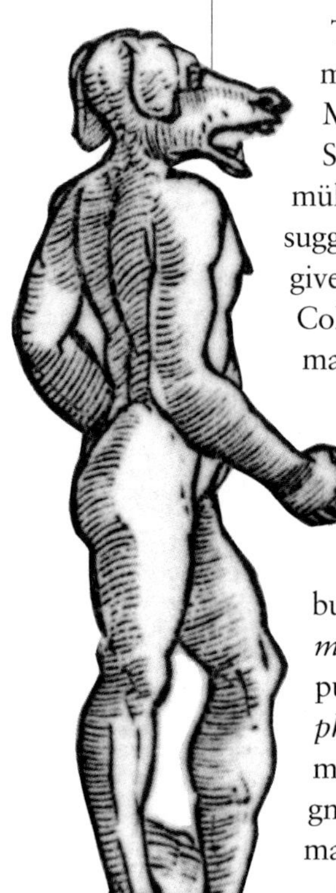

A monstrous inhabitant of Sebastian Münster's Asia.

The two leading figures in German mapmaking, however, were Martin Waldseemüller and Sebastian Münster. Waldseemüller is most often remembered for suggesting that the name America be given to the new lands explored by Columbus and others, but he made many other substantial contributions to cartography, including two world maps (1507 and 1516) and a large-scale map of Europe (1511). Münster produced a number of maps but is chiefly known for his *Cosmography,* a book that was first published in 1544. The *Cosmography* contained many woodcut maps lavishly illustrated with vignettes and fanciful creatures; these maps were widely reprinted and imitated. Münster also started the practice of making separate maps for each continent.

During the 17th century, German mapmakers were overshadowed by the Dutch, but in the 18th century, when Dutch cartography began to decline, the Germans became more active in map and atlas publishing. The Homann family of Nürnberg produced a number of atlases, including several atlases of Germany and the *Atlas Major* (1780), which contained more than 300 maps and covered the entire known world. High-quality maps continued to be published in Germany throughout the 19th and 20th centuries. In addition, German scholars have taken a leading role in the study of ancient and medieval cartography.

SEE ALSO

Apian, Peter; Behaim, Martin; Cusanus, Nicolaus; Etzlaub, Erhard; Germanus, Dominus Nicolaus; Martellus, Henricus; Münster, Sebastian; Ptolemy; Waldseemüller, Martin

Germanus, Dominus Nicolaus

GERMAN MONK AND MAPMAKER

- *Active: 15th century, Italy and Germany*

DOMINUS NICOLAUS GERMANUS was a Benedictine monk from Reichenbach, Germany, who drew the maps for several editions of Ptolemy's *Geography* that were published in the 15th century. In 1466 he was working in Florence, Italy, where he produced an edition of Ptolemy for the duke of Ferrara; other mapmakers imitated his work in preparing their own versions of Ptolemy.

Germanus edited another edition of Ptolemy in 1482. It is called the Ulm edition because it was produced in Ulm, Germany. The Ulm edition was the first edition of Ptolemy to be printed outside Italy. It featured maps printed from woodcuts. All of the maps described in Ptolemy's book were reproduced, but Germanus also added five new maps, one of which was the first known printed map of Scandinavia.

Global positioning system (GPS)

A GLOBAL positioning system is an instrument that can tune in to satellite signals and use them to pinpoint its location anywhere on earth. Global positioning systems are used on many airplanes and oceangoing ships. Once too bulky and expensive for most private owners, they are now fairly common among boaters, off-road drivers, and researchers or adventurers who make long treks into the wilderness. These travelers need never fear being truly lost, for a GPS will give them a readout of their latitude and longitude.

Globe

A GLOBE is a map on a spherical surface. A globe of the earth is called a terrestrial globe. Because the globe is a sphere like the earth, it is the only type of map that does not suffer some distortion. Only on globes can the sizes, shapes, and directions of geographic features be shown in their true relation to one another. Globes of the moon, planets, and constellations have also been made; these are called celestial globes.

Like other kinds of maps, globes can be made to communicate many different types of information: political boundaries, land cover, topography, or some combination of these.

Globes are known to have existed among the ancient Greeks, but no specimens of these ancient globes have survived. The oldest existing globe was made by Martin Behaim in 1492. Globes of all sorts are manufactured today for various uses. Some are decorative objects (including popular reproductions of antique globes) or novelties (such as beach balls with maps printed on them). Most globes, however, are used in classrooms and libraries.

The manufacture of globes has changed over the years. In Behaim's time and for centuries afterward, globes were made by gluing a drawn or printed map onto a wooden sphere. The map itself consisted of a number of tapering strips called gores, designed to fit together neatly on a spherical surface. Some globes are still made this way, although the gores are now usually pasted onto spheres of plastic, chipboard, or heavy cardboard. Most globes, however, are mass-produced, printed onto sheet aluminum or chipboard, which is then molded into shape by hydraulic presses. Relief globes, generally made of molded plastic, feature raised plateaus and mountains and indented canyons and seabeds.

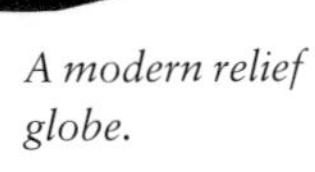

A modern relief globe.

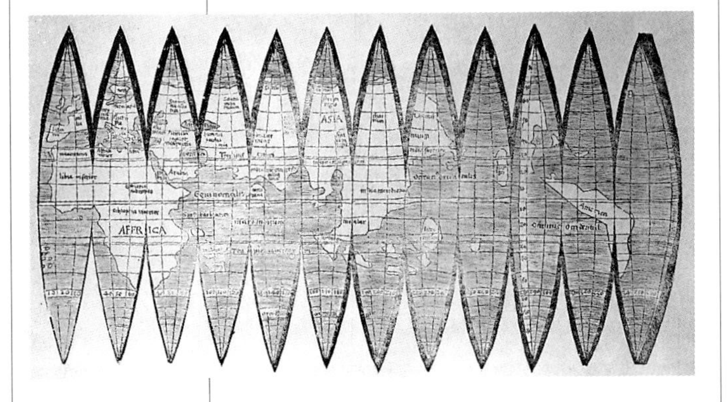

Gores for a globe of Martin Waldseemüller's 1507 world map. America is at the far right.

SEE ALSO
Behaim, Martin; Relief

Gough map

THE GOUGH MAP is a map of Great Britain that was made around 1335. England is very accurately mapped; Wales is somewhat less accurate; and Scotland is quite inaccurate. Despite its shortcomings, the Gough map was the best map of Britain that had been made up to the early 14th century—the best one of that period that survives today, at any rate.

The map includes some interesting features designed to make it easy to use. Straight lines between towns are marked

The Gough map is oriented to the east. Scotland is at the left; the cluster of large buildings near the upper right is London.

with the distances between the towns, and different symbols are used for walled and unwalled towns. Color is used to convey information; for example, the roads are red.

The maker of the Gough map is unknown. The map is called the Gough map because it was brought to public notice by the British historian Richard Gough in 1780. It is also sometimes called the Bodleian map because it is kept in the Bodleian Library at Oxford University in England.

Great circle

IMAGINE A huge, flat plane passing through the center of the earth like a razor blade through a pea. The line around the earth's surface where the earth is bisected by the plane is a great circle. In other words, a great circle is any line that goes completely around the earth, with the center of the earth at the middle of the circle.

The equator is a great circle. Every meridian of longitude forms a great circle with its opposite meridian. But a great circle can also bisect the earth at any angle, as long as its center is the center of the earth.

Great circles are important to transportation because the shortest line between any two points on earth forms part of a great circle. Airplane routes that are intended to cover long distances as quickly as possible are called "great circle routes." But because most people who are not navigators or pilots are not accustomed to visualizing the world in terms of great circles, these routes often seem contrary to common sense. For example, the great circle route—and thus the shortest flight path—from Washington, D.C., to Beijing, China, does not cross the Pacific Ocean at all, but instead passes over the north pole.

NORTH POLE

SOUTH POLE

A great circle can encompass the earth at any angle.

It may be easier to understand great circles by looking at a globe than by studying maps. Pull a piece of string taut between Washington and Beijing, or any two other points on the globe, and you have found the great circle route between those places. If you extend the piece of string all the way around the globe, you have traced a great circle.

Great Trigonometrical Survey of India

IN THE 18th century, the East India Company—the merchant firm that ran Great Britain's colony in India—launched the Survey of India, a program intended to map every part of the huge Indian subcontinent. Several decades later, after a number of maps had been produced, responsibility for the survey passed to an army engineer named William Lambton, who decided to start all over and make the survey more rigorously correct. Lambton wanted not just to map routes and communities but to make a map of India that would put each point in its exact latitude and longitude. To do this, he planned to use a surveying method called triangulation, based on the mathematical principles known as trigonometry. If he could establish a baseline of points whose position was carefully measured, he could work outward from there, relating each new survey point to a point that had already been surveyed. He would need to make thousands of sightings in an intricate pattern of interlocked triangles crisscrossing the subcontinent, but first he had to lay out a baseline all the way from the southern tip of India to the northern frontier of Kashmir in the Himalaya Mountains.

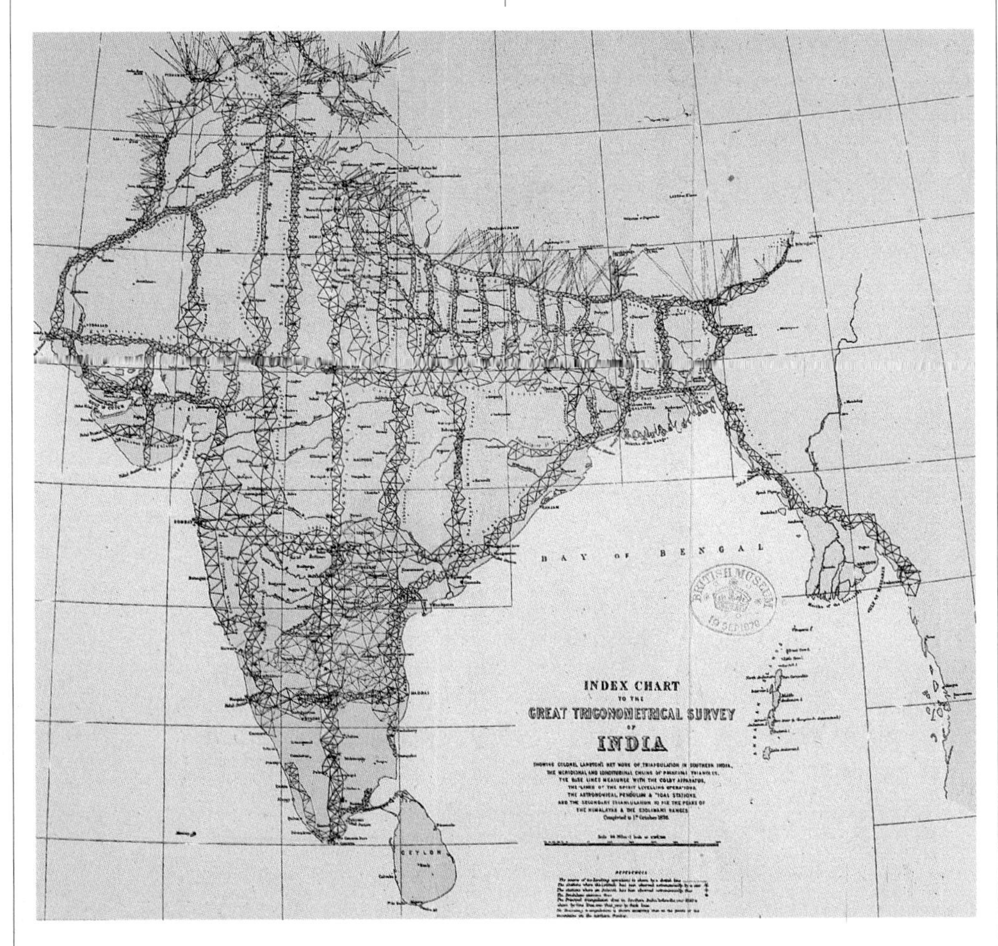

The British survey of India was begun in the 1790s and completed in 1876. Each tiny triangle on the survey grid represents a series of painstaking measurements, often carried out in grueling conditions.

Lambton's first task was to obtain a theodolite, an instrument for measuring horizontal and vertical angles at a distance. He ordered a special theodolite from London. This precious instrument had many adventures. The ship carrying it to Lambton in India was captured by the French, who were then at war with the British—but the French commanders courteously sent the instrument on to Lambton, who began taking sightings in 1806. In the years that followed, the theodolite was wheeled in carts or carried on poles across thousands of miles of mountains, deserts, rivers, and mountain gorges. It was hauled by teams of sweating workers up scores of observation towers so that Lambton could squint through it at distant flags or fires on other towers. Disaster fell on one such occasion, when a rope broke and the theodolite smashed into a wall. Refusing to give up on his project, Lambton took the ruined instrument into his tent and rebuilt it with his own hands in six weeks.

Lambton worked with single-minded determination, hoping against hope to finish his project before he died. But when he died in his survey tent in 1823 at the age of 70, the project was not even half completed. Fortunately, his chief assistant, George Everest, was equally single-minded. Everest took over the project and pushed the line of triangles northward. He had to overcome many obstacles: snakes, tigers, jungle, heat (and cold, in the mountains), and disease. Politics, too, could be tricky. Local rulers were often hostile to the British and suspicious—with good reason—of their intentions. Everest did what he could to disarm them, resorting to ploys like abandoning the usual survey flags for flags of plain green, the color of Islam, when he entered Muslim territories.

Everest retired in 1843, having guided the survey into the foothills of the Himalayas. No one knows, however, whether he ever saw the cloud-wrapped peak that rises on the border of Tibet and Nepal. The people of Nepal called it Sagarmartha, the Goddess Mother of the Earth, but the surveyors who determined in 1849 that it was the tallest peak in the world named it Mount Everest in George Everest's honor.

The Great Trigonometrical Survey of India was completed in 1876 by Andrew Waugh. It was the single largest mapping effort that had ever been undertaken—and one of the most arduous, carried out under extremely trying conditions. It was also one of the great success stories of cartography. When the survey was done, the larger task of mapping all of India was not yet complete, but a framework had been laid in place. It set a new standard of perseverance and excellence in surveying and cartography.

SEE ALSO

Rennell, James; Triangulation

Great U.S. Exploring Expedition

THE GREAT U.S. Exploring Expedition spent the years 1838–42 exploring and mapping the Pacific Ocean. It was an official, government-sponsored mission, voted upon by Congress and directed by the U.S. Navy. The expedition had several purposes: to gather scientific data about one of the least-known parts of the world; to bolster Americans' feelings of national pride and importance; and to give the United States a foothold on territorial claims and economic opportunities in the Pacific.

Charles Wilkes

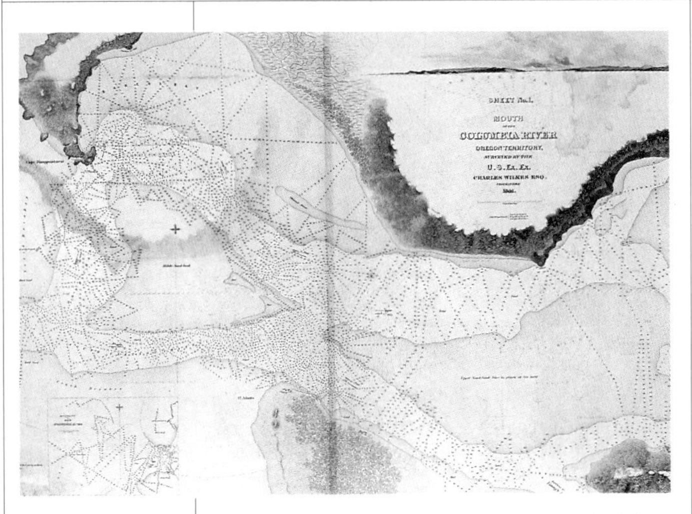

Wilkes's chart of the mouth of the Columbia River, in the Oregon Territory. Wilkes was far more gifted as a surveyor than as a leader.

The expedition was commanded by Lieutenant Charles Wilkes (1798–1877), a young officer with some training in mathematics and science. It included nine civilian scientists and artists. Although these individuals gathered 160,000 specimens and made thousands of pages of notes and drawings on the botany, zoology, geology, and anthropology of the places they visited, Wilkes insisted that the expedition's foremost goals were surveying and chart making. He gave his officers extensive training in the arts of taking sightings and drawing charts, and whenever they made landfall at a Pacific island, he kept them busy at these tasks from dawn until dusk.

In some ways, the Wilkes expedition was an epic of ineptitude, desertion, and bloodshed. Wilkes proved to be tyrannical, moody, and a poor leader of men. But he also proved to be a superb maker of charts. He and the officers under his supervision produced two atlases containing 241 maps. These covered the northwest coast of North America, the Columbia River, and 280 islands in the Pacific Ocean. So complete and accurate were Wilkes's charts of these Pacific islands that they were used by Allied forces during World War II (1939–45). Wilkes also produced a map of 1,500 miles (2,414 kilometers) of the coast of Antarctica. His expedition was the first to map a significant stretch of the Antarctic coastline, proving that the land sighted at far southern latitudes was a continent and not simply a collection of islands. Part of coastal Antarctica is still called Wilkes Land in his honor.

SEE ALSO
Antarctica, mapping of; Pacific Ocean, mapping of

FURTHER READING

Stefoff, Rebecca. *Scientific Explorers: Travels in Search of Knowledge.* New York: Oxford University Press, 1992.

Viola, Herman J., and Carolyn Margolis, eds. *Magnificent Voyagers: The U.S. Exploring Expedition, 1838–1842.* Washington: Smithsonian Institution Press, 1985.

Wolfe, Cheri. *Lt. Charles Wilkes and the Great U.S. Exploring Expedition.* New York: Chelsea House, 1991.

Greek mapmakers

SEE Ancient and medieval mapmakers

Greenwich, England

GREENWICH, ENGLAND, now a borough of London, was the site chosen for Great Britain's Royal Observatory in 1675. It was a center for the study of astronomy and navigation, subjects closely connected with mapmaking.

Greenwich—specifically the Royal Observatory—was used by the British Admiralty as the prime meridian, the starting point of zero degrees for the measurement of longitude. Latitude was standardized from the beginning because the starting point in the measurement of latitude is the equator, fixed midway between the poles. But the prime meridian, starting point for measuring longitude, could be anywhere that a mapmaker wanted to put it; there is no geographic

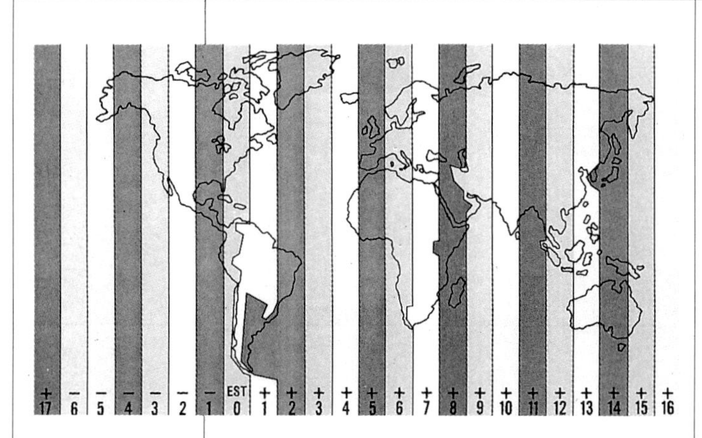

Time zones of the world. The Greenwich prime meridian provided a standard global time system as well as standard longitude.

reason for it to run through one place rather than another. Any meridian of longitude could be designated zero degrees—and many of them had been, at one time or another.

As far as the Spanish explorers were concerned, the prime meridian ran through Madrid, Spain, or the Canary Islands (claimed by Spain), and they reckoned their longitude accordingly. Around 1800 the French declared that the prime meridian ran through Paris. At about this time Pierre-Simon Laplace, a French mathematician, suggested that the world adopt a single prime meridian so that the maps of all nations would have a standard frame of reference and geography would become a standardized science. But 80 years later, at least 14 different prime meridians were still being used by the mapmakers of various nations. Mostly for patriotic reasons, cartographers ran the prime meridian through such places as Brussels, Belgium; Istanbul, Turkey; Lisbon, Portugal; Beijing, China; and Tokyo, Japan. Greenwich, however, was the most commonly used prime meridian, because the British Empire covered much of the world in the 19th century and British Admiralty charts were used by navigators of many nations.

In 1881, representatives of 29 nations met in Vienna, Austria, for the Third International Geographical Congress. They agreed that they needed a standardized, uniform system of meridians for measuring longitude, but they could not agree on where the prime meridian should run. Another conference was held in Washington, D.C., in 1884 to settle that question. After much debate and a few quarrels, representatives of the nations finally agreed to accept the Greenwich meridian as the prime meridian. From then on, longitude has been measured east and west of Greenwich. The world has also been divided into 24 time zones, with the Greenwich meridian as the center of the first time zone. Thus, the establishment of the Greenwich prime meridian made it possible for people around the world to share not only a standardized mapmaking system but standardized timekeeping as well.

SEE ALSO
Longitude

Gutiérrez, Diego

SPANISH MAKER OF CHARTS AND INSTRUMENTS

- *Born: 1485, Seville, Spain*
- *Died: 1554*

DIEGO GUTIÉRREZ served briefly as the pilot major, the official in charge of sea charts, for the Casa de la Contratación de las Indias in Seville, a government bureau that supervised voyages of exploration and trade. He was also the royal cosmographer to the king of Spain. His position at court meant that Gutiérrez heard the latest news from Spanish explorers and colonists around the world. He incorporated his chartmaking skills and geographic knowledge in several maps: one of the Atlantic in 1550, one of the world in 1551, and one

The Amazon River winds through exaggerated bends on Gutierrez's 1562 map of South America, made two decades after Francisco de Orellana became the first European to cross the continent.

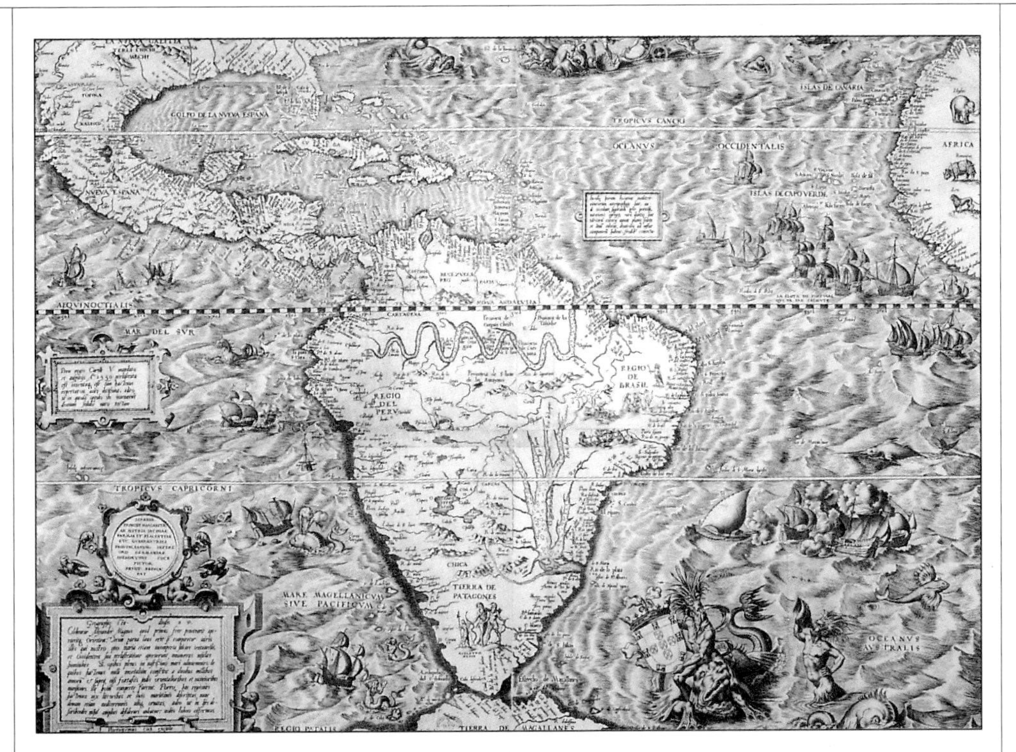

of America in 1562. His was the first printed map of America to name California.

SEE ALSO
Casa de la Contratación de las Indias

Gyogi maps

GYOGI MAPS are maps of Japan that are modeled on those made in the 8th century by Gyogi Bosatsu (670–749), a Korean Buddhist priest and cartographer who went to Japan as a young man. In addition to teaching the arts of building roads, canals, dams, and bridges, Gyogi also advanced the art of mapmaking. He is said to have made the first general map of Japan.

None of Gyogi's own maps survive, but a similar map from 1305 is preserved in the Temple of Ninna-ji, near Kyoto, Japan. It is the oldest known map in the Gyogi style. A map that appeared in a Japanese encyclopedia published between 1596 and 1614 is thought to be a reproduction of Gyogi's original map. His mapmaking style is evident in the work of cartographers who produced their own Gyogi maps.

Gyogi maps were sketches of Japan's various provinces and the roads that connected them. South was generally at the top of the map. The Gyogi maps did not give an accurate shape to the islands of Japan; instead they were concerned with the relationships of provinces to one another and the position of cities in the provinces. They were the dominant form of Japanese map until the 16th century.

SEE ALSO
Japanese mapmakers

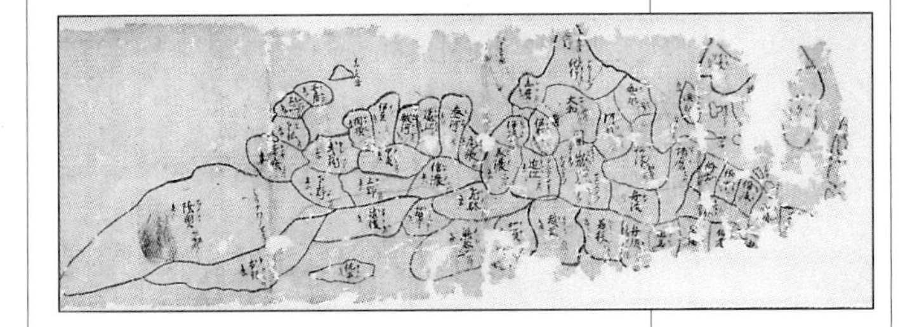

Part of a Gyogi map of Japan made in 1305. South is at the top; the borders are those of provinces.

Hachure

ONE WAY that cartographers show relief is by hachuring. Hachures are lines that show the direction and steepness of slopes. Steep slopes are indicated by short, thick lines drawn close together; gentler slopes are shown by fine, long lines more widely spaced.

Skillful hachuring can reveal the topography of hilly or mountainous regions. The technique was used on many 19th-century topographic maps. One artist of hachuring was a cartographer named G. H. Dufour, who used hachures in his atlas of Switzerland, published in 1833–63. Dufour's maps reveal not only the versatility of hachuring but also its biggest drawback: in very mountainous country, such as the Swiss Alps, the map is so covered with hachures of various lengths and thicknesses that it becomes quite dark, and the place names are almost impossible to read.

In the 20th century, hachuring has largely been replaced by other methods of illustrating relief.

SEE ALSO

Relief

Hakluyt, Richard

BRITISH GEOGRAPHER

- *Born: about 1552, in or near London, England*
- *Died: Nov. 23, 1616, London, England*

RICHARD HAKLUYT, orphaned at an early age, was brought up by a cousin who was a lawyer and was acquainted with a number of London merchants, travelers, and explorers. Even when quite young, Hakluyt was fascinated by books about geography and travel. He

A detail from G. H. Dufour's atlas of Switzerland. He used hachures to portray the steep terrain of the Alps.

wrote that he read "whatsoever printed or written discoveries and voyages" he could find.

Hakluyt went to college at Christ Church in Oxford. Later he gave lectures in geography there. His greatest pleasure, however, was getting to know sea captains, explorers, mapmakers, and others connected with travel and exploration. He devoted himself to promoting expeditions and colonial ventures, including the establishment of Sir Walter Raleigh's Virginia colony and the search for the Northwest Passage. He also sought out and published travel narratives, by both English and foreign explorers. In 1582 he published a book about the first voyages to the Americas, but his best-known work was *The Principall Navigations, Voiages, and Discoveries of the English Nation* (1589). In 1846, 230 years after Hakluyt's death, the Hakluyt Society was founded in London to publish scholarly editions of the travel and geography books of bygone days. Since that time, the Hakluyt Society has issued hundreds of volumes, keeping alive the works of travelers from ancient times to the recent past.

Harrison, John

BRITISH INSTRUMENT MAKER

- *Born: 1693, Wragby, England*
- *Died: 1776, London, England*

UNTIL THE late 18th century, sailors had a serious problem. They could not determine their longitude—that is, how far east or west they were of a known point. Since ancient times they had been able to measure their latitude, which is the distance north or south of the equator, but judgments about longitude were based on dead reckoning and guesswork. Much time was lost on long voyages, and sometimes ships were lost with all their crew and cargo because captains could not be sure whether they had gone too far or not far enough for their destinations.

In 1598 the king of Spain offered a large reward to anyone who could come up with a reliable method of measuring longitude. Portugal, Venice, and the Netherlands also offered rewards. Many scientists experimented with "the longitude," as the problem was called; among these was Galileo Galilei, who discovered Jupiter's satellites and thought of using them to determine longitude. By comparing the moment when a satellite passed in front of Jupiter to a table of that satellite's movements at known points of longitude, a navigator ought to be able to figure out how far east or west he was from a known point. Jean Dominique Cassini of France later refined this idea into a system for measuring longitude by observing the moons of Jupiter.

The timekeeper with which John Harrison finally solved the problem of measuring longitude at sea.

But the accurate measurement of longitude by this method required large telescopes, which were almost impossible to use effectively on the pitching deck of a ship at sea. It also required accurate, precise timekeeping. Better timekeepers than hourglasses and sundials were needed to time the moons of Jupiter and other heavenly bodies in their dance. When the pendulum clock was invented in the mid-17th century, it seemed to offer just what the navigators needed. Unfortunately, however, neither pendulum-driven clocks nor the newer spring-wound clocks worked well enough. Each clock told time a little differently, and the problem was magnified at sea, where the pitching motion of the ship affected the clock's mechanism. In addition, changes in climate during long voyages caused the metal to expand and contract, further addling the instrument. The measurement of longitude called for a standardized, reliable timepiece that would work for long periods at sea. This timepiece was finally produced by John Harrison.

Harrison was a self-taught mathematician and watchmaker. When he was 21 years old, the British Parliament offered a reward of £20,000—a very large fortune at that time—to anyone who could devise a method of determining longitude at sea. A few years later, Harrison began working on a marine chronometer, a clock designed for use at sea. It took him seven years to build his first chronometer. The government tested the device and demanded a smaller, still more accurate version. Harrison spent years perfecting the chronometer, making several versions that did not quite satisfy Parliament's Board of Longitude. Finally, in 1765, Harrison's "Number Four" chronometer passed a sea test. It seemed that the long-standing problem of longitude had at last been solved.

Harrison's problems were not over, however, for the board paid him only part of the reward, insisting that Harrison build two more chronometers before receiving the rest of the money. Harrison turned in "Number Five" when he was 78 years old, and still Parliament would not pay him. Harrison took his case to King George III, who is said to have exclaimed indignantly, "By God, Harrison, I'll see you righted!" Finally, in 1773, Harrison received the rest of his reward.

The benefits of Harrison's chronometer were felt immediately. Captain James Cook used it on his second and third Pacific voyages and reported that it worked perfectly. At last, navigators on the seas had a method of determining the exact longitude of their positions. Together with their latitude, this information told them not only where they were but how to reach their destination. Mapmaking took a great step forward, too, for the locations of islands and coastlines could now be plotted properly on charts.

SEE ALSO

Dead reckoning; Greenwich, England; Longitude; Navigation

FURTHER READING

Sobel, Dawn. "Longitude." *Harvard Magazine,* March–April 1994, pp. 44-52.

Hayden, Ferdinand Vandiveer

AMERICAN GEOLOGIST AND MAP SURVEYOR

- *Born: Sept. 7, 1829, Westfield, Massachusetts*
- *Died: Dec. 22, 1887, Philadelphia, Pennsylvania*

FERDINAND VANDIVEER HAYDEN was the first head of the United States Geological and Graphical Survey of the Territories, one of several government-sponsored organizations that mapped the frontier lands of the American West in the 19th century. Hayden's group mapped Colorado and the Yellowstone region; indeed, it was partly because of the public interest in Hayden's reports that Yellowstone was made the first national park. In 1879, Hayden's group was merged with three other surveys to form the U.S. Geological Survey (USGS), the principal mapmaking body in the United States since that time.

Ferdinand Hayden (center, hatless) and his survey team on a meal break in Wyoming. Before becoming a government surveyor, Hayden roamed the West as a fossil hunter.

Hayden believed that his greatest responsibility as a territorial geologist was to provide people with information about the location of useful resources such as coal, water, timber, and grazing land. His reports and maps emphasized this practical approach and were snapped up by businessmen, ranchers, and railroad developers.

SEE ALSO
Americas, mapping of

Hearne, Samuel

BRITISH EXPLORER OF CANADA

- *Born: 1745, London, England*
- *Died: 1792, London, England*

AS A YOUNG man, Samuel Hearne served in the British navy. In 1766 he joined the Hudson's Bay Company (HBC), the trading firm whose agents, in their search for beaver pelts, were taking a leading role in the British exploration of Canada. Hearne was assigned to an HBC trading post on the shore of Hudson Bay. In Hearne's time this outpost was called the Prince of Wales's Fort; today the city of Churchill, Manitoba, stands on the site. From there, Hearne was sent by his superiors on a mission to explore the region north and west of the fort. He was expected to find the way to the Arctic Sea and especially to look for any rivers or straits that might prove to be the long-sought Northwest Passage, the waterway that was believed to run through the northern part of North America, connecting the Atlantic and Pacific oceans. He was also supposed to keep an eye out for any commercial possibilities along the way, such as good beaver-hunting grounds, prime locations for whale fishing, or valuable minerals such as gold.

From Native Americans who visited the HBC fort, Hearne had heard rumors of a copper mine somewhere in the interior of Canada. He made two attempts to reach it, but both times he was forced to turn back when his food ran out or his Indian guides deserted him. His third attempt, in 1770–72, was a success. Guided by Matonabbee, a leader of the Chipewyan people, Hearne traveled

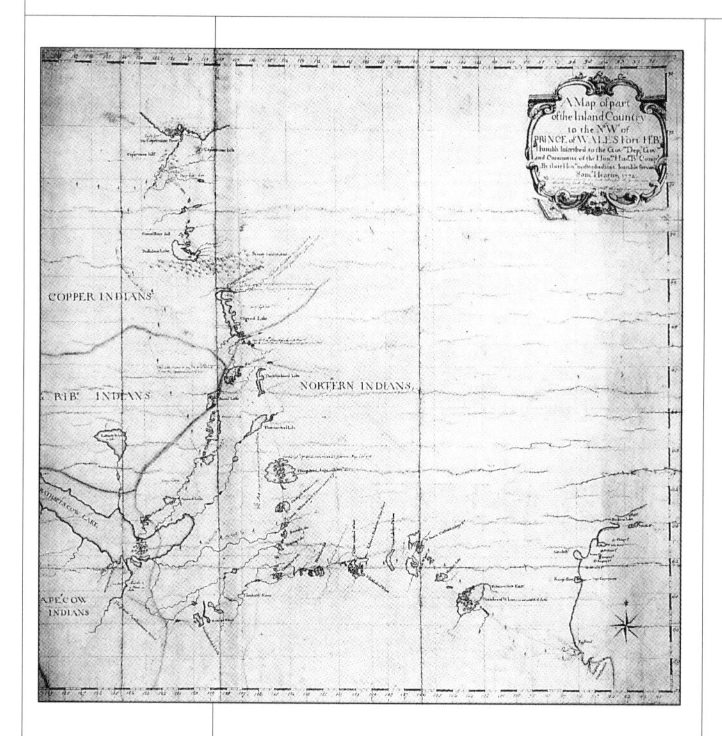

Part of Samuel Hearne's 1772 map, showing the route of his long journey across Canada to the Arctic coast (upper left).

overland into the heart of the Canadian wilderness. They found a river, which Hearne named the Coppermine, and followed it north until it emptied into the Arctic Ocean at a place Hearne called Coronation Gulf. Hearne realized that the barren, bleak land and the ice-choked sea did not offer a practical route to the Pacific: "This is no North West Passage," he wrote dejectedly in his journal. "It is a rocky suburb of Hell." Still, he located the Indians' copper mine and headed south to Great Slave Lake and then east, reaching the Prince of Wales's Fort after a journey of 3,500 miles.

Hearne returned to London in 1787 and died there five years later. His account of his journey—including a map he made of his route and his discoveries along the way—was not published until 1795, but his expedition had been much discussed by explorers and HBC officials since 1772. Hearne had not only proved that there is no Northwest Passage leading westward from Hudson Bay, but he had also explored more of the interior of northern Canada than any other European of his time.

SEE ALSO
Arctic, mapping of; Northwest Passage

FURTHER READING

Cooke, Alan, and Clive Holland. *The Exploration of Northern Canada, 500–1920.* Toronto: Arctic History Press, 1978.
Hearne, Samuel. *A Journey from Prince of Wales's Fort in Hudson's Bay to the Northern Ocean.* Edited by Richard Glover. Toronto: Macmillan, 1958.
Newman, Peter. *Company of Adventurers: The Story of the Hudson's Bay Company.* New York: Penguin, 1985.

Hecateus

SEE Ancient and medieval mapmakers

Hemisphere

A HEMISPHERE is half of a sphere. As applied to geography and cartography, the term can refer to any half of the earth. It is common to speak of the Eastern Hemisphere—consisting of Europe, Asia, Africa, and Australia—and the Western Hemisphere, consisting of the Americas. The world can also be divided into the Northern and Southern hemispheres, which are separated by the equator.

Henry, Prince of Portugal

PORTUGUESE PATRON OF EXPLORATION

- *Born: 1394, Porto, Portugal*
- *Died: 1460, Sagres, Portugal*

PRINCE HENRY (Henrique) of Portugal was neither an explorer nor a mapmaker. He sponsored the work of many explorers and mapmakers, however, and helped launch the era of European seagoing exploration. A Portuguese court chronicler reported that Henry's fate was foretold in the stars. At his birth an astrologer claimed that the prince "was bound to engage in great and noble conquests, and above all was he bound to attempt the discovery of things which were hidden from other men, and secret."

Henry was made governor of the Algarve, Portugal's southern province. There, on Cape St. Vincent, a promontory that juts into the Atlantic, he built a residence in a village called Sagres. He gathered a library of maps and books about travel and geography, and he invited geographers, astronomers, cartographers, instrument makers, sea captains, and merchant traders from all parts of Europe and the Mediterranean to visit Sagres and share their knowledge. He also encouraged local shipwrights to design vessels suitable for open-sea sailing. Henry's court became a center of geographic activity.

Henry is portrayed here as a martial prince, but symbols of scholarship—books and cartographic instruments—are shown at the upper left.

From Arab traders and travelers, Europeans had learned of the existence of Guinea, as the African lands south of the Sahara Desert were called. Caravans had crossed the desert from North Africa to Guinea for centuries, but Henry wanted to find a sea route to Guinea. Beginning around 1419, he sent a series of captains—many of them young nobles of his household—south into the Atlantic to probe the African coast. It took many years and many voyages before the Portuguese navigators made it around the great, dry, westward bulge of upper Africa and began exploring, trading, and slave raiding along the Guinea coast. In the year Henry died, his navigator Pedro de Sintra reached the coast at what is now Sierra Leone, the southernmost point that had yet been reached by a European explorer. Henry's captains had also explored the lower Gambia and Senegal rivers and explored and colonized the Madeira, Azores, and Cape Verde island groups in the eastern Atlantic.

Henry is sometimes called Prince Henry the Navigator. He was not a navigator, and he did not hold that title during his lifetime; it was coined by a British historian in the 19th century. It is also sometimes said that Henry launched the search for a seaway around Africa to India, but this is not strictly true. There is no evidence that he ever thought about rounding Africa. He did, however, want Portugal to explore West Africa and the Atlantic. His motives were probably mixed: the desire for trade and exploitation, the missionary urge of a devout Catholic prince to see Christianity spread in the world, and geographic curiosity.

Indirectly, though, Henry did set in motion the events that led Vasco da Gama around Africa to India at the end of the 15th century. The rest of the Portuguese royal family was sufficiently impressed with the discoveries made by Henry's captains that after the prince's death, the kings of Portugal continued to sponsor voyages of exploration southward along the flank of Africa. Mariners such as Diogo Cão and Bartolomeu Dias doggedly extended the mission of exploration that Henry had begun, until at last the continent was rounded and the sea road to India and the East lay open.

SEE ALSO

Cão, Diogo; Dias, Bartolomeu; Gama, Vasco da

FURTHER READING

Divine, David. *The Opening of the World: The Great Age of Maritime Exploration.* New York: Putnam, 1973.
Scammell, G. V. *The First Imperial Age: European Overseas Expansion, 1400–1715.* London: Unwin Hyman, 1989.
Stefoff, Rebecca. *Vasco da Gama and the Portuguese Explorers.* New York: Chelsea House, 1993.

Hereford map

ONE OF the most spectacular surviving medieval maps is preserved in Hereford Cathedral in England. It is a *mappa mundi,* or world map, painted around 1280 on vellum, a form of split calfskin that was especially prized as a writing surface. The Hereford map is a T-and-O map, in which the earth is shown in the form of a circle. It measures 5 feet (1.5 meters) across. The sea is green—except for the Red Sea, which is red. Rivers are blue.

An inscription around the map's border says, "The measurement of the world was begun by Julius Caesar," following a medieval tradition that Caesar had ordered the first survey of the world. Like other medieval world maps, the Hereford map is thought by scholars to be based in part on Roman Empire maps that are now lost. Its geography is rather poor: even familiar places such as Spain, Italy, and Scotland are not very accurately drawn. Routes of pilgrimage to the Holy Land (present-day Israel) are shown on the map, as are major cities such as Rome and Paris.

But the Hereford map reflects far more than geography. It also reflects a view of the world shaped by myth, legend, and Christian dogma. The influence of Christianity placed Jerusalem at the center of the map and the east—the direction of paradise—at its top. Among the pictures on the map are such biblical scenes as Christ with a band of angels, the Garden of Eden, Noah's ark, and the Tower of Babel. Mythical creatures are also present: mermaids, unicorns, and the bizarre races of people described by the imaginative 3rd-century geographer Solinus.

The Hereford map conveyed as much information about Christian dogma as about geography. Such maps were visual aids that helped drive home biblical lessons in an age when most churchgoers could not read. Jerusalem is at the map's center; above the map, Christ sits flanked by angels.

SEE ALSO

Mappa mundi; Solinus, Gaius Julius; T-and-O map

Herodotus

SEE Ancient and medieval mapmakers

Hipparchus

SEE Ancient and medieval mapmakers

Historical maps

HISTORICAL MAPS are maps that help us understand historical periods and

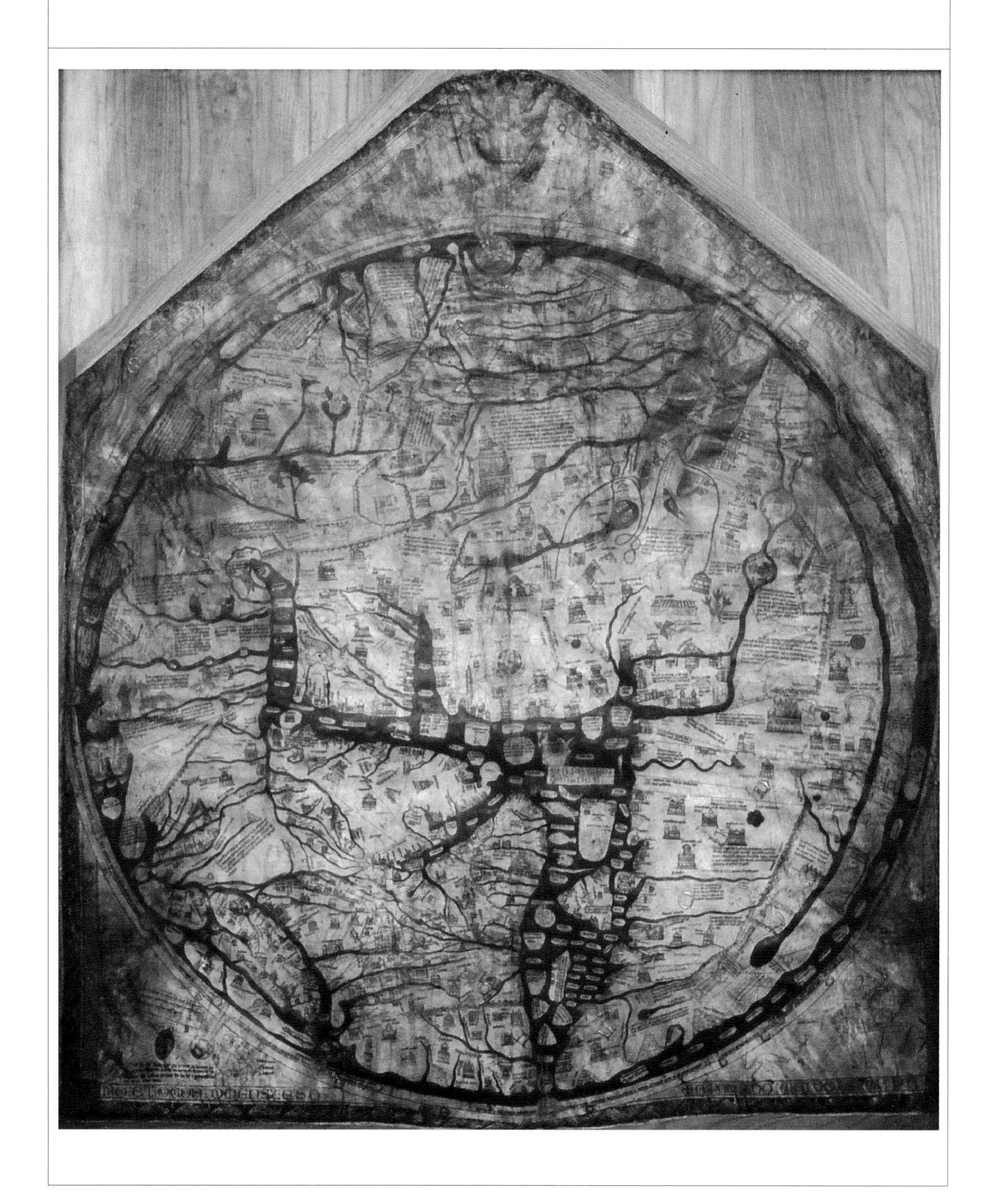

events. They are not necessarily old maps, dating from the historical period in question, although such maps can help us interpret history by showing us how people viewed their world during the time in question. But most historical maps are modern maps designed to reveal certain facts about the geography of bygone periods. Classrooms and textbooks are full of historical maps about everything from the provinces of the Roman Empire to the battles of the American Civil War.

Some historical maps show how part of the world looked not just at one point in history but over a period of time, by using contrasting colors or different kinds of line. For example, the same map could show the extent of the Roman Empire in blue or with a solid line, and the extent of the Byzantine Empire in red or with a dotted line.

Historical maps are produced for educational purposes or to accompany works on history. There are also historical atlases, which are collections of historical maps, usually with accompanying text that relates the events of history to the places in which those events occurred. Some atlases show multiple aspects of specific events or periods; for example, there are atlases of various wars and of different centuries.

FURTHER READING

Atlas of American History. New York: Facts on File, 1991.
Atlas of World History. Chicago: Rand McNally, 1983.
Harper Atlas of World History. New York: HarperCollins, 1992.
Historical Atlas of Canada. Toronto: Toronto University Press, 1987.
McEvedy, Colin. *The Penguin Atlas of Ancient History.* New York: Penguin, 1967.
———. *The Penguin Atlas of Medieval History.* New York: Penguin, 1961.
———. *The Penguin Atlas of Modern History.* New York: Penguin, 1972.

Hokusai, Katsushika

SEE Japanese mapmakers

Homem family

PORTUGUESE MAPMAKERS

THE FIRST of the Homem family of mapmakers was Lopo Homem, who died in 1565. Court records establish that he was working in Portugal as a geographer by 1517; judging from the number of payments he received from royal accounts, he must have been busy. In 1524 he served on a royal commission to determine the border between Spanish and Portuguese territorial claims in the Americas and Asia. Several of his maps survive, including one made in Florence, Italy, in 1554.

Lopo Homem's son Diogo Homem was exiled from Portugal in 1547, accused of involvement in a murder. Details of his life are unknown, but he made many maps and globes between 1557 and 1576. Some scholars believe that the first world atlas, Abraham Ortelius's 1570 *Theatrum Orbis Terrarum,* included some of Homem's work. A cartographer named Andrée Homem was active in Paris around 1565, but his identity and his relationship to Lopo and Diogo, if any, are not known.

SEE ALSO
Ortelius, Abraham

Homolographic projection

SEE Projections

Jodocus Hondius (right) and Gerardus Mercator, two of the leading figures in Dutch mapmaking. The Mercator-Hondius atlas combined the best work of the two cartographic families.

Hondius family

DUTCH MAPMAKERS

THE HONDIUS FAMILY was one of the major cartographic dynasties in the Netherlands during the golden age of Dutch mapmaking, from the late 16th to the late 17th century. The founder of the family business was Jodocus Hondius, born in Flanders in 1563. Hondius learned drafting and engraving and concentrated on making maps, globes, and charts. He did not draw all of his own maps; rather, he "borrowed" the work of other mapmakers—a common practice at the time—and re-engraved it to make a handsome product. He made the largest globes that had yet been constructed and also helped engrave the charts for *The Mariners Mirrour,* an influential sea atlas.

Hondius set himself up in business in 1595, and less than 10 years later he took advantage of an outstanding business opportunity. He bought from the Mercator family the copper plates that had been used to print Gerardus Mercator's atlas. Hondius reissued the atlas in 1606, using the Mercator plates and adding about 50 maps of Africa, Asia, and the Americas from other sources. The atlas was issued again in 1607, and that year a pocket-size edition was also published.

Jodocus Hondius died in 1612. The business of publishing the atlas was carried on by his widow, Coletta van der Keere, and his two sons, Jodocus II (1594–1629) and Henricus (1597–1651). After the death of Jodocus Hondius II, Henricus Hondius sold 34 of the Mercator-Hondius map plates to Willem Janszoon Blaeu, which were incorporated into a series of atlases published by the Blaeu family. Later,

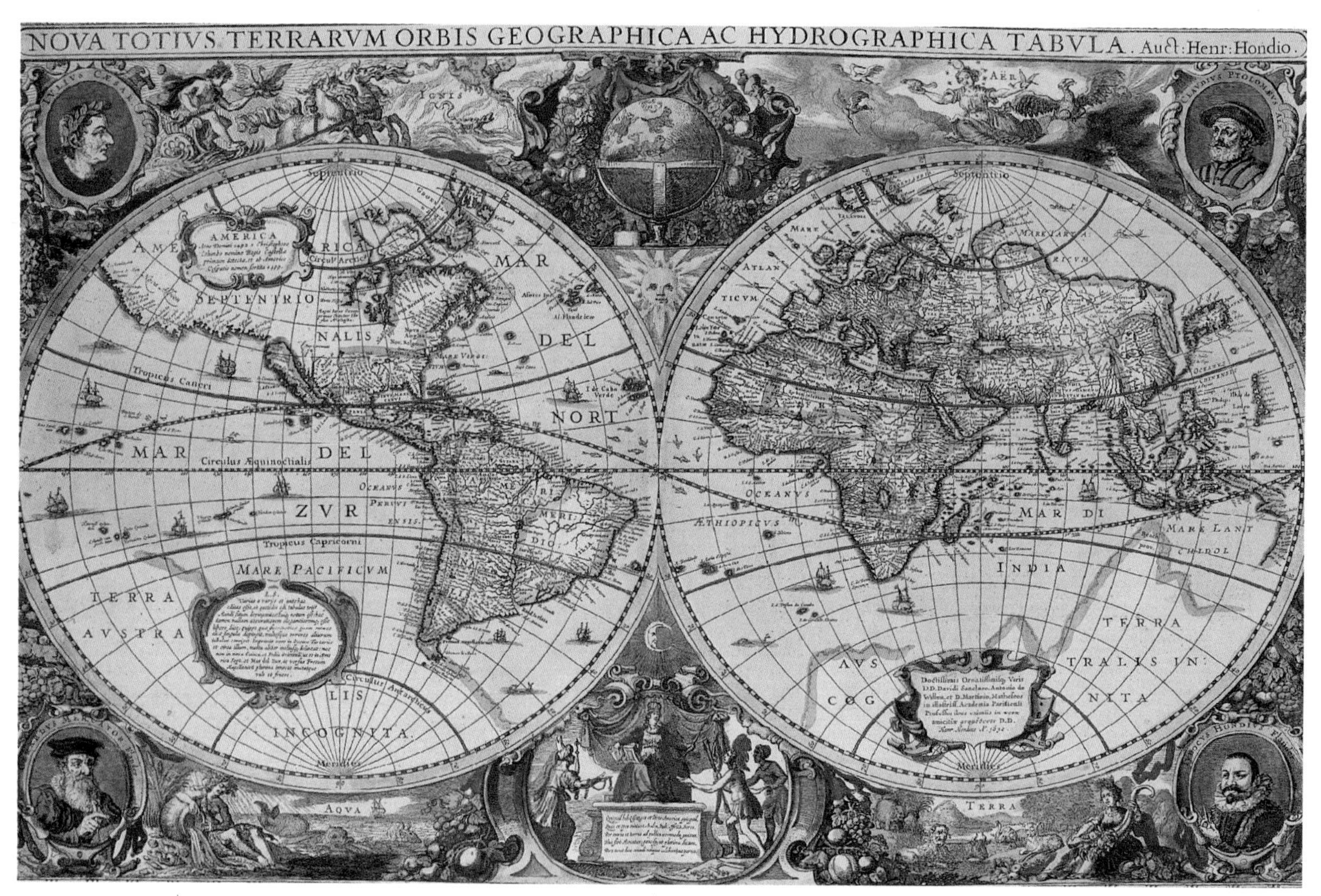

Henricus Hondius's 1630 world map is flanked by images of Julius Caesar (upper left), Ptolemy (upper right), Gerardus Mercator (lower left), and Jodocus Hondius.

however, Henricus Hondius went into partnership with his brother-in-law, Jan Jansson, to compete with the Blaeu atlases. Using the remaining Mercator-Hondius plates as well as maps from other sources, Hondius and Jansson reissued several versions of the Mercator-Hondius atlas. Jansson continued publishing the atlas after Hondius died.

SEE ALSO

Blaeu family; Jansson, Jan; Mercator family

Hudson, Henry

ENGLISH NAVIGATOR

- *Born: date and place unknown*
- *Died: probably 1611, Hudson Bay, Canada*

IN 1607 Henry Hudson was hired by an English merchant company to find the Northwest Passage. He sailed to Greenland and Spitsbergen, reaching a point about 700 miles (1,126 kilometers) from the north pole—the northernmost point that had been reached by any European.

In 1609 Hudson was hired by a Dutch company to sail through the Northeast Passage, from Europe to Asia through the waters north of Russia. But when his crew objected to the intense cold of the far northern region, Hudson turned around and set course for North America instead. On this trip he explored the east coast of America from present-day North Carolina to New England. He not only entered Chesapeake Bay and Delaware Bay but also sailed up the Hudson River as far as the site of present-day Albany, New York. Because he was sailing for the Dutch, Hudson's explorations became the basis for the Dutch claim to New York, which for a time was a Dutch colony called New Netherland.

Hudson's last voyage began in 1610. Again he sought the Northwest Passage. He entered a huge bay on the coast of eastern Canada, believing it would lead into the Pacific Ocean. After a cold, hungry winter in the bay he tried to press westward, still searching for an opening to the Pacific, but his crew mutinied. The rebellious crewmen set Hudson, his son, and a handful of loyal sailors adrift in a small boat before fleeing back to England. Hudson's fate is unknown; no trace of the marooned party has ever been found. The mutineers who returned to England saved themselves from being executed by producing Hudson's charts, which were considered so valuable that the mutineers' lives were spared.

For some time after the mutiny, geographers believed that Hudson really had found the Northwest Passage. In 1612, a mapmaker named Hessel Gerritz, working for the Dutch East India Company, drew a chart showing "the Great Sea first discovered by Mr. Hudson." Today, that "Great Sea" is known as Hudson Bay. Although it is a vast body of water, it is separated from the Pacific Ocean by all of central and western Canada.

SEE ALSO

Arctic, mapping of; Northeast Passage; Northwest Passage

FURTHER READING

Brown, Warren. *The Search for the Northwest Passage.* New York: Chelsea House, 1991.

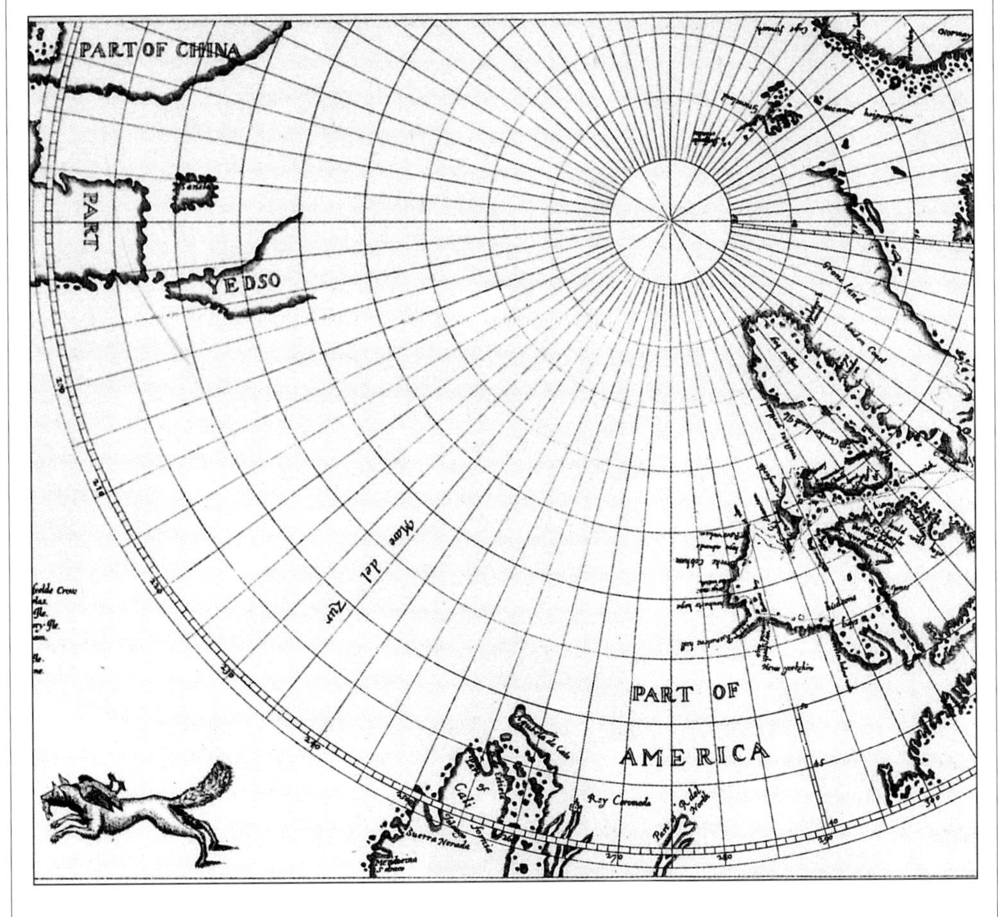

Hudson Bay appears near the lower right of this 1631 English map. Despite Hudson's failure to find a Northwest Passage out of the bay that bears his name, the map shows a possible waterway leading off to the west.

Lehane, Brendan, and Time-Life editors. *The Northwest Passage.* Alexandria, Va.: Time-Life Books, 1981.
Morison, Samuel Eliot. *The Great Explorers: The European Discovery of America.* New York: Oxford University Press, 1978.

Humboldt, Friedrich Wilhelm Heinrich Alexander von

GERMAN SCIENTIST AND EXPLORER

- *Born: Sept. 14, 1769, Berlin, Germany*
- *Died: May 6, 1859, Berlin, Germany*

ALTHOUGH HE came from a wealthy, aristocratic family, Alexander von Humboldt chose a life of work and action. He studied the sciences, especially geology, and was trained as a mining engineer. From 1799 to 1804 he and a French botanist named Aimé Bonpland traveled in Spanish America, covering more than 6,000 miles (9,656 kilometers) in Venezuela, Colombia, Ecuador, Peru, and Mexico.

Although much of this ground had been covered by earlier explorers, Humboldt and Bonpland made several significant contributions to geography.

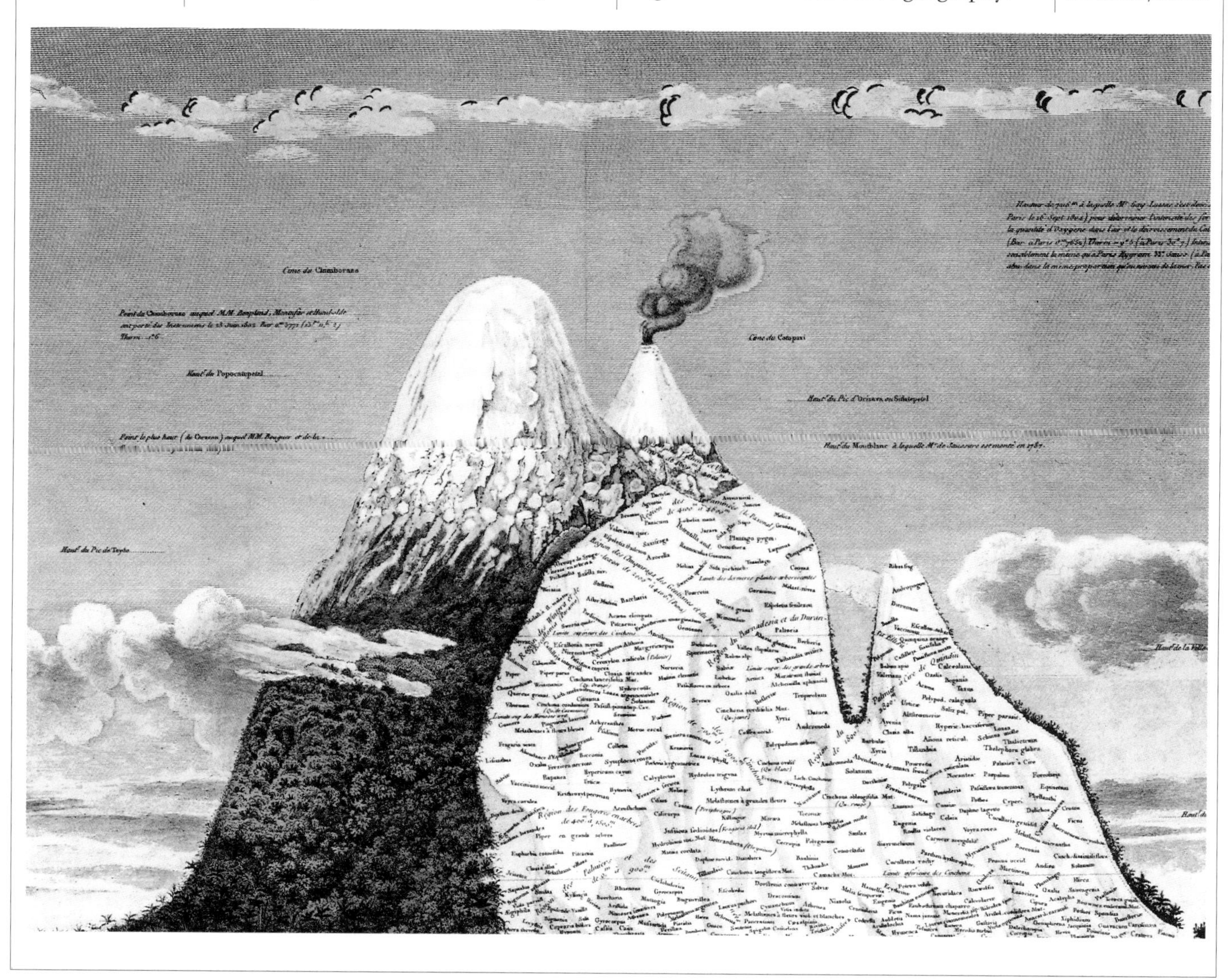

Humboldt's map of Chimborazo, an Ecuadoran volcano, shows what kinds of vegetation grow on the mountain. In Humboldt's hands, the map became a versatile scientific tool.

They explored and mapped a unique topographic feature called the Casiquiare Canal, a natural channel that links the Orinoco River of Venezuela with Brazil's Amazon River network. They also climbed Mount Chimborazo, a volcano in Ecuador, to a height of 19,286 feet (5,878 meters), setting a world mountain-climbing record that stood unbroken for 36 years.

A meticulous scientist, Humboldt amassed enormous amounts of data during his travels. His interests were wide-ranging: animal and plant life, weather, magnetism, ocean currents, and geology. Upon his return to Europe, he spent more than 20 years in Paris, writing books about his finds and supervising their publication.

Humboldt was responsible for a number of cartographic innovations. By drawing simplified maps and cartograms to illustrate such things as the geographic relationships of ancient Mexican cities, or the types of plants that grow at different altitudes on the sides of mountains, he pioneered the use of charts and diagrams to convey a great deal of information in an efficient, easy-to-read way. He also invented the isothermic map, a form of weather and temperature map widely used today. It was Humboldt's notion that a map could be used to convey much more than just topography: it could present social, meteorological, agricultural, or geological information—the whole "diversity of form or feature on the surface of our planet," as he wrote in his 1807 book *Essay on the Geography of Plants.*

SEE ALSO

Cartogram; Isogram

FURTHER READING

Botting, Douglas. *Humboldt and the Cosmos.* New York: Harper & Row, 1973.
Gaines, Ann. *Alexander von Humboldt: Colossus of Exploration.* New York: Chelsea House, 1991.

Hydrography

HYDROGRAPHY IS the study of the earth's bodies of water. A more specific use of the term refers to the act of charting navigable seas, lakes, and rivers. Hydrographic charts show the contours of seafloors, lake bottoms, and riverbeds, as well as the depth of the water at various points and the speed and direction of currents. People who prepare such charts are called hydrographers. Their work has many practical applications in addition to assisting navigators. Hydrographic studies are a necessary part of building and operating dams, bridges, canals, harbor works, jetties, and other engineering projects that involve water.

Hypsometric shading

HYPSOMETRIC SHADING is one of several techniques that mapmakers use to show relief on the flat surface of a map.

Transparent washes of color—sometimes called tints—are applied to the map, with each shade representing an area of a particular average elevation. For example, a series of greens may be used to show levels below sea level or close to it, while yellowish browns and then reddish browns are used for increasingly higher elevations; the highest points of all are white. The same technique can be used on maps of the oceans, with the darkest shades of blue covering the deepest regions, and lighter shades used for shallower areas. The term *hypsometric* comes from the hypsometer (derived from the Greek words *hypsos,* meaning "height," and *metron,* meaning "mea-

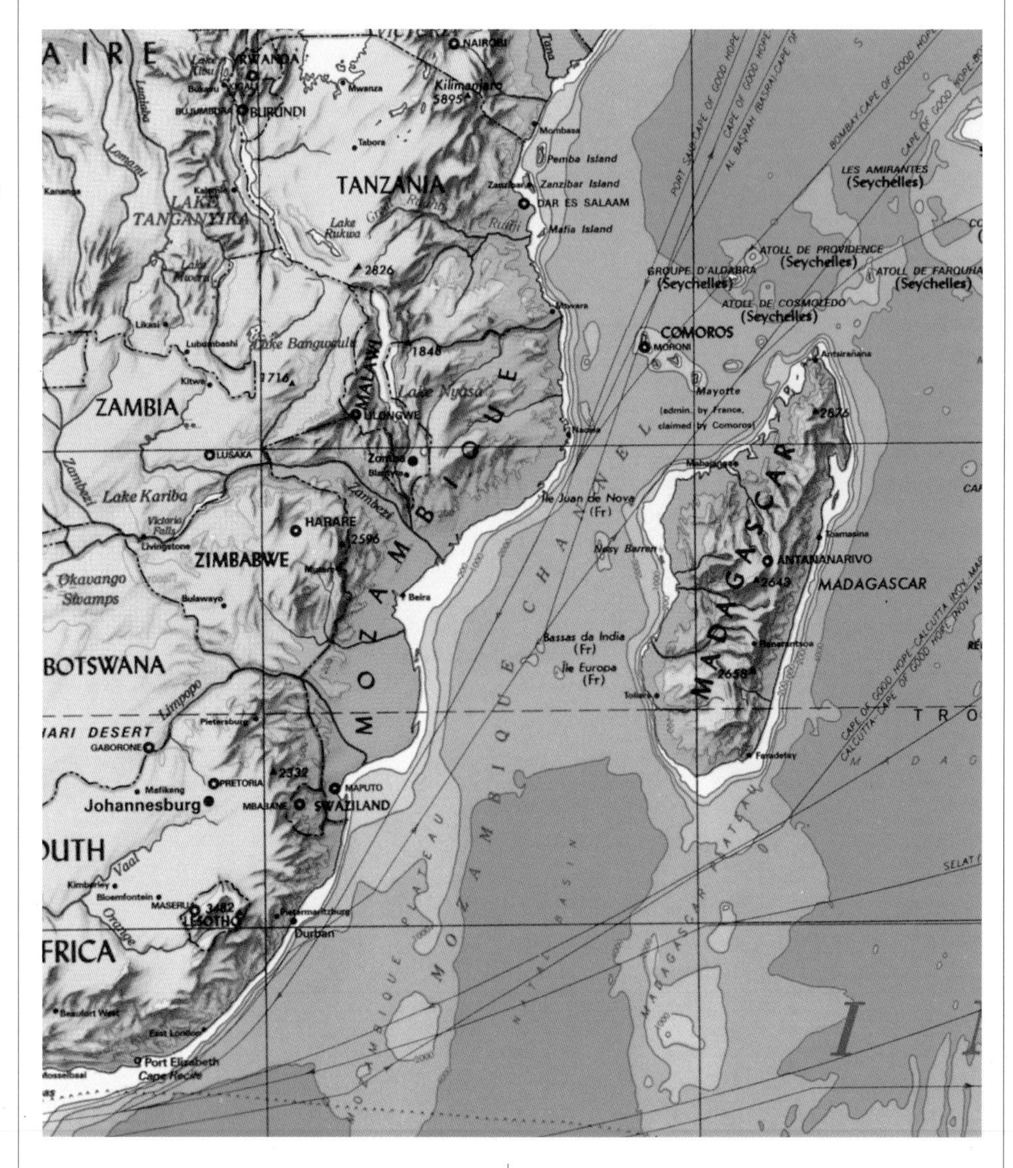

A hypsometric-bathymetric map of southeast Africa, Madagascar, and part of the Indian Ocean uses color shadings and contour lines to show the elevation of the land and the depth of the sea.

sure"), an instrument used to measure altitude. Information about altitude is thus hypsometric.

Ibn Battuta

MOROCCAN TRAVELER

- *Born: 1304, Tangier, Morocco*
- *Died: 1368, Morocco*

IBN BATTUTA (Sheikh Abu Abdallah Muhammad ibn Abdallah ibn Muhammad ibn Ibrahim al-Lawati) was born to an aristocratic family. He was trained as an Islamic scholar and judge, and his learning made him welcome in Muslim communities all over Africa and Asia in the 14th century.

Ibn Battuta's journeys began in 1325 with a pilgrimage to Mecca, during which, he said, he was filled with a desire "to travel through the world." He spent three decades traveling, accompanied by a retinue of wives and attendants. In Asia he visited Persia, Afghanistan, India, Sri Lanka, Sumatra, and China; in Africa he ranged from Kilwa and other Arabic trading ports on the east coast to the Niger River and the kingdom of Mali in West Africa. He was not an explorer in the strict sense of the word, for he generally traveled with caravans or merchants along well-known trade routes, but he was a keen observer of geography.

Ibn Battuta is thought to have been the most widely traveled person of his time, having covered more than 75,000 miles (120,700 kilometers). *Rihlah* (*Travels*), a book he wrote about his journeys, is a treasure trove of information about medieval Islam. Its details about Mali and other black African kingdoms are particularly valuable to modern scholars. In Ibn Battuta's own time, too, this information was eagerly absorbed by Europeans, who heard stories about Ibn Battuta's travels from Arabic-speaking traders and travelers. Few Europeans knew anything at all about the sub-Saharan African realms until after Ibn Battuta's account of them had begun to circulate. A European map called the Catalan atlas, made in the late 14th century, shows that its maker, Abraham Cresques, knew something about the Islamic world—in particular, the map shows the Saharan caravan routes and the gold-crowned black king of Mali described by Ibn Battuta. A map made by Portuguese cartographer Gabriel Vallseca in 1438 also shows the Atlas Mountains of North Africa and, south of them, the Sahara Desert threaded by trade routes and dotted with prosperous-looking kingdoms. Ibn Battuta's description of the wealth of Mali and Guinea helped turn the attention of the Portuguese to Africa. It thus contributed to the series of voyages of exploration along the West African coast that were sponsored by Prince Henry of Portugal in the 15th century.

SEE ALSO

Islamic geographers and mapmakers

FURTHER READING

Abercrombie, Thomas J. "Ibn Battuta: Prince of Travelers." *National Geographic*, December 1991, 2–49.

Ibn Battuta. *Travels, A.D. 1325–1354*. Translated by H. A. R. Gibb. 3 vols. Cambridge, England: Cambridge University Press, 1958–71.

Idrisi, Abu Abdullah Muhammad

ARAB GEOGRAPHER

- *Born: 1099, possibly in Ceuta, Morocco*
- *Died: about 1166*

IDRISI IS thought to have studied at Cordova, a center of Muslim learning in Spain, before traveling through North Africa and the Middle East. In the mid-12th century he was invited to join the court of King Roger II of Sicily, where he was treated as an honored guest and invited to share his geographic knowledge.

Legend says that Idrisi made an engraved silver globe or map of the world for Roger, although this treasure was later destroyed. But Idrisi's geographic book, which took 15 years to complete, has survived. Idrisi gave it a flowery Arabic title that can be translated as "Delight of him who wishes to traverse the regions of the world," but it is more commonly known as the *Kitab Rujua* (*Book of Roger*). It appeared in 1154 and included a map of the world on 70 sheets.

Idrisi's map, like many Arab maps, was oriented so that south was at the top. It was based on Ptolemy's geography but also drew on the contemporary lore of sailing instructions and sea charts. A slightly abridged version of the map, possibly also prepared by Idrisi, appeared in 1161. Idrisi's map was copied many times over the succeeding centuries

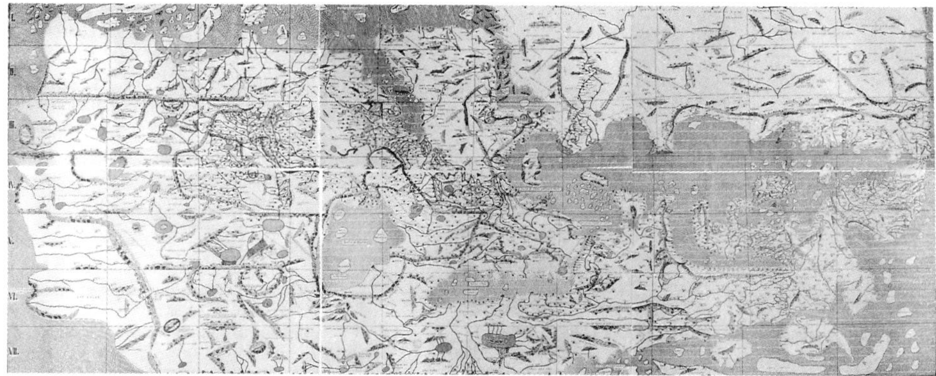

Oriented with south at the top, Idrisi's world map echoes practical sailing guides as well as ancient geographers. Arabia and the Middle East are the most accurate portions of the map.

by cartographers and historians in Europe and the Islamic world alike. One such copy was made as late as 1620 by Petrus Bertius, a Belgian mathematician and geographer who found employment at the court of France. Bertius recreated Idrisi's view of the world in a map called "New Map of the World, from the Nubian Geography" (Idrisi's geography text had been translated into Latin under the title *Geographica Nubiensis,* or *Nubian Geography;* Nubia refers to part of the Nile River Valley.) Although the Americas were being explored in Bertius's time, he did not show them on the map, limiting himself to the three continents—Europe, Asia, and Africa—that had been known to Idrisi. Bertius made the map as an exercise in scholarship rather than for any practical purpose, for by the early 17th century Idrisi's geography was out of date.

Imaginary places

SOME WELL-KNOWN maps depict places that have never existed outside the imagination. These are not maps of the world that show mythical or theoretical places mixed in with real locations, but rather maps that are deliberate works of fiction and imagination.

Most maps of imaginary places were made to accompany stories. Maps not only make it easier for the reader to follow the activities of characters who are moving through a fictional landscape but also add an air of "realness" to their adventures. One famous example is Robert Louis Stevenson's map of Treasure Island. Stevenson did not draw the map to accompany the book, as one might suppose. Instead, he made the map in 1881 as a game to amuse his stepson, but before long the map suggested a story, which eventually became the novel *Treasure Island.* The map that was printed in the book was quite realistic, with a compass rose, rhumb lines, and topographic markings; it looked very

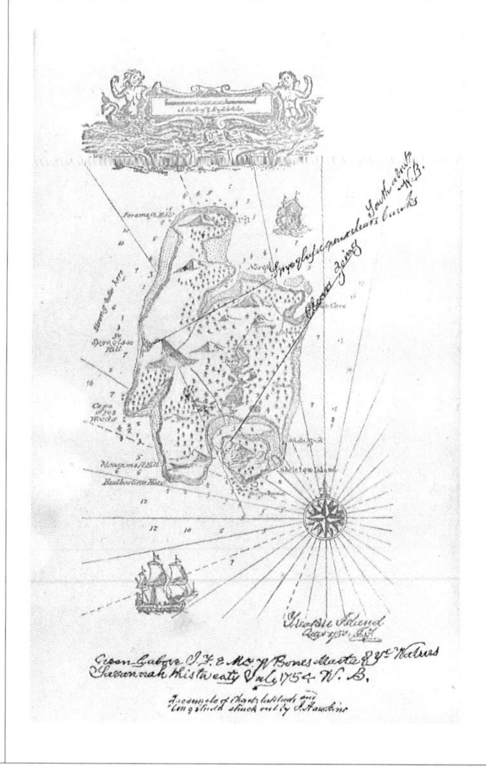

Robert Louis Stevenson's map of Treasure Island shows landmarks such as Spyeglass Hill and Skeleton Island.

much like the nautical maps that many 19th-century readers had seen.

Both before and after Stevenson, a number of fictional places have been mapped by their authors. Sir Arthur Conan Doyle, the creator of Sherlock Holmes, drew a map of the imaginary South American realm of Maple-White-Land to illustrate his 1912 adventure novel *The Lost World.* J. R. R. Tolkien, who drew maps of his fictional world, Middle Earth, to illustrate *The Hobbit* (1937) and *The Lord of the Rings* (1954–55), set the fashion for contemporary fantasy novels, most of which contain maps.

FURTHER READING

Manguel, Alberto, and Gianni Guadalupi. *The Dictionary of Imaginary Places.* New York: Harcourt, Brace, Jovanovich, 1987.

Indies

"THE INDIES" is a geographic term that has fallen out of use in modern times. It still appears on some maps, however, as well as in discussions of exploration and colonization. In the 13th and 14th centuries, at the dawn of the age of European expansion, "the Indies" referred to a vague, almost unknown region south and east of the known world that was the source of the spices, gems, and silks that reached Europe along Arab trade routes. In the following centuries, as Europeans began learning more about the world, the term "Indies" became rather flexible. It almost always included India but could also include China, Southeast Asia, Japan, the Spice Islands (the Moluccas, in what is now Indonesia), and the other islands that surround the East Asian mainland.

When Christopher Columbus bumped into the Americas, he thought he had reached the Indies. He called the Native Americans Indians, a mistake that became common usage. When geographers realized that Columbus had actually landed somewhere other than Asia, they began referring to his landfall as the West Indies to differentiate it from the East Indies, his original goal. These terms remained in use for centuries, throughout the colonial era, and are still used occasionally.

International date line

THE INTERNATIONAL date line is an imaginary line on the earth's surface that marks the point at which one day ends and the next begins. When it is midnight at the international date line, one day changes to the next—for example, Friday night becomes Saturday morning.

The date line is the meridian of longitude that lies halfway around the world from the prime meridian, the meridian that runs through Greenwich, England, and is considered to be zero degrees of longitude. When it is midnight on the date line, it is high noon on the prime meridian. Thus the date line is the 180th degree of longitude, which runs through the Pacific Ocean. The line does not follow the meridian straight north and south from the equator, however; it zigzags in several places so that island groups will not be divided between two days. For example, the line takes a pronounced westward jog off the coast of southern Alaska so that it will always

Each new day officially begins in the middle of the Pacific Ocean.

be the same day in the Aleutian Islands as in the rest of Alaska. Most world maps and globes, as well as many maps of the Pacific region, show the international date line.

SEE ALSO
Longitude

International Geophysical Year (IGY)

THE INTERNATIONAL Geophysical Year (IGY) was actually two and a half years: the 30-month period from July 1, 1957, to December 31, 1959. During this time, 30,000 scientists from 70 countries took part in an international research effort organized by the International Council of Scientific Unions to study the earth and its environment.

The main areas of study were the upper atmosphere, weather, the oceans, glaciers, and geology. Maps were involved in every step of the process, from planning individual projects to finding sites for research stations to communicating the data gathered by researchers. In addition, many IGY projects were directly concerned with mapping. Observers in many stations made new determinations of latitude and longitude so that existing maps could be corrected and new ones prepared. Airplanes, balloons, and rockets were also used to take pictures that helped cartographers refine their maps.

The polar regions received close attention during the IGY. Antarctica was the scene of much surveying activity, and the shape of the continent beneath its permanent mantle of ice began to be known for the first time. The IGY paved the way for international cooperation on a host of later scientific projects, including many cartographic efforts.

SEE ALSO
Antarctica, mapping of

International Map of the World

IN 1891, the Fifth International Geographical Congress was held in Bern, Switzerland. At the conference, a young Swiss geography professor named Albrecht Penck proposed that the nations of the world should work together to create a uniform International Map of the World. Every part of the world would be mapped on the same scale and the same projection, so that any two sheets of the map could be joined together or easily compared. In addition, the same colors and symbols would be used throughout.

Penck's idea was the most ambitious cartographic project ever proposed. For years afterward, the nations talked about it but took no action. When the International Geographical Congress met again in 1904, Penck displayed samples of maps made according to his plan. The world's geographers agreed that a uniform world map would be immensely useful, but there was much debate over how the project was to be carried out. Finally, in 1913, the International Conference on the International Map of the World met in Paris and agreed on standards for the map sheets. Historians of cartography agree that at this meeting the science of mapmaking attained a new level of maturity.

The representatives who attended the conference agreed that the map would show topographic features but not political boundaries, and that place names would be shown in the local spelling of each country. The scale of each map would be 1:1,000,000, or one to a million; a centimeter on the map would equal 10 kilometers on the ground, and an inch on the map would equal 15.78 miles on the ground. Because of this choice of scale, the International Map of the World is sometimes called the Millionth Map or the 1/M map.

Work on the map has been slow and erratic, interrupted by two world wars. It took 25 years, for example, to make the series of maps covering Hispanic America. If it is ever completed, the International Map of the World will consist of 2,122 sheets showing the entire land area of the globe. These will have been prepared by various national cartographic surveys as well as by other mapmaking bodies, such as the American Geographical Society, a private organization that produced the maps of Hispanic America for the International Map. A 1973 report by the United Nations, which coordinates the work of the individual mapping agencies, said that "about half the land area of the globe" remained to be mapped by the project's standards. As of the early 1990s, some 750 sheets of the International Map of the World had been published—but some of them are known to be inaccurate and out of date. In recent years, some cartographic specialists have suggested that the International Map of the World need not be completed in accordance with the original plan because it has been made obsolete by satellite mapping and other new cartographic technologies.

Interrupted projection

SEE Projections

Isidore's simple outline provided the structure used in more elaborate medieval maps, such as the Hereford and Ebstorf maps.

Isidore of Seville

SPANISH ARCHBISHOP, THEOLOGIAN, AND HISTORIAN

- *Born: about 560, probably in Cartagena, Spain*
- *Died: 636, probably in Seville, Spain*

ISIDORE OF SEVILLE, who was made a saint of the Roman Catholic church in 1598, became archbishop of Seville in 600. He was a scholar who addressed himself to many branches of learning, including grammar, history, theology, and science. His most important work was an encyclopedia in 20 sections in which he summed up everything that he had learned from his study of earlier writers. This book became one of the leading medieval reference works. More than a thousand hand-written copies of it have survived. It included a simple T-and-O map showing the division of the world into the continents of Asia, Africa, and Europe. This symbolic map was widely copied and became one of the standard ways the world was pictured by Europeans in the Middle Ages. It was drawn in hundreds of manuscript versions; the earliest known printed version of a T-and-O map dates from a 1482 edition of Isidore's work.

SEE ALSO

Ancient and medieval mapmakers; Ebstorf map; Hereford map; T-and-O map

Islamic geographers and mapmakers

WHEN THE Roman Empire came to an end in the late 5th century A.D., science and scholarship were obscured for a time in Europe, and much of the knowledge accumulated by the ancient Greeks and Romans was neglected or forgotten by Europeans. The Arabs, however, continued to read, use, and revise the classical works of geography and astronomy. This knowledge became part of the Islamic culture that arose between the 7th and 12th centuries, when Islam spread from Arabia throughout the Middle East, North Africa, Spain, and parts of South Asia. Much of what medieval Europeans learned about science and medicine came from the Arab world, either passed on from the works of the Greeks and Romans that had been translated into Arabic or discovered independently by Muslim scientists and scholars.

Among other things, the Arabs preserved the *Geography* of Ptolemy, which was lost to the Western world for centuries. Arab cartographers made maps based on Ptolemy's geography and added their own discoveries to these maps. In particular, the Arabs knew a great deal about the Red Sea, the Persian Gulf, the Indian Ocean, and the west coast of India because they had been carrying on trade between Arabia, Persia (now Iran), and India since ancient times. Arab merchants also ventured as far afield as Korea and China. In 1490, Ahmad Ibn Majid, a pilot on the Indian Ocean, summed up his knowledge in a book that, he said, described "many unknown places and things." By that time the Arabs were skilled navigators, able to lay courses using compasses. They also used sextants to determine latitude.

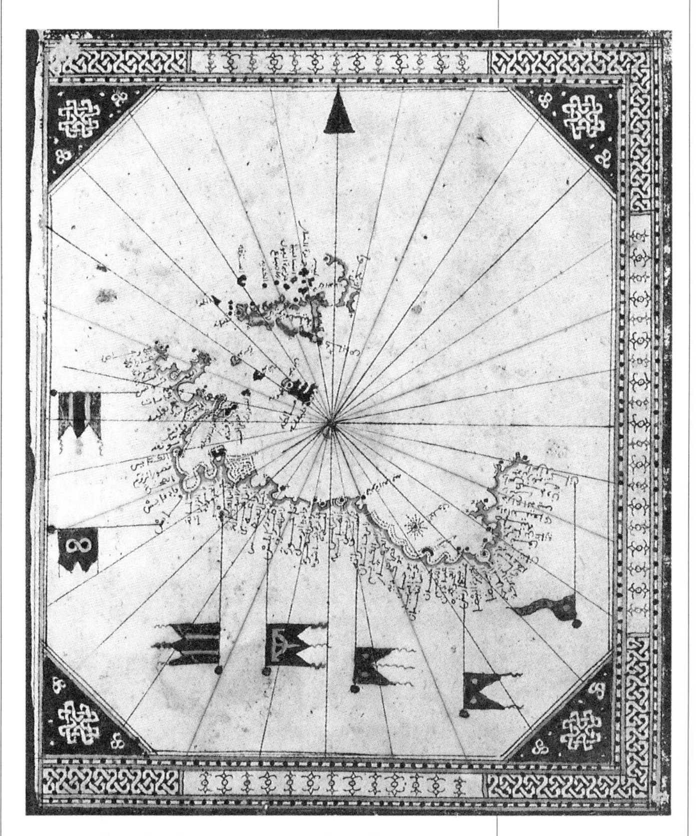

Part of the Italian coastline, from a 16th-century portolan atlas made by the al-Sifaqsis, a Tunisian mapmaking family. South is at the top.

Another source of geographic information was the pilgrimage. Pious Muslims were required to visit Mecca, an Islamic holy city in Arabia. Many pilgrims extended their journeys to see other parts of the Islamic Empire. One such was Ibn Battuta, who journeyed through much of Asia and part of Africa in the 14th century, becoming the farthest-ranging of all medieval Muslim travelers. He and other Islamic pilgrims and wanderers added greatly to the geographic knowledge the Arabs had acquired from the ancients.

The Arabs were skilled in astronomy, which led naturally to geography and mapmaking because the stars were used as reference points to fix the locations of places on earth. One of the most important Arab geographers was al-Biruni, who was born in Central Asia in 973 but spent his adult life in Persia, Turkey, and Iraq. Al-Biruni speculated that the Indian and Atlantic oceans were connected by a sea passage south of Africa, an idea that happened to be true, even though it was contrary to Ptolemy. Al-Muqaddasi, a Syrian geographer who died in the year 1000, wrote a book called *The Best Description for an Understanding of All Provinces,* which covered the geography of the whole Islamic Empire.

Although Islam was regarded as a threat to Christianity, the eminence of Arab geographers was recognized by many Europeans. Arab maps circulated in the courts and universities of Europe during the late Middle Ages. In the 12th century, King Roger II of Sicily brought the geographer Idrisi to his court, and there Idrisi produced a map that combined Arab and European knowledge of the world.

In the 13th century, the Persian scholar Qazwini developed a geography of the whole earth based on Ptolemy and on his own travels. Yet the Islamic cartographers did not simply rehash the works of the ancients with details of the Arab world thrown in; they also paid close attention to new geographic developments in the West and added European discoveries to their own maps. The Piri Reis map, made in Turkey in 1513, may have been based on Christopher Columbus's own charts of his voyages to the Americas.

SEE ALSO

Ibn Battuta; Idrisi, Abu Abdullah Muhammad; Piri Reis map; Ptolemy

Isogram

The isogram (sometimes called the isoline) is one of the mapmaker's most useful symbols. It is nothing more than a line, but it can be used to convey a great variety of information.

An isogram is a line that connects points whose values are equal (*iso* means "equal" in Greek). A mapmaker can draw isograms that wander across the landscape, linking places that are alike in some way. The first isogram was developed by the British scientist Edmund Halley, who in 1701 and 1702 published maps of the Atlantic Ocean and the world on which lines connected points of equal magnetic variation. In other words, because compass needles point to magnetic north rather than true geographic north, Halley was able to chart the difference between geographic north and the magnetic compass readings in many locations.

By connecting with a line each point where the compass needle pointed 20 degrees west of true north, for example, or 10 degrees east, Halley created a map that revealed the global pattern of magnetic variation. By revealing the pattern, Halley allowed people to make reliable estimates of the degree of magnetic variation that their compasses would show at all points within the pattern, even if a compass reading had not yet been taken at those points. Knowledge of this pattern in variation, clearly set forth on an easy-to-use world map, made compasses much more reliable. With it, navigators could add or subtract the necessary number of degrees from their compass readings to bring the readings into line with true north; thus, they could stay on course with regard to their charts.

The idea of a curve-line connecting points of equal value quickly caught on

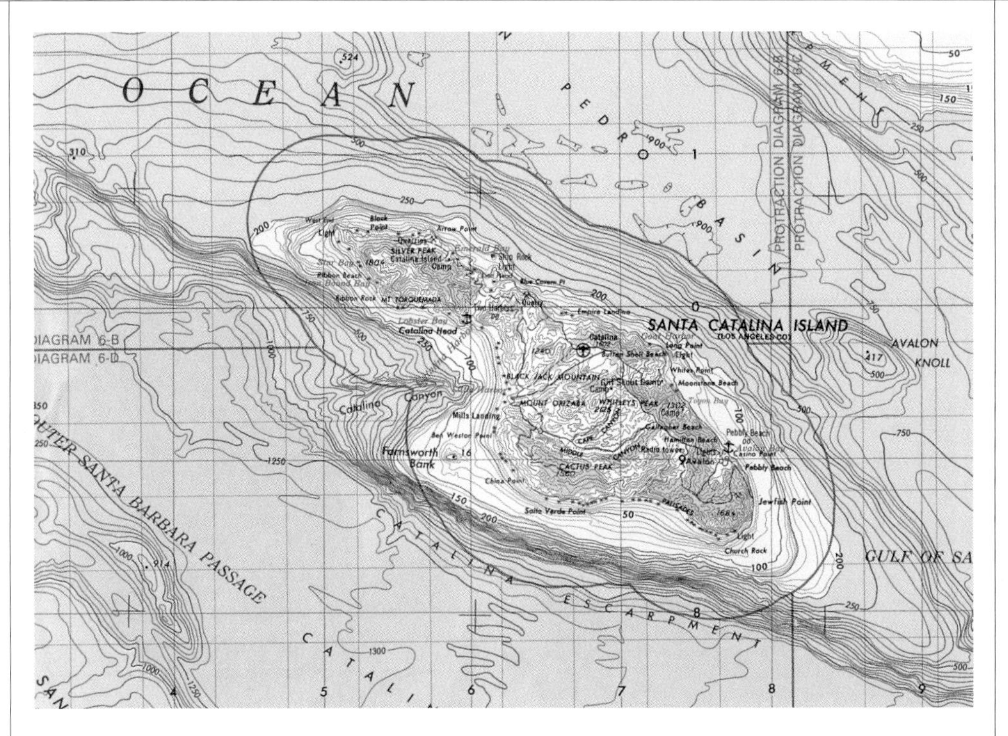

Closely spaced isobaths show the seafloor dropping steeply on the south side of Santa Catalina Island, off the California coast.

among other mapmakers, who began using isograms to communicate a wide variety of information. One very common type of isogram is the contour line. On relief maps, contour lines connect points that are the same distance above sea level; thus, they give a picture of the hilliness or flatness of the terrain. Similar to the contour line is the isobath, which is used on nautical charts and ocean maps to connect points of equal depth, thus creating a relief map of the undersea world.

Various specialized kinds of isograms have been developed to convey specific facts. An isotherm, for example, is an isogram that connects points that have the same temperature. The German explorer and scientist Alexander von Humboldt invented the isothermic map in 1817, and today every weather report on television or in the newspaper uses a form of it, often with color added between the isotherms so that the coldest areas, for example, are dark blue and the warmest ones are red-orange. Isobars are lines that connect points of equal air pressure. Weather maps that show high-pressure and low-pressure systems use isobars.

There are as many kinds of isograms as there are qualities that can be measured. The isanemone connects points of equal wind velocity. Isanthesic lines connect places where plants of the same species bloom at the same time. The isohaline traces the different degrees of saltiness in the ocean; the isohyet outlines places that receive equal rainfall.

These and other isograms are used on meteorological and agricultural maps, as well as on maps for many other special purposes. By connecting points of equal value, isograms reveal the patterns in natural phenomena, making them clear and easy to grasp. For example, if you read a list of the average annual rainfall in a thousand towns in India, you might not realize from the rows of place names and figures that parts of India are drenched by seasonal rains and parts are almost desert. But if you look at a map of India with isograms showing rainfall, you will immediately see that some areas receive a lot of rain and how the amount of rainfall shades away little by little from these wet regions; you will also identify the driest regions at a glance. Isograms are a form of visual shorthand. They allow a single map to communicate information that might take many pages if it had to be spelled out in words.

SEE ALSO
Contour lines

Italian mapmakers

ITALY WAS the center of the mapmaking trade in the late 15th century, at the time of the major European voyages of exploration. The revival of Ptolemy's worldview began in Italy; the first new editions of Ptolemy's *Geography* since ancient times were printed in Italy between 1477 and 1490.

Mapmaking came naturally to Italy, which was the commercial center of the Mediterranean and a crossroads for news from Europe, North Africa, and the Middle East. Italian maps incorporated the reports of Marco Polo and other 13th- and 14th-century Italian travelers to Asia. And as a maritime nation, Italy had a tradition of chart making that was adapted to general cartography in the 16th century. Island atlases such as those of Benedetto Bordone were a specialty of Italian mapmakers, reflecting the country's maritime heritage.

The Italian cartographers tended to make individual maps rather than atlases, although Giacomo Gastaldi, the foremost Italian mapmaker of the 16th century, did produce a small "pocket atlas" of the world in 1548. Gastaldi's maps are fine examples of the best Italian copperplate engraving, with a clean style

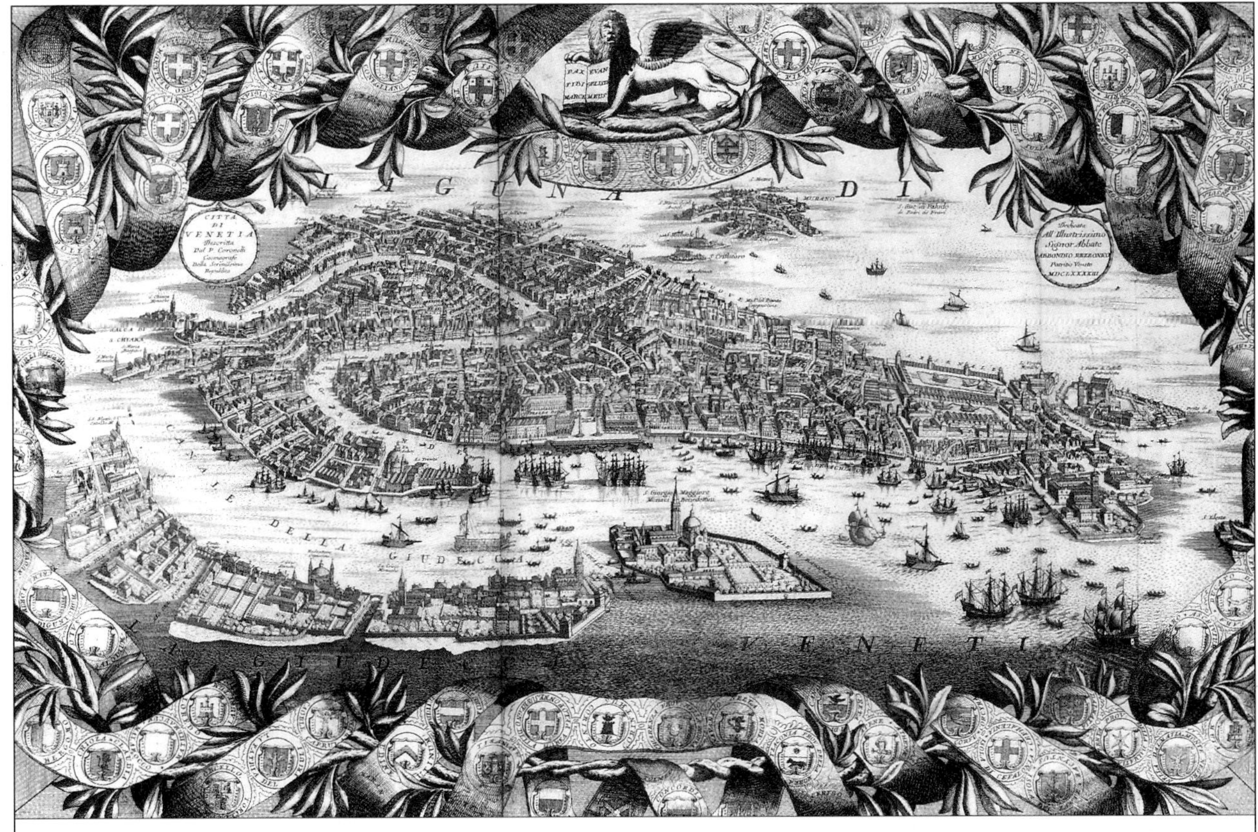

A scene from Vincenzo Coronelli's four-volume atlas of Venice. The atlas contains more than 300 maps and sea charts, as well as many pictures.

in which monsters and other whimsical illustrations are limited to the oceans.

Italy's next great mapmaker was Vincenzo Coronelli, who dominated the nation's cartography at the end of the 17th century. He made several globes, including a pair of painted globes each 15 feet (4.5 meters) across for King Louis XIV of France. He also published an atlas of globe gores in 1693 and the *Atlas of Venice* in four volumes in 1690–96, complete with pictures of the ships, princes, and buildings of Venice. Despite Coronelli's importance, however, Italian dominance in mapmaking gave way to Dutch supremacy during the 17th century.

In the late 18th century, several Italian map publishers issued world atlases. A three-volume atlas by Giovanni Maria Cassini, issued from 1792 to 1801, contains maps of many Pacific islands, drawn from the reports of James Cook's voyages.

SEE ALSO

Bordone, Benedetto; Gastaldi, Giacomo; Globe; Ptolemy

Jaillot family

SEE French mapmakers

Jansson, Jan

DUTCH MAPMAKER

- *Born: 1596, Arnhem, Netherlands*
- *Died: 1664, Amsterdam, Netherlands*

JAN JANSSON was a maker of maps and globes in Amsterdam during the 17th century, the period of Dutch supremacy in mapmaking. He married a daughter of the Hondius map publishing family and went into business with his brother-in-law, Henricus Hondius, in 1630. Together they published various editions of the Mercator-Hondius atlas, using copper plates that the Hondius family had bought from the family of Gerardus Mercator. After Henricus Hondius died, Jansson continued publishing the atlas in ever larger and more ambitious editions. An edition in Dutch, published 1647–62, ran to 10 volumes; the Latin edition had 11 volumes. These and other Jansson atlases would probably have been regarded as the highest expression of the Dutch mapmaker's art if not for the rival Blaeu family of cartographers, who had also acquired some of the original Mercator-Hondius plates. The Blaeus produced a sumptuous multivolume atlas that somewhat overshadowed the work of Jansson.

SEE ALSO

Blaeu family; Hondius family; Mercator family

Japanese mapmakers

JAPANESE MAPMAKING originated in the 7th century and was chiefly concerned with mapping Japan; little attention was given to exploring or mapping the outside world. The first known Japanese maps are surveys of population and land, made to aid tax collectors and administrators. The earliest known Japanese surveyor was Ise-No-Kimi, who published a collection of surveys of the provinces in 683, although none of his works survive. The oldest remaining Japanese maps, which date from the 8th century, are sketches of estates, showing buildings, cultivated land, streams, forests, and mountains on each landholder's property.

The first map of Japan in its entirety was made by Gyogi Bosatsu (670–749),

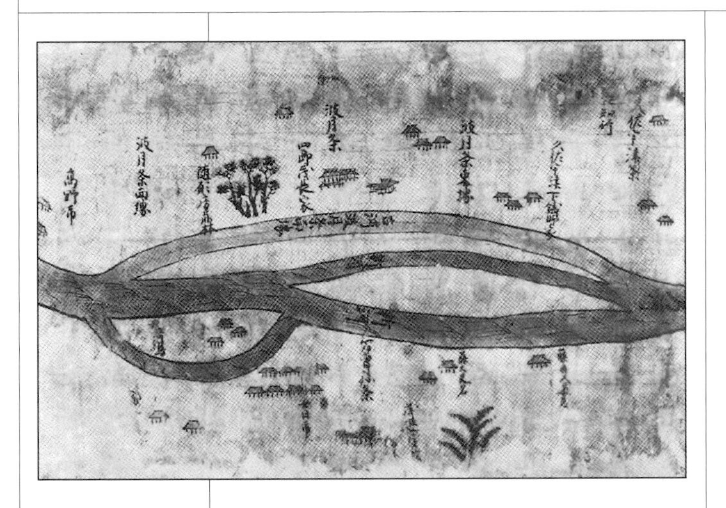

A 13th-century map of a feudal estate owned by two warrior brothers; their manor houses appear on opposite sides of the river.

a Buddhist priest. Although the map was not geographically accurate—the overall shape of Japan was badly distorted—it depicted the relationships between cities and provinces and the routes between them. This style of mapmaking was used by later cartographers, whose works are called Gyogi maps.

Although the Japanese made few maps of the world outside their own country, one world map from 1364 survives. Called the Gotenjikuzu (Map of Five Regions), it is a blend of geography and cosmography, showing the influence of traditional Indian, Tibetan, and Chinese Buddhist art. The Himalaya Mountains, the birthplace of Buddhism, are at the center of the map; mythical mountains from Buddhist legends appear just above them. China, Korea, and Japan are recognizable.

Japan's contact with the West began in 1534, when European ships first sailed into Japanese waters. For a century, traders and missionaries carried European influences into Japan. Among these influences were maps—especially the world maps made by European cartographers, who were in the midst of an explosion of world mapping. The works of cartographers such as Abraham Ortelius influenced some Japanese cartographers, who made their own world maps based on European models.

After 1643, the Japanese government embarked on a policy of strict isolationism. Foreigners were not permitted to mingle with the Japanese, and Japanese people were not allowed to leave their homeland. For the next two centuries, Japanese mapmakers turned away from the world and focused again on their own country, producing local maps, many of which were works of fine art. Ishikawa Toshiyuki (active from 1688 to 1713) made town plans that are notable more for their beauty than their geographic accuracy. Nagabuko Genshu (also called Sekisui) produced a map of Japan that was not only fairly accurate but also used latitude and longitude lines, a legacy of the Europeans. The government sponsored many careful land surveys, but these were not available to the public. However, a wealthy brewer named Ino Tadataka Chukei prepared more than 200 regional maps in the years 1800–16. These maps were widely circulated and remained in use into the 20th century.

Another tradition in Japanese mapmaking was the highway map—maps of the historic routes that crossed mountains and plains to link Japan's provinces. Some of these maps were of primarily ornamental value, decorated with paintings of beautiful views and other sights along the highway. But many were practical, filled with information useful to travelers and pilgrims, such as the distances between towns and landmarks and the locations of inns and shrines. One such map was made in 1690 by Hishikawa Moronobu, who mapped the Tokaido highway. Katsushika Hokusai (1760–1849), an engraver, artist, and printer who is described in one chronicle as "an old man mad about painting," printed a woodcut map of the Kiso highway in 1819; then, at the age of 81, he made a bird's-eye or panoramic map of China.

Japanese maps were generally made on silk or paper, although in the 19th century a few maps were made on porcelain plates—for decorative, not practical, use. The first printed maps were made in the 17th century; a printed Gyogi map from 1651 is preserved in the Imperial University in Kyoto. Most printed maps were woodcuts, printed from carved wood blocks. In 1792, Shiba Kokan published the first Japanese map printed from engraved copper.

SEE ALSO
Gyogi maps

Jode family

SEE Dutch mapmakers

Jolliet, Louis

FRENCH-CANADIAN EXPLORER AND HYDROGRAPHER

- *Born: Sept. 1645, Quebec, Canada*
- *Died: 1700, Quebec, Canada*

LOUIS JOLLIET was born in New France, the French colony in Canada. He attended a school run by the Jesuits, members of a Roman Catholic religious order called the Society of Jesus. There he excelled in mathematics, drawing, and mapmaking.

In 1668, Jolliet became a fur trader. He spent several years living and traveling in the wilderness around the Great Lakes. Then, in 1672, the governor of the French colony chose Jolliet for an important mission: to find a large river that, according to the Indians, flowed somewhere west of the lakes. A Jesuit priest, Jacques Marquette, was selected to accompany Jolliet. They made a good team. Jolliet knew how to live in the wilderness and could make good maps of the territory they explored; Marquette was interested in geography and took notes about the things they saw.

Jolliet and Marquette succeeded in locating the mysterious river, called "Father of Waters" by the Native Americans. They then sailed south almost the whole length of the Mississippi, turning back about 400 miles (644 kilometers) from the river's mouth because they did not want to trespass on territory claimed by Spain. They returned to New France by way of the Illinois River valley.

Both men are credited with making the French discovery of the Mississippi River (Spanish explorers had already found the river's mouth on the Gulf of Mexico). Marquette died in 1675 in the Illinois wilderness. Jolliet lost most of his notes and maps before reaching Quebec but was still able to recall many details of the journey and to redraw his maps from memory. In 1697, Jolliet was made the first official hydrographer of New France. He held that position until his death three years later.

SEE ALSO
Americas, mapping of; Hydrography; Marquette, Jacques

King, Clarence

AMERICAN GEOLOGIST AND SURVEYOR

- *Born: Jan. 6, 1842, Newport, Rhode Island*
- *Died: Dec. 24, 1901, Phoenix, Arizona*

TRAINED AS a geologist at Yale University, Clarence King joined a team of

Clarence King climbing in the Sierra Nevada of California. He combined a surveyor's skill with a passionate appreciation of nature's beauty.

government surveyors mapping California in the 1860s. He explored the southern Sierra Nevada range and some of the southwestern deserts. In 1867 he was made director of the United States Geological Exploration of the Fortieth Parallel, the first systematic exploration of a region along a line of latitude. He and his team spent a decade surveying, mapping, and studying a 100-mile-wide (160 kilometers) strip of land running from Cheyenne, Wyoming, through the Rocky Mountains to the Sierra Nevada. The results of this survey were published in eight volumes (including an atlas) in 1878 under the title *Systematic Geology*, setting a new standard for scientific reporting.

The following year King's survey and three other government surveying organizations were merged into a single body, the U.S. Geological Survey (USGS), of which King was the first director.

SEE ALSO
U.S. Geological Survey

La Cosa map

THE LA COSA map is the first known world map to show the Americas. It was made in 1500 by a Spanish navigator named Juan de la Cosa, who accompanied Christopher Columbus on at least one voyage.

Not much is known about La Cosa's life. He may have served as the master of the *Santa Maria* on Columbus's very first western voyage, in 1492. He certainly sailed on Columbus's second voyage, in 1493–94, and later he made several other journeys to the New World. He was killed in a battle with Indians in South America in 1509.

Sometime in the early years of the 16th century, La Cosa prepared a map, drawing upon everything that he knew about world geography. Painted in bright colors on an ox hide, the map is a portolan, or sailing guide, with coastal points and compass directions clearly marked. It shows the sailing tracks of Columbus's first, second, and third voyages. It depicts Puerto Rico, Jamaica, and Cuba, as well as other islands in the Caribbean Sea; it also shows Labrador, the part of Canada that was sighted by John Cabot in 1497. La Cosa's treatment of

Part of the La Cosa map, showing the Americas. Flags mark European landfalls along the coastline, with the Caribbean islands in the center.

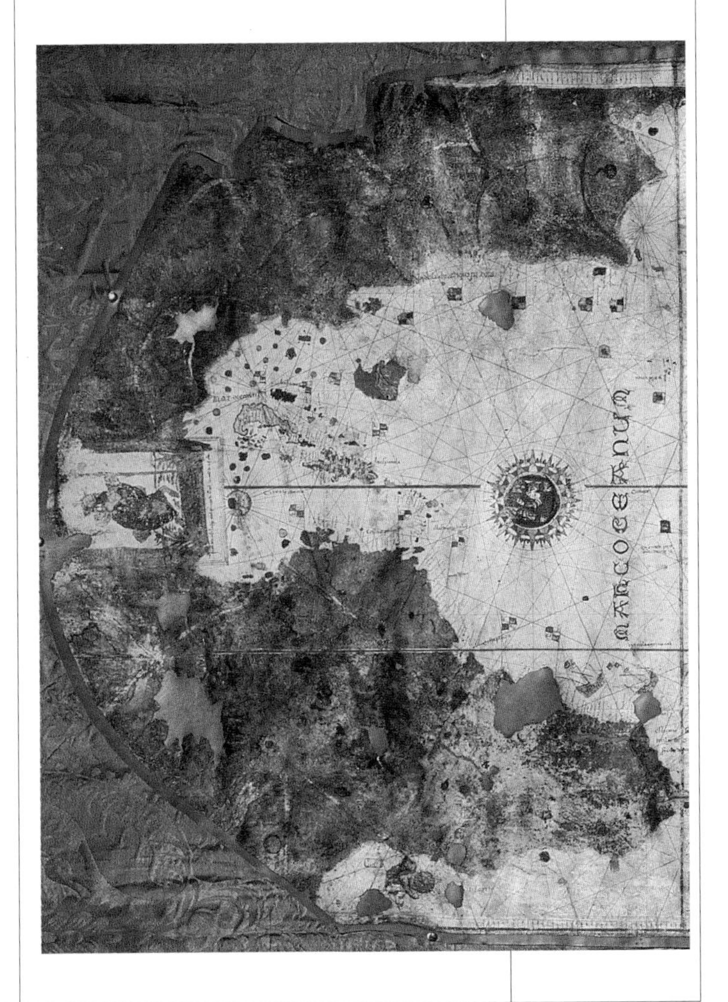

Cuba is particularly interesting because Columbus, insisting that he had landed upon part of the Asian mainland, had forced his entire crew to swear that Cuba was part of a continental landmass, not an island. La Cosa knew better, however, and showed Cuba as an island upon his map, regardless of his forced pledge to Columbus. Although La Cosa mapped various islands and coastal points of the Americas fairly accurately, he does not seem to have realized that the Americas were continents separate from Asia. But by the time of his death, that fact was becoming generally known.

Other parts of the La Cosa map prove that the mapmaker was familiar with the explorations and geography of his day. The Mediterranean region and the west coast of Africa are shown quite accurately; East Africa and Asia are more speculative, however, as geography gives way to hearsay and legend. One of the many illustrations on the map is the fleet of Portuguese explorer Vasco da Gama rounding the Cape of Good Hope at the southern tip of Africa.

SEE ALSO

Americas, mapping of; Columbus, Christopher; Portolan

Land-use maps

LAND-USE MAPS are special-purpose maps that show how land is being used: what parts of it are cropland, forest, city, range land, and so on. Most land-use maps today are based on satellite scans of vegetation and water cover, combined with geologic and topographic data. Patterns of color and shading are used to indicate areas of different land cover or use and are superimposed onto the outlines of countries, states, or smaller political units.

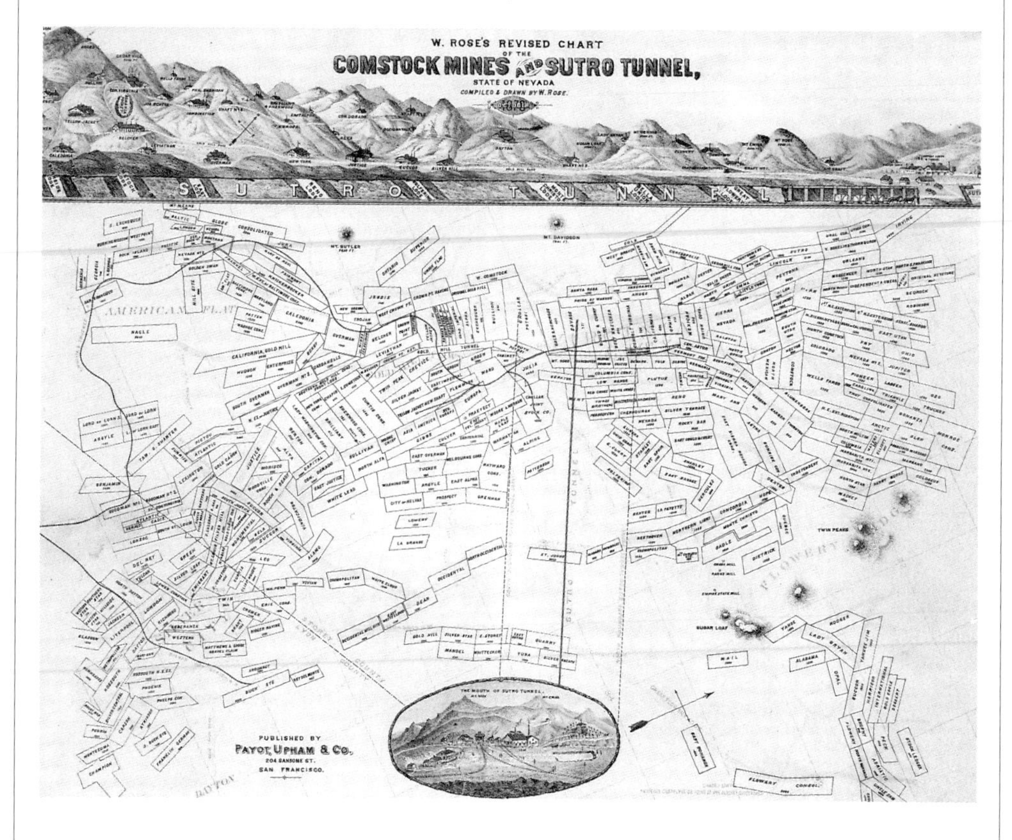

An 1878 map of mining claims on the Comstock silver lode beneath Virginia City, Nevada. At the top is a cross section of the tunnel built to drain water from the mines.

Land-use maps, which show how land is used now, are vital to decisions about how it should be used in the future. Such maps are tools for building developers and environmentalists—two groups that often use the same maps to argue dramatically different points of view—and also for farmers, foresters, and government planning agencies such as zoning boards and the administrators of public lands. Most land-use maps are produced by government agencies such as Energy, Mines, and Resources Canada or the U.S. Geological Survey.

Closely related to the land-use map is the map of natural resources. Whenever a valuable resource is found, maps are made to help people make the most of that resource. A 17th-century map shows the location of salt mines in Poland that had been worked since the 11th century. The 19th century saw the production of many resource maps: maps to guide whaling ships to the parts of the world's oceans where whales congregated in greatest numbers, maps to guide hopeful prospectors to the goldfields of Australia and the Yukon, and maps of the oil fields of western Pennsylvania, where the petroleum industry was born. Modern resource maps take many forms, from the complex geologic surveys of the oil companies to hand-sketched maps photocopied by birdwatchers to show other birdwatchers where a rare species has been sighted.

La Salle, René-Robert Cavelier, Sieur de

FRENCH EXPLORER OF NORTH AMERICA

- *Born: Nov. 1643, Rouen, France*
- *Died: Mar. 1687, Texas*

RENÉ-ROBERT CAVELIER was born in France. He studied to become a priest but decided instead to become a fur trader in New France, the French colony in Canada. He arrived in New France in 1667 and spent most of the rest of his life, except for a few visits to France, trading and exploring in North America. Giving himself a nobleman's title—without waiting for the king's permission—he began calling himself the Sieur de La Salle.

La Salle explored the region around the Great Lakes and built a number of forts in present-day Michigan and Illinois. His biggest contribution to geography was a journey he made in 1681–82, sailing south along the entire length of the Mississippi River and claiming the land around the river for France. La Salle called this region Louisiana, after King Louis XIV of France.

Mapmaking—or, more accurately, map trickery—helped bring about La Salle's second expedition to the Mississippi. La Salle wanted to establish a colony at the mouth of the Mississippi, where the river flows into the Gulf of Mexico. He went to France to seek the king's support for this scheme, and he swayed the king by arguing that the colony would be a good launching point for attacks against Spain's settlements and silver mines in Mexico. To support this argument, he drew a map that deliberately distorted the course of the Mississippi. Instead of showing the river flowing almost straight south, La Salle made it veer far to the west, emptying into the Gulf of Mexico in what is now Texas. This trick made it seem that the mouth of the Mississippi was much closer to the Spanish colony than it really was. Perhaps the phony map did the trick; at any rate, the king gave his support to the project.

The expedition was a disaster. With 4 ships and about 300 men and women,

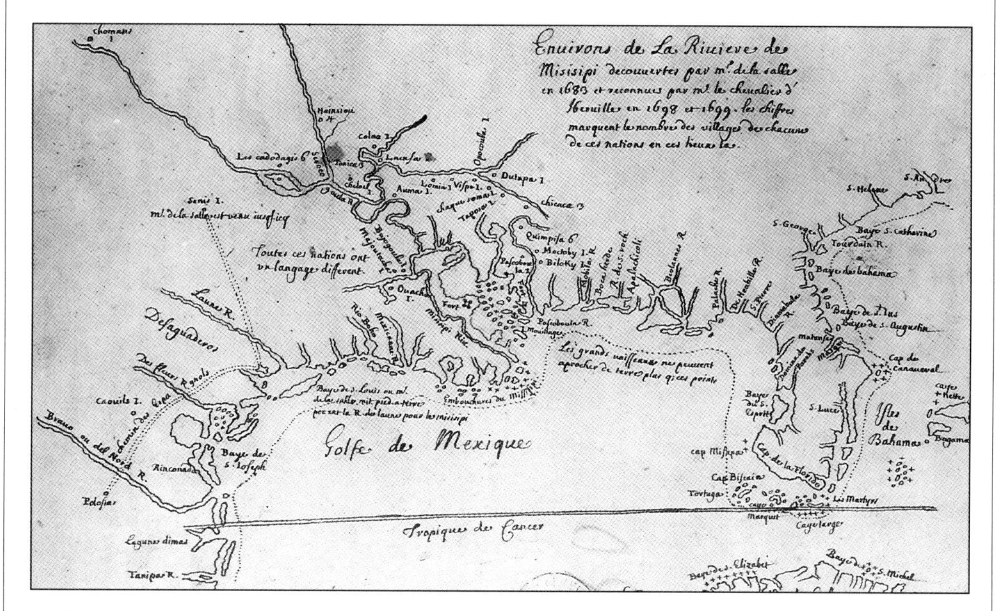

The Mississippi Delta, as sketched by French explorer Pierre Le Moyne in 1699. A few years earlier, La Salle had faked a map of the region—and then died there.

La Salle tried to reach the Mississippi Delta by sea from the Gulf of Mexico. Ironically—in view of the fact that he had faked the location of the Mississippi to win the king's backing—La Salle was unable to find the river. He missed it by about 400 miles (644 kilometers) and came ashore at Matagorda Bay, Texas, in 1685. Two years later, with the colony greatly reduced by disease and starvation, La Salle tried to lead a small group overland to the river to seek help. Along the way, several members of the party mutinied and murdered La Salle.

Although La Salle failed to establish a colony in Louisiana, he did manage to establish France's claim to a huge region that included all or part of 14 present-day states. This territory more than doubled the size of the United States when President Thomas Jefferson bought it in the 1803 Louisiana Purchase.

The confusion created by La Salle over the true location of the Mississippi's mouth persisted for some years after his death. Many maps showed the river emptying through Texas, as La Salle had drawn it on his trick map. Not until 1699 did a French explorer named Pierre Le Moyne produce a reasonably accurate map of the Gulf of Mexico and the Mississippi Delta. In 1718, the French cartographer Guillaume De L'Isle made a map of the Louisiana Territory with a spot marked "Here was killed Mr. La Salle." La Salle's other legacy to cartographers, his journeys of exploration around the Great Lakes, were also recorded by the mapmakers of New France and Europe.

SEE ALSO

Americas, mapping of; L'Isle family

FURTHER READING

Coulter, Tony. *La Salle and the Explorers of the Mississippi.* New York: Chelsea House, 1991.

Latitude

LATITUDE IS the measurement of a location's distance north or south from the equator. It is measured in degrees. There are 90 degrees of latitude in the Northern Hemisphere and 90 degrees in the Southern Hemisphere. The equator is at zero degrees, and the north and south poles are at 90 degrees north and south. Thus, the phrase "the high latitudes" refers to the cold regions approaching the poles, and "the low latitudes" refers to the warmer regions surrounding the equator.

A degree equals 1/360th of the distance around the world, or about 69 miles (111 kilometers). Each degree is subdivided into 60 minutes, and each minute is further subdivided into 60 seconds. The symbol for a degree is °; a minute is '; and a second is ".

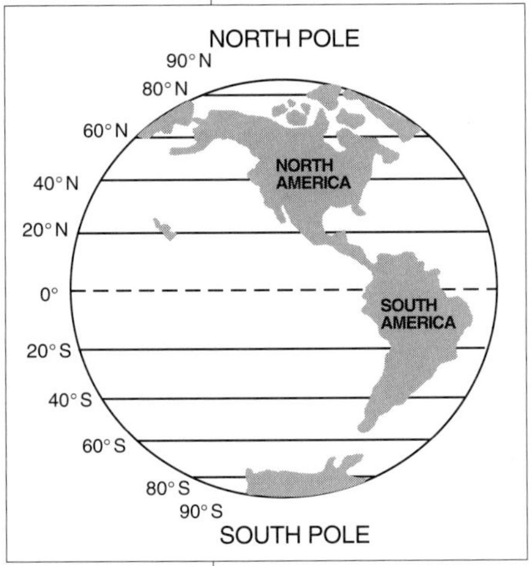

Latitude is indicated on maps by a series of imaginary lines that circle the earth, running east and west parallel to the equator. These lines of latitude are called parallels. The scale of each map determines which parallels will be shown on that map. For example, a map of Connecticut may need only three parallels: 41°N, 41°30'N, and 42°N. A map of China, however, will be on an entirely different scale and may show eight parallels, from 15°N to 50°N in 5-degree intervals.

Latitude and climate are closely connected. Places that have the same latitude are the same distance from the equator, which is the part of the earth's surface that is closest to the sun. Low latitudes, being close to the equator, have the hottest climates; as latitude increases, either north or south of the equator, the climate grows cooler. Ancient geographers recognized the link between latitude and climate. Ptolemy, who wrote about world geography in the 2nd century A.D., divided the world along horizontal lines, or lines of latitude, into seven climate zones. A similar view of the world holds true today: when we talk about polar and subpolar zones or tropical and subtropical regions, we are talking about both climate and latitude.

SEE ALSO
Coordinates; Longitude; Navigation

Legend

A MAP'S legend (sometimes called its key) is a summary of the information that a user will need to interpret the map. It may consist of nothing more than the map's scale, usually printed in the margin of the map or enclosed in a frame, or cartouche, within the map. In more elaborate cases, the legend consists of an example of each of the different symbols used on the map, together with a note that explains what each symbol means: a star for national capitals, perhaps, or circles of various sizes to show cities of different population sizes. Every symbol that appears on the map should be explained in the legend, and so should the map's scale and the use of colors to illustrate elevation or land cover. Some atlases have a single large legend that explains all the symbols used on all the maps, so that individual maps do not need legends.

SEE ALSO
Scale

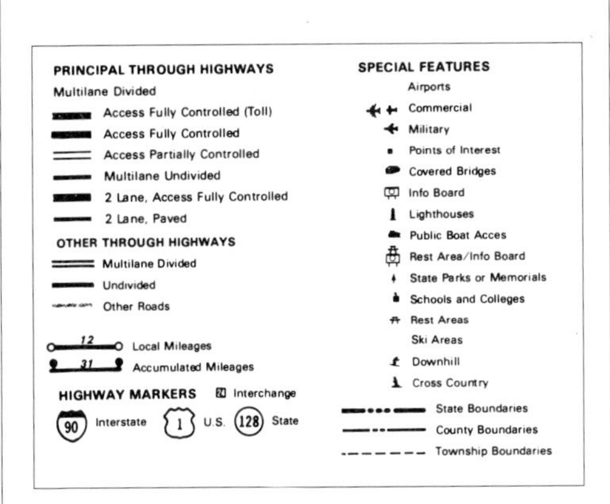

A typical road-map legend. It identifies nine kinds of roadway.

Lewis and Clark Expedition

THE LEWIS and Clark Expedition was the first American exploring expedition into the western part of North America. Under orders from President Thomas Jefferson, army officers Meriwether Lewis and William Clark led a party of about 40 people from St. Louis to the Pacific Ocean and back in the first overland crossing of U.S. territory.

In 1803, the United States bought from France all the territory drained by the Mississippi River and its tributaries: in effect, the land west of the Mississippi River to the summit of the Rocky Mountains. This immense tract, called the Louisiana Purchase, was almost entirely unknown to white Americans. Jefferson assigned Lewis and Clark to explore it and also to search for a route from the Missouri River to the Pacific Ocean. Along the way, they were to take special note of places that might be good for settlement. Jefferson wanted to make good the U.S. claim to the newly acquired land because other nations had already staked claims to North American territory: the British had claimed the Pacific Northwest in 1792, and the Spanish held claims to California and Florida.

The Lewis and Clark Expedition set out from St. Louis, Missouri, in April 1804. Ascending the Missouri River, the explorers spent the winter of 1804–5 with the Mandan and Hidatsa, Native American peoples of what is now North Dakota. In the spring they pushed west and managed to cross the Rocky Mountains in present-day Idaho before the winter snows closed the passes. Coming down out of the mountains, they continued west toward the Pacific on a series of rivers: the Clearwater, the Snake, and the Columbia. They reached the Pacific at the mouth of the Columbia in November 1805 and spent the winter in a fort there, near the site of Astoria, Oregon.

In the spring of 1806 they set out on the return trail, following their earlier path except in Idaho and Montana, where the expedition split into several parties to explore alternative routes. By the time the expedition reached St. Louis in September 1806, it had been gone for 28 months and had covered 8,000 miles (12,874 kilometers) of territory, most of it previously unknown except to Native

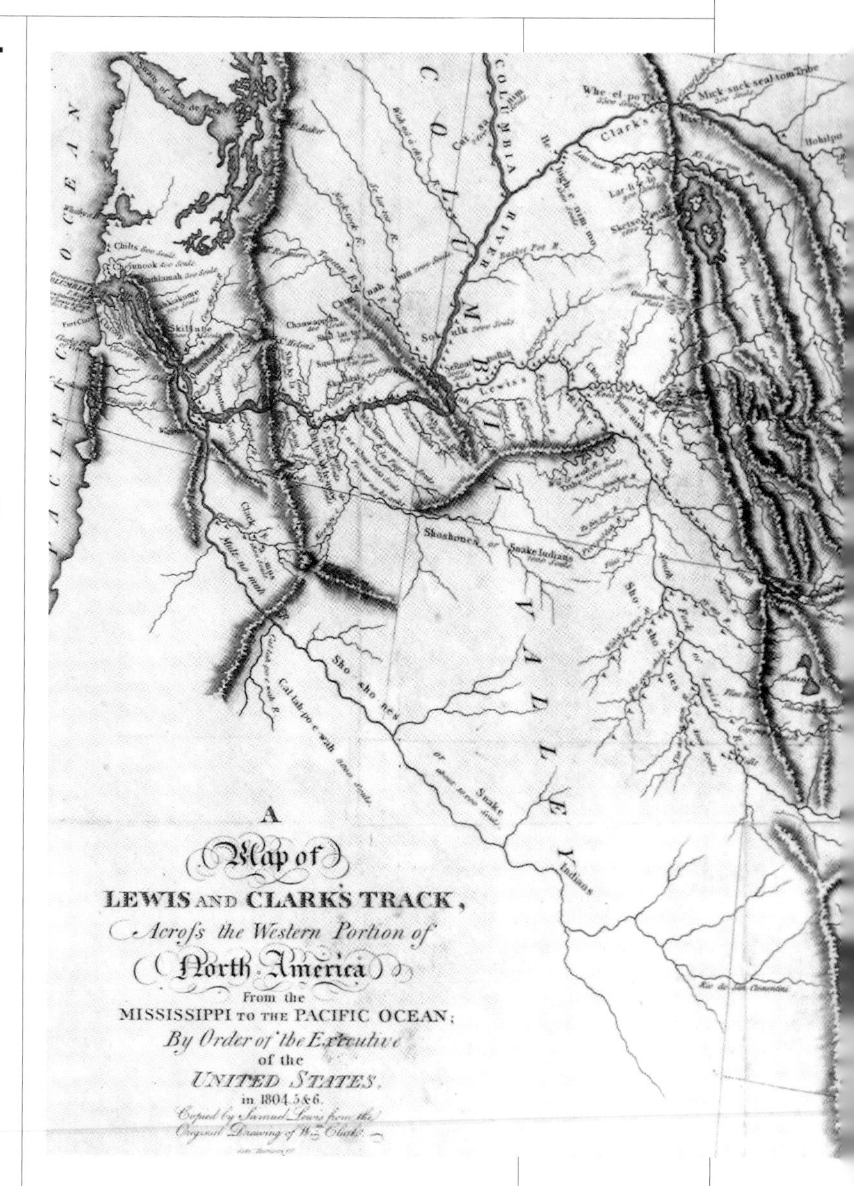

William Clark's 1814 map of the country crossed by the Lewis and Clark expedition. This was the first map to show the link between the Missouri and Columbia river systems; it served as a blueprint for American expansion into the West.

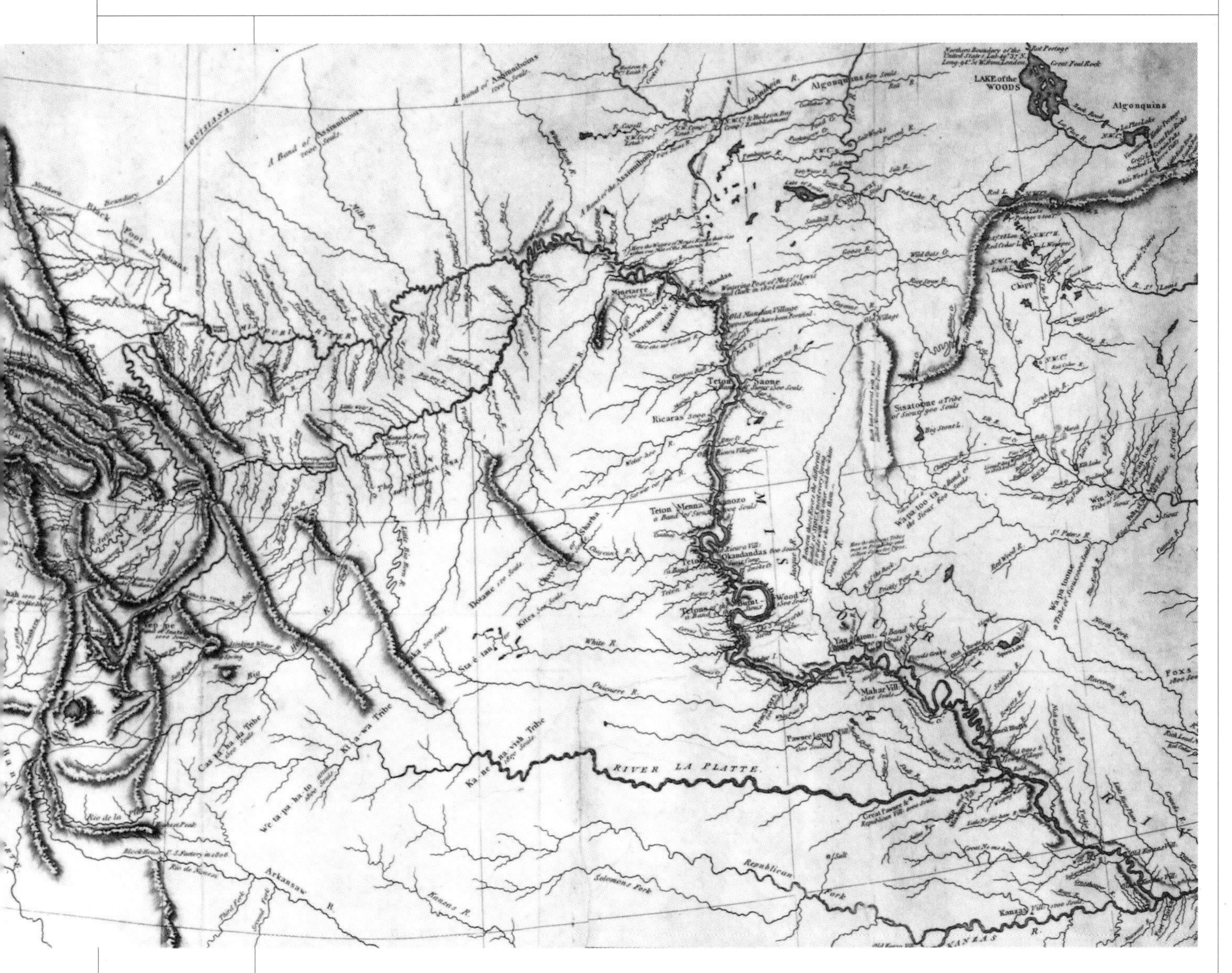

Americans. Only one man had died—a victim of appendicitis. There had been remarkably little dissension and almost no suffering, and the two leaders remained in harmony throughout. All in all, it was one of the best-managed major exploring expeditions in history.

It was also one of the most successful in terms of bringing back information. The notebooks of Lewis and Clark bulged with observations about the geography, weather, resources, inhabitants, and natural history of the land through which they had passed. The explorers had discovered and described many new species of animal life, including prairie dogs and horned toads. They had also compiled a wealth of information about the organization, customs, and languages of the Native American peoples they had encountered.

The Lewis and Clark Expedition is a good example of the profound effect that maps can have on the world. Before the expedition, the geography of western North America was a mystery, as shown on an 1804 map of North America by British cartographer Aaron Arrowsmith. This large map, intended for use as a wall chart, outlines the continent of North

America fairly accurately. The eastern region, between the Atlantic Ocean and the Mississippi River, is crowded with details and place names. So is Mexico, well known after centuries of Spanish colonization. Eastern Canada, colonized by the French in the 17th century, contains a fair amount of detail, and Arrowsmith filled in some of western Canada's lakes, rivers, and mountains, based on the discoveries made in the Canadian northwest by Alexander Mackenzie in the late 18th century. A trickle of place names runs down the west coast of America from Alaska to California, for mariners had been sighting and naming landmarks on that shore for hundreds of years. But the region between Canada and Mexico, between the Mississippi River and the Pacific Ocean, is empty, a tantalizing blank space broken only by a line of mountains—the Rockies, which were known from Indian accounts but had not yet been explored. This was what the world knew about the American West in the year that Lewis and Clark set out on their historic journey.

The next important map of the region is William Clark's own, published in 1814. The difference is remarkable. The area along the expedition's track is filled with detail: mountains, passes, rivers and streams, locations of Indian settlements, and the broad Columbia Valley. For the first time, a map connected the eastern part of the United States with the continent's west coast. That map was like a road that led people westward. It guided thousands of American settlers across the Rocky Mountains and into the Oregon Territory, which eventually became part of the United States, just as Jefferson had hoped. By recording details of the western landscape on his map, Clark set in motion a process that populated that landscape—and thus transformed it forever.

SEE ALSO

Clark, William; Lewis, Meriwether; Mackenzie, Alexander

FURTHER READING

Bergon, Frank, ed. *The Journals of Lewis and Clark*. New York: Penguin, 1989.
Cavan, Seamus. *Lewis and Clark and the Route to the Pacific*. New York: Chelsea House, 1991.
DeVoto, Bernard, ed. *The Journals of Lewis and Clark*. Boston: Houghton Mifflin, 1953.
Duncan, Dayton. *Out West: An American Journey*. New York: Viking, 1987.
Eide, Ingvard H., ed. *American Odyssey: The Journals of Lewis and Clark*. New York: Rand McNally, 1969.
Lavender, David. *The Way to the Western Sea: Lewis and Clark Across the Continent*. New York: Harper & Row, 1988.
Moulton, Gary E., ed. *Journals of the Lewis and Clark Expedition*. 11 vols. Lincoln: University of Nebraska Press, 1983– .

Lewis, Meriwether

AMERICAN EXPLORER OF THE WEST

- *Born: Aug. 18, 1774, Charlottesville, Virginia*
- *Died: Oct. 11, 1809, near Nashville, Tennessee*

MERIWETHER LEWIS served as a militiaman and then joined the U.S. army, which sent him to the Ohio River Valley to serve in the Indian wars. In 1801 he was chosen by President Thomas Jefferson, a friend of the Lewis family, to make the first American exploring expedition into the lands beyond the Mississippi River.

With the help of William Clark, a friend from his army days, Lewis led the expedition from St. Louis, Missouri, to the Pacific coast of Oregon and back again in 1804–6. By bringing back a wealth of information about the geography, inhabitants, resources, and natural history of the American West, the explorers launched the great wave of exploration and settlement that carried Americans westward throughout the 19th century.

Lewis and Clark were heroes upon their return from the West. Jefferson made Lewis the governor of the Missouri Territory, but Lewis ran into problems. He found himself unable to finish the account of his journey that Jefferson expected him to write, and he also had money troubles. On his way to Washington, D.C., he died of a gunshot wound along the Natchez Trace. Many historians believe that he killed himself.

SEE ALSO

Clark, William; Lewis and Clark Expedition

Linschoten, Jan Huyghen van

DUTCH GEOGRAPHER

- *Born: 1563, Haarlem, Netherlands*
- *Died: 1610*

JAN HUYGHEN van Linschoten was a traveler who spent several years in India and also sailed with Willem Barents on two voyages in search of the Northeast Passage. In 1596 he published a book called the *Itinerario* (*Itinerary*) that was full of information about trade and travel in Asia. The book was illustrated with many maps engraved by Peter Plancius and others. Based on Linschoten's own cartographic researches in India and elsewhere, these maps were extremely useful to navigators because they contained a wealth of data drawn from Portuguese sea charts that had never been published. The *Itinerario* also contained pictures and charts of islands in the Atlantic; a plan of the Portuguese settlement at Goa, India; and the most detailed map yet made of the Molucca Islands in Indonesia, called the Spice Islands. A later edition of the book contained a map of Barents's voyage to the Arctic.

SEE ALSO

Barents, Willem; Plancius, Peter

L'Isle family

FRENCH MAPMAKERS

THE L'ISLES (also called the De L'Isles or Deslisles) were two generations of cartographers and geographers. Claude De L'Isle (1644–1720) studied geography under Nicolas Sanson, a renowned geographer and cartographer. L'Isle produced some maps, including a historical atlas (1654), but his accomplishments were later overshadowed by those of his sons.

Guillaume De L'Isle (1675–1726) studied under the astronomer Jean Dominique Cassini, who taught him to rely on mathematical certainty and direct observation, not on speculation and imagination. The hundred or so maps, globes, and atlases that Guillaume De L'Isle published between 1700 and 1726 reveal his passion for mathematical accuracy. They introduced a new note of sober science into French cartography. In particular, his world map of 1700, based only on verifiable observations, eliminated many errors drawn from Ptolemy's *Geography* that had been passed from map to map for centuries.

Guillaume De L'Isle's son-in-law, Philippe Buache, continued to publish maps into the 1770s. L'Isle's two brothers, Joseph Nicolas and Louis De L'Isle, also made contributions to cartography. They visited Russia at the invitation of Czar Peter the Great. Louis died in 1741

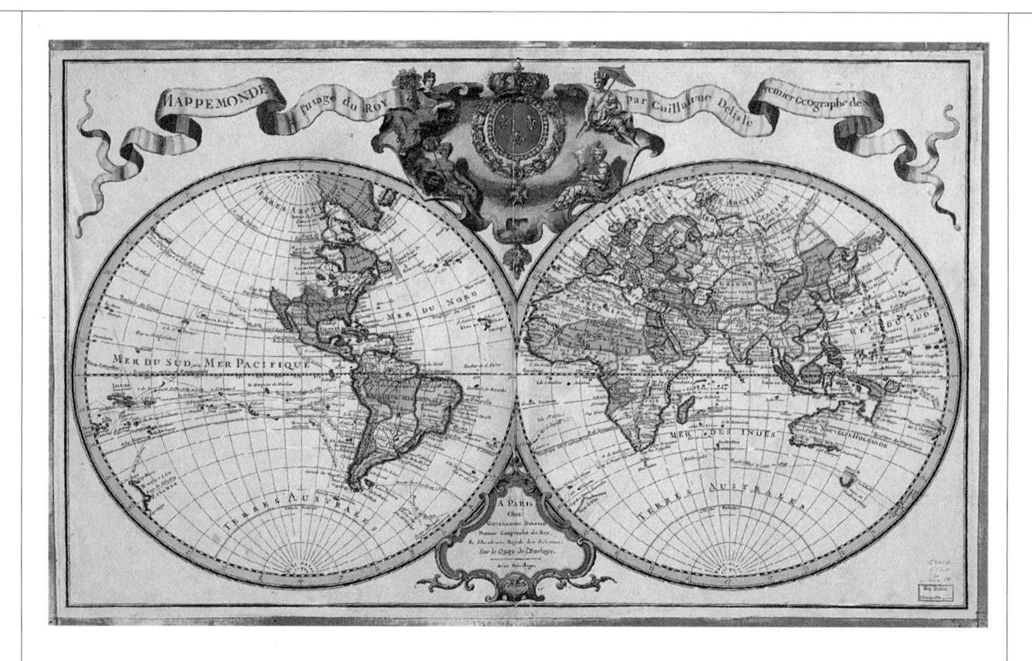

A 1720 world map by Guillaume De L'Isle, the leading French mapmaker of his time. L'Isle began drawing maps at the age of 8 and published his first major works at 24.

on an expedition through Siberia with the explorer Vitus Bering; Joseph Nicolas produced the first printed atlas of Russia in 1745.

SEE ALSO

Buache, Philippe; Cassini family; French mapmakers; Sanson family

Lithography

LITHOGRAPHY IS a method of printing from a flat surface that has been treated with some form of grease to which ink will stick. Originally, smooth tablets of polished stone were used as the printing surfaces; this gave the process its name, from the Greek words *lithos* (stone) and *graphein* (write). In lithography, the image that is to be reproduced is not carved or engraved into the printing surface. Instead it is drawn onto the surface with a substance that will hold ink. The surface is then inked and rinsed; the ink stays on the treated parts, which form a print when the surface is pressed to a sheet of paper.

Lithography was invented in Germany in 1798 but was guarded as an industrial secret for several decades. After about 1825 it quickly became the popular way to print illustrations and maps. Lithography was cheaper than copperplate engraving for several reasons: printing stones were less costly than copper, and drawing the image was far less time-consuming than engraving it. Lithography thus offered printers a way to meet the growing demand for inexpensive mass-produced maps. By 1860 printers had developed color lithography. Lithography was not only cheaper than engraving, it was also faster. In 1866, George Washington Bacon of London published a color lithograph map of the United States showing not just the results of the most recent western explorations but also the restoration of the union after the Civil War—which had ended just a few months earlier—and the new railroads that were being built across the land. The states were printed in only four colors—blue, yellow, pink, and green—which were arranged so that no neighboring states had the same color. The practice of using a limited number of carefully arranged col-

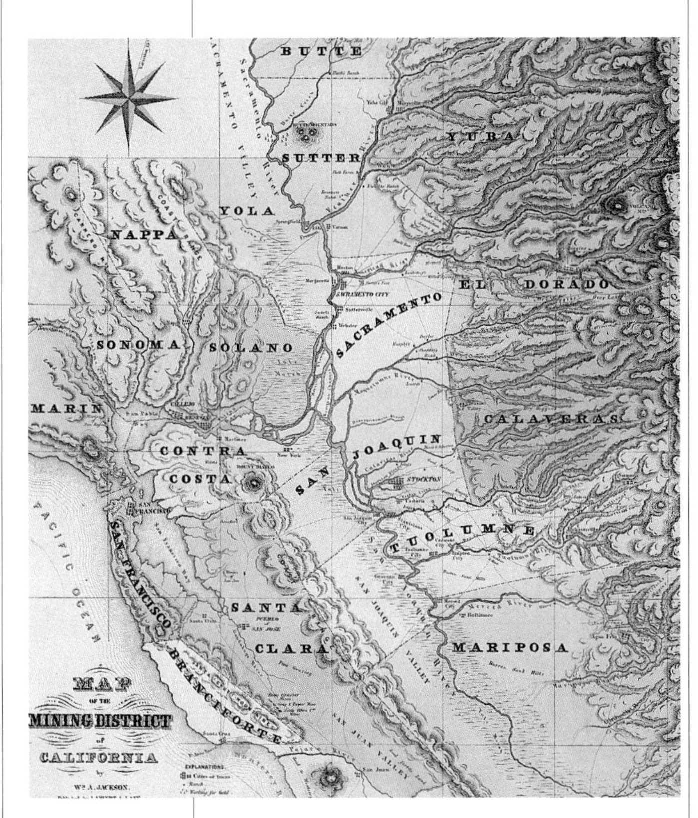

An 1851 lithograph map of the mining district around Sutter's Mill, California, where the gold rush began in 1848. This map was hand-painted; color lithography came into use a few years later.

ors soon became standard, and hand-colored maps, like engraved ones, were no longer produced after the mid-19th century. Today, lithography has been replaced in mapmaking by photographic reproduction, in which maps are printed from various kinds of photographic plates and films, and by computer-assisted cartography, in which maps are generated by special printers from computerized databases. Lithography is, however, still used by artists.

Livingstone, David

SCOTTISH MISSIONARY AND EXPLORER OF AFRICA

- *Born: Mar. 19, 1813, Blantyre, Scotland*
- *Died: May 1, 1873, Africa*

DAVID LIVINGSTONE earned credentials as a doctor and a minister with the goal of becoming a missionary. In 1841, the London Missionary Society sent him to southern Africa. Livingstone was interested in natural history, and soon he developed a longing to explore the unknown reaches of the continent. He believed that British trade and commerce should be spread throughout Africa to "civilize" the Africans and especially to stamp out the trade in slaves, which was carried out in the interior by Arab and African slavers. Exploration became the focus of Livingstone's life. "I shall open a way to the interior," he said, "or perish."

Livingstone made many expeditions, but his biggest contribution to geography came in 1853–56, when in a series of journeys he crossed Africa from west to east, becoming the first European known to traverse the continent. He charted the course of the Zambezi River and made the European discovery of a great waterfall, which he named Victoria Falls. This trip made Livingstone a popular hero in Great Britain, but the rest of his journeys were less successful.

In 1865 the Royal Geographical Society sent Livingstone to look for the source of the Nile River in East Africa. When no word of him had reached the outside world for several years, newspaperman Henry Morton Stanley led an expedition to "rescue" Livingstone. The two men met on the shore of Lake Tanganyika in 1871, and Stanley uttered one of the most famous phrases in the history of exploration: "Dr. Livingstone, I presume." Although he was in poor health, Livingstone refused to leave Africa with Stanley. He continued his search for the source of the Nile and died in present-day Zambia in 1873.

Livingstone played an important role in the exploration and mapping of Africa. His measurements of latitude and longitude were unusually careful, which meant that the maps based on them were highly accurate. He also collected geologic, zoologic, and botanic information and specimens. The books he wrote about his trips—as well as his immense popularity—aroused public interest in African exploration and encouraged others to follow in his footsteps. These volumes contained carefully engraved maps that drew on the explorer's own notes and sketch maps.

SEE ALSO

Africa, mapping of; Stanley, Sir Henry Morton

FURTHER READING

Huxley, Elspeth. *Livingstone and His Journeys.* New York: Saturday Review Press, 1974.
Ransford, Oliver. *David Livingstone: The Dark Interior.* New York: St. Martin's, 1978.

Longitude

LONGITUDE REFERS to the distance of a given point either east or west of the prime meridian, an imaginary line that runs from the north pole to the south pole, passing through the Royal Observatory at Greenwich, England. Longitude is measured in degrees (shown by the symbol °). There are 360 degrees of longitude on the earth's surface: 180 west and 180 east of Greenwich. Each degree is subdivided into 60 minutes (') and each minute is subdivided into 60 seconds (").

The lines that depict longitude on maps are called meridians. Unlike parallels of latitude, which remain parallel to one another all the way around the globe, the meridians of longitude meet at two locations, the north and south poles.

The term *meridian* has entered everyday English in the terms A.M. and P.M., used to mean "morning" and "afternoon" or "night." A.M. stands for "ante meridian" (before meridian) and refers to the hours from midnight to noon, before the sun crosses the meridian. When the sun crosses a meridian, it is at the highest point overhead during that day. P.M. stands for "post meridian" (after meridian) and refers to the hours from noon to midnight—after noon.

Locations on the earth's surface are given in terms of latitude and longitude. Any place can be pinpointed if its latitude and longitude are known. Of the two measurements, however, longitude was far more difficult for early explorers and mapmakers to determine. Although many geographers of the ancient world—in Egypt, Greece, and elsewhere—understood the concept of mapping the world with a grid of latitude and longitude coordinates, they were unable to put this idea to practical use. They had several fairly accurate ways of measuring latitude, but no standard, reliable way to measure and express longitude. For centuries, astronomers and navigators searched for a way to measure longitude. The solution to this puzzle was one of the most important milestones in the history of mapmaking.

Baselines are needed to measure both latitude and longitude. In the case of latitude, the baseline is fixed, determined by geography: it is the equator, the midpoint of the earth. In the case of longitude, however, there is no fixed baseline. Any meridian can be used as a baseline, and over the centuries many of them have been so used. The standard for modern maps is the prime meridian, the meridian that runs through Greenwich, England.

Once a baseline is determined, the longitude of any point on earth can be obtained by using the difference in clock time between that point and the prime meridian. For example, to determine the longitude of an unexplored island, a navigator needed to know the time in

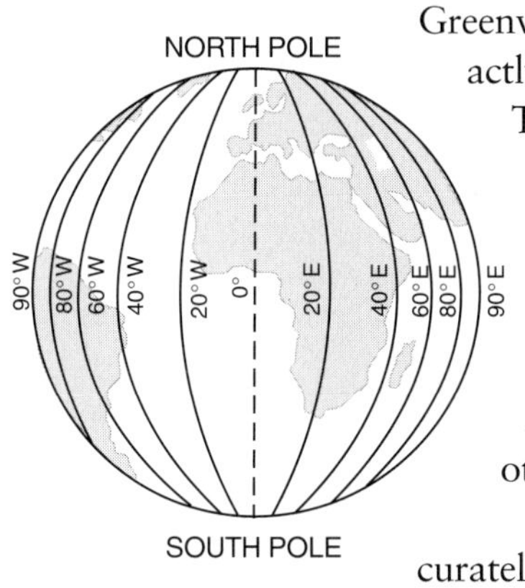

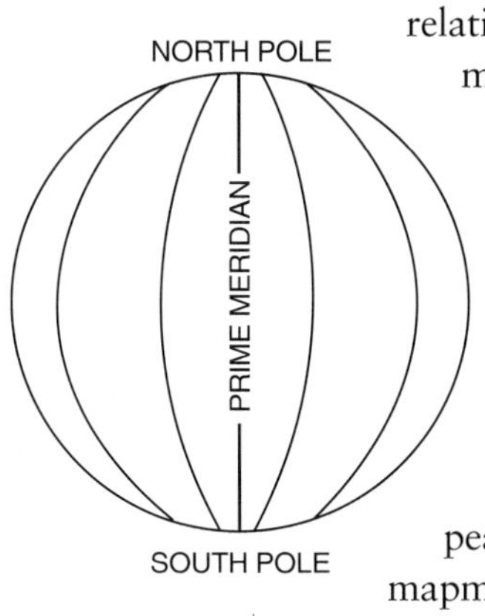

Greenwich when the sun was exactly overhead on the island. The difference in time between the two places enabled the navigator to calculate distance in degrees east or west of Greenwich for the island's location—in other words, its longitude.

Longitude may be accurately measured by using the sun's position overhead and time relative to the standard, or prime, meridian. Therefore, longitude involves time as well as space. In order to measure longitude precisely, people needed accurate, reliable chronometers—timekeepers such as clocks and watches. Yet during the heyday of European exploration and mapmaking, timekeepers were crude and not very precise. Water clocks, sundials, and the like could tell approximately what time of day it was, but they could not be used for precise scientific measurements. By the 16th century, mechanical clocks driven by springs and pendulums had appeared, but the early clocks were somewhat erratic in their accuracy. And even the later, more accurate clocks failed to perform well at sea, where their mechanisms were thrown out of order by waves, storms, humidity, and temperature changes. In order to measure longitude accurately on long voyages, navigators needed a clock that would remain true to their baseline time for months and years, so that they would always know what time it was on their prime meridian.

Throughout the 17th century, scientists, mariners, and instrument makers strove mightily to invent a way of determining longitude at sea. Various systems were developed. Some of these were extremely complex, involving not only a great deal of mathematics but also such operations as the timing of the movements of Jupiter's moons. These methods of determining longitude were either unreliable or impossible to perform effectively on a ship's deck.

Finally, in the mid-18th century, the long-standing puzzle of longitude was solved by a British instrument maker named John Harrison, who produced a sealed, spring-driven chronometer that kept accurate time on long sea voyages. At last, longitude could be measured with certainty. High noon at any point on the globe could be compared with the time on the prime meridian, and then a simple calculation would yield the position in degrees east or west of that meridian.

Harrison's chronometer revolutionized both navigation and mapmaking. Reliable measurements replaced the cautious guesswork of earlier eras. For the first time, maps and charts could truly be counted on. Longitude was determined as accurately as latitude, and a new level of precision was reached. Maps became more uniform and more universally useful, especially after Greenwich was selected as the prime meridian. The decision to use the prime meridian as the international standard made it possible for the nations of the world to agree on standardized time zones, on an international date line, and on standards for international maps.

Today, navigation and global survey grids use the Greenwich, or prime, meridian as the basis for measuring both time and longitude. However, when it is a question of determining one's actual location, explorers today rely on high-technology tools such as computers, satellites, gyroscopes, and global positioning systems to find their longitude as well as latitude.

SEE ALSO

Celestial navigation; Chronometer; Coordinates; Greenwich, England; Harrison, John; International date line; Latitude; Navigation

Loxodrome

SEE Portolan

Mackenzie, Alexander

SCOTTISH EXPLORER OF CANADA

- *Born: 1764, Stornoway, Scotland*
- *Died: Mar. 11, 1820, Pitlochry, Scotland*

ALEXANDER MACKENZIE was born in Scotland but was taken to North America by his father when he was a child. He became a fur trader and a partner in the North West Company, a rival of the powerful Hudson's Bay Company. From his base at Fort Chipewyan on the shore of Lake Athabasca, Mackenzie made two long journeys into western Canada.

In 1789 he followed a large river north from the Great Slave Lake in northwestern Canada all the way to the Arctic Ocean. Because he had hoped the river would lead to the Pacific, he named it the River of Disappointment; mapmakers, however, renamed it the Mackenzie, the name it bears today. It is Canada's longest river. Mackenzie did reach the Pacific on his second journey, in 1792. He went west from Lake Athabasca through the towering Canadian Rockies and the jumbled coastal ranges on their western flanks, arriving at the Pacific coast of what is now British Columbia. Mackenzie was the first white person known to have crossed the North American continent north of Mexico. His journeys were immediately reflected in new maps, notably the "Map of America" made by Aaron Arrowsmith of London in 1804; Arrowsmith's map shows the Mackenzie River flowing into the Arctic Ocean. Reports of Mackenzie's journeys and maps showing his discoveries did much to hasten European settlement of northwestern North America. Mackenzie's continental crossing also inspired President Thomas Jefferson of the United States to launch the Lewis and Clark Expedition.

SEE ALSO

Americas, mapping of; Lewis and Clark Expedition

FURTHER READING

Bancroft, Hubert H. *The Voyage of Alexander Mackenzie across Canada, 1789–1793.* Seattle: Shorey Book Store, 1968.

Daniells, Roy. *Alexander Mackenzie and the North West.* New York: Barnes & Noble, 1969.

Mackenzie, Alexander. *Voyages from Montreal: First Man West—Alexander Mackenzie's Journal of His Voyage to the Pacific Coast of Canada in 1793.* Edited by Walter Sheppe. Westport, Conn.: Greenwood, 1976.

Madaba map

THE MADABA MAP dates from the 6th century and is one of the oldest known Christian maps. It illustrates Christian teachings and originally formed part of the decoration of a church. Discovered in 1896, the map is a mosaic, a picture made of many tiny bits of colored stone, in the floor of a church near the town of Madaba, Jordan. The mosaic was made sometime between A.D. 542 and 565. It once measured 50 by 20 feet (15.2 by 6.1 meters) and showed the Holy Land and parts of Egypt, Arabia, and the Mediterranean, but much of it has been destroyed. One remaining part that is of

A detail from the Madaba map of Jerusalem. The rows of columns near the center represent streets (or possibly a marketplace) built when Justinian (483–565) ruled the Byzantine Empire.

special interest to historians is a map of the city of Jerusalem as it appeared in the 6th century.

Magellan, Ferdinand

PORTUGUESE NAVIGATOR

- *Born: about 1480, Portugal*
- *Died: Apr. 27, 1521, Mactan Island, Philippines*

FERDINAND MAGELLAN spent his youth at the Portuguese royal court, where he was educated in navigation and geography. In 1505, soon after Vasco da Gama made the first sea passage from Europe to India, Magellan was sent to the Indian Ocean as part of Portugal's colonizing army. For eight years he sailed Asian seas, perhaps venturing as far east as the Philippines, although records of his travels during this period are fragmentary and unclear.

After a falling-out with the Portuguese court, Magellan moved to Spain in about 1516. Ever since Columbus's voyages, Spain had been seeking a westward passage through the Americas to Asia. Magellan was given a fleet of 5 ships and 240 men and told to find that passage. He left Spain in 1519, crossed the Atlantic Ocean to Brazil, and spent the winter on the South American coast. After quelling a mutiny, Magellan pushed southward, probing every river mouth and bay in the hope of finding a passage to the west. Eventually, he found it by accident when several ships were driven by a storm into what looked like a dead-end bay but proved to be the mouth of a long, winding channel. That channel, which today is called the Strait of Magellan, led past glaciers, cliffs, and scores of rocky islands, finally emptying into the South Pacific Ocean. Magellan thus became the first European mariner to sail from the Atlantic Ocean to the Pacific.

The Pacific world was entirely new to the Europeans of the 16th century, and Magellan vastly underestimated the ocean's size. The crossing took 90 days, far longer than he had expected. The crew's sufferings from hunger and thirst

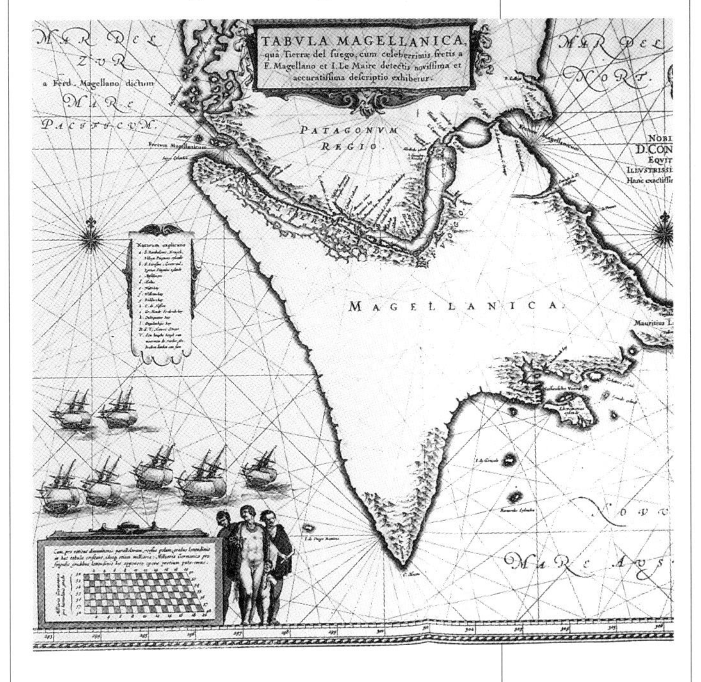

A Dutch map published in 1640 shows the Strait of Magellan, the narrow passage separating Magellanica (Tierra del Fuego) from the South American mainland.

were nightmarish. Finally, the fleet fetched up in the Philippine Islands, which Magellan claimed for Spain. He was killed there in 1521 in a battle with the native inhabitants, but one of his ships, the *Victoria,* continued south and west, returning to Spain in 1522 with 18 survivors.

Those 18 men were the first people to travel all the way around the world. As soon as they arrived back in Europe, their voyage was reflected in new maps, including one made by Antonio Pigafetta, an Italian traveler who had joined Magellan's expedition and survived the trip home aboard the *Victoria.* Diego Ribeiro, a Portuguese cartographer working for Spain, traced Magellan's track across the Pacific in a chart he drew in 1529. The Strait of Magellan, too, soon appeared on many important maps, and the southern part of South America was called Magellanica for years.

SEE ALSO

Pacific Ocean, mapping of; Pigafetta, Antonio

FURTHER READING

Cameron, Ian. *Magellan and the First Circumnavigation of the World.* New York: Saturday Review Press, 1973.
Stefoff, Rebecca. *Ferdinand Magellan and the Discovery of the World Ocean.* New York: Chelsea House, 1990.

Magnus, Olaus

SWEDISH CHURCHMAN AND MAPMAKER

- *Born: 1490, Linköping, Sweden*
- *Died: 1558, Rome, Italy*

OLAUS MAGNUS, the archbishop of Uppsala, Sweden, made one of the earliest maps of Scandinavia. Printed in Venice in 1539 from wood blocks, it covered Norway, Sweden, Finland, Denmark, and Iceland on nine sheets. This is the first known large-scale map of any part of Europe.

Sea monsters chomp and crush ships off the coast of Norway on Olaus Magnus's map. A vessel is shown sinking into a whirlpool, which, unlike the monsters, was a real hazard.

The Olaus Magnus map is crude in design, but its bold black-and-white lines give it a striking, lively appearance. The map is of special interest because it includes more than 100 small pictures of sea monsters, ships at sea, people, and animals. These decorative images were reused and copied by other mapmakers; they were also used as illustrations in travel books.

Only one copy of the original Olaus Magnus map is known to have survived. It belongs to the Munich State Library in Germany. A smaller version of the map was printed from copper plates in Rome in 1572, and seven copies of this edition survive. The map also appears in Olaus Magnus's *Historia,* a book that was printed in Switzerland in 1567.

Map

A MAP is a drawing that shows where various things are located in relation to one another. (The term *map* is a shortened form of the Latin phrase *mappa mundi,* meaning "map of the world.") In other words, maps express the locations and relationships of things. Maps of things on earth are called terrestrial maps; maps of the stars and other astronomical objects are called celestial maps.

Maps need not be concerned only with geography. Today, scientists are "mapping" many fields of knowledge. Neurobiologists map the areas of the brain where different functions occur. Geneticists map the chromosomes that carry our genetic heritage from one generation to another. Physicians use scanning equipment such as magnetic resonance imagers (MRIs) to make maps of patients' bodies, showing bones, organs, and anomalies such as tumors. Physicists are busy mapping the largest and smallest domains of science: the expanding universe and the insides of atomic particles.

Most of the time, however, the word *map* is used in a geographic sense. When we speak of a map, we are usually referring to a flat surface on which the world or some part of it is portrayed. Over the centuries, cartographers, or mapmakers, have invented many techniques to depict the curved, three-dimensional surface of the earth on flat, two-dimensional maps. Some maps—called globes—are on spherical surfaces, which permit more accurate representations of the earth.

There are countless kinds and styles of maps. Maps have been made on every surface on which people could make marks: stone, wood, bark, animal skins, paper, metal, glass, plastic, and more. The oldest known maps are bits of sunbaked clay covered with ancient scratchings; the newest are streams of digitized data flowing onto computer screens from disks and databases. A map can cover an area as small as a garden plot or as large as a world. It can show the physical features of the land, the political divisions of society, or both—and a host of other kinds of information, as well. Even time can be mapped.

Paleobotanists have made maps showing the locations of lakes and forests millions of years ago, when earth's landscape was young; and geologists have constructed maps to show us how the very continents have drifted across the earth's surface, joining and parting, over the ages. Ways of making and of using maps have been evolving since the dawn of human culture, and the future will undoubtedly bring both new maps and new uses for them.

SEE ALSO
Cartography; Geography; Globe

Map collecting

MAP COLLECTING has probably been around for almost as long as there have been maps. Until the late 15th century, however, only kings and wealthy scholars could afford to collect maps, which were hand-drawn and thus rare and expensive. After map printing was introduced in 1477, maps became more widely available, and collectors were able to indulge their passion. In the time of Great Britain's Queen Elizabeth I, there were many who "liketh, loveth, getteth and useth maps, charts, and geographical globes," reported her royal astronomer and geographer Dr. John Dee (1527–1608). A few decades later, Samuel Pepys of London, whose diaries have given historians a priceless glimpse of 17th-century life, mourned the loss of his atlas, which perished in the Great Fire that ravaged London in 1666. Many mapmakers and map publishers were also map collectors; one such was Dutch cartographer Abraham Ortelius, who produced the first modern atlas in 1570. In more modern times, map collecting was an interest of the Swedish explorer Adolf E. Nordenskiold, who traveled through the Northeast Passage in 1878–79 and who contributed to the study of cartography by publishing reproductions of the rare maps in his collection.

Some map collectors specialize in antique maps, and there are a number of dealers and stores around the world devoted to the trade in these rare items; *The Map Collector* is a journal published four times each year in Great Britain for collectors of antique maps. But original maps produced before 1850—especially the work of well-known cartographers—are beyond the budget of many people who might enjoy collecting maps.

If you are interested in maps, there are many ways to build a collection without spending a fortune. For example, you could collect inexpensive modern reproductions of classic old maps—many books of such reproductions have been published. Or you could focus on contemporary maps of your own city, state, or country, trying to obtain as many different maps as you can, starting with road maps, historical maps, topographic maps, and census maps. You could collect maps of places you have visited, or you could select a part of the world in which you are interested and acquire maps of that area.

Another way to approach map collecting is to concentrate on a particular theme, such as military, historical, or geologic maps, and try to get as many such maps of different parts of the world as you can. Mail-order map services (see Appendix 2) offer some possibilities for finding a wide variety of maps. Collectors interested in novelty maps, such as map puzzles, cartoons, T-shirts, stamps, or matchbooks, can share information about this aspect of map collecting through *Cartomania,* the journal of the Association of Map Memorabilia Collectors.

SEE ALSO

Antique maps; Appendix 2: Organizations and Publications Related to Maps

FURTHER READING

Makower, Joel, ed. *The Map Catalog.* 3rd ed. New York: Vintage, 1992.

Potter, Jonathan. *Antique Maps.* New York: Chartwell, 1988.

Tooley, R. V. *Tooley's Handbook for Map Collectors.* Chicago: Speculum Orbis, 1985.

Mappa mundi

MAPPA MUNDI is a Latin phrase meaning "map of the world"; generally the term is limited to maps made during the Middle Ages (A.D. 400–1450) that reflect cosmographies or geographic theories as well as geographic facts. Many *mappae mundi* seem to have been meant to serve decorative purposes. They are covered with pictures of towns, churches, kings, exotic animals (some real, such as the elephant, and some mythological, such as the hippogriff), biblical scenes, and other curiosities. Brilliant colors were used to render them beautiful. The Hereford map and the Ebstorf map, both dating from the 13th century, are good examples of the *mappa mundi.* Although they reflect some geographic knowledge, neither would be of much practical use to a traveler.

The term *mappa mundi* comes from the Latin words *mappa* (napkin or towel) and *mundi* (world). Apparently, some early *mappae mundi* were drawn or painted on pieces of cloth, and the phrase remained in use even when the mapmakers shifted to animal skins and paper. The modern word *map* is a shortened form of *mappa mundi.*

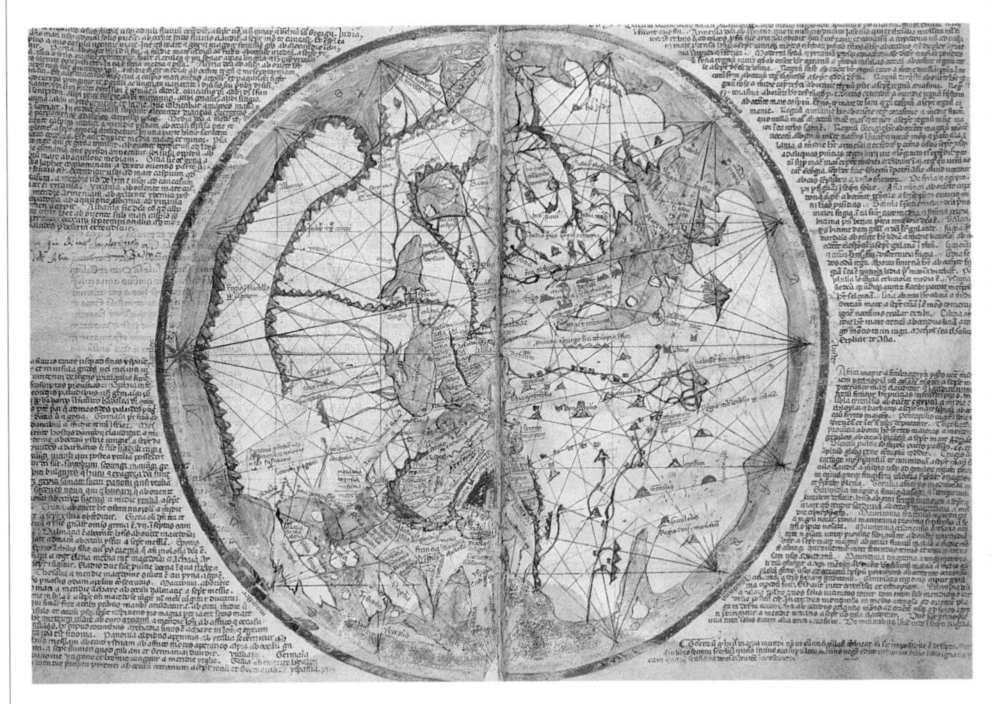

The mappae mundi *encompassed the known world—and a generous amount of myth and legend. This Italian* mappa mundi *dates from 1321.*

SEE ALSO

Ancient and medieval mapmakers; Ebstorf map; Hereford map; Mauro, Fra

Mapparium

VISITORS CAN see the world from the inside in a one-of-a-kind globe in the Christian Science Publishing Society building in Boston, Massachusetts. The Mapparium is a glass globe 30 feet (9 meters) in diameter, made up of 608 stained-glass panels, each covering 10 degrees of latitude and longitude. Visitors can enter the globe and walk across a bridge that spans its center. When the Mapparium is illuminated by the 300 electric lights that surround the sphere, people inside are surrounded by the brilliant colors of the lands and seas.

The Mapparium was built between 1932 and 1935 and shows national boundaries as they existed in the early 1930s. Although many countries' borders and even their names have changed since then, the Mapparium has not been changed because it is considered an original work of art. The First Church of Christ, Scientist, which operates the Christian Science Publishing Society, encourages people to visit the Mapparium and says, "The Mapparium is a reminder that wherever you live in the world, we are all neighbors." The Mapparium is pictured at the front of this book.

Marquette, Jacques

FRENCH EXPLORER OF NORTH AMERICA

- *Born: June 1, 1637, Laon, France*
- *Died: May 18, 1675, Michigan*

JACQUES MARQUETTE, often called Father Marquette, was a missionary priest in the Roman Catholic Society of Jesus, the Jesuits. In 1666 he arrived in Quebec, France's North American colony. Six years later he was invited by the explorer Louis Jolliet to take part in an expedition in search of a mighty western river of which the Indians had spoken. That river was the Mississippi. Jolliet and Marquette found the river and traced much of its course, returning with the first accurate European information about one of the world's major rivers. The two explorers were not only the first Europeans to provide information about the course of the Mississippi but they also were the first to map it. They made several maps, but all have been lost except one, which is preserved in a Jesuit archive in Quebec. It was made by Marquette and shows his route from what is now Wisconsin south along the Mississippi to the mouth of the Arkansas River.

SEE ALSO

Americas, mapping of; Jolliet, Louis

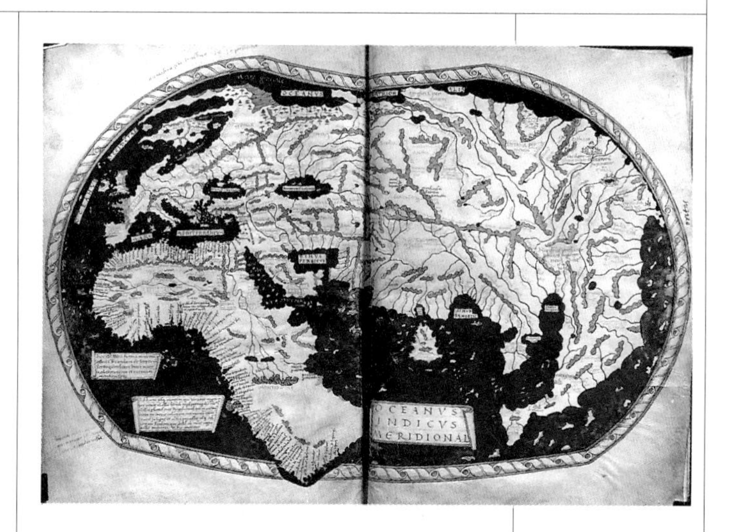

Henricus Martellus's 1490 world map reflects the worldview of Europeans on the eve of Columbus's fateful voyage. The west coast of Africa was known in considerable detail after decades of exploration by the Portuguese.

Martellus, Henricus

GERMAN GEOGRAPHER AND MAPMAKER

• *Active: 1480–96*

HENRICUS MARTELLUS worked in Florence, Italy, during the late 15th century. Around 1490 he published a book of geography that contained a world map. Martellus's world map is based partly on traditional notions of geography drawn from Ptolemy and partly on the recent explorations of Portuguese mariners. Departing from Ptolemy, Martellus shows open sea south of Africa and labels the southern tip of that continent "the true modern form of Africa from the description of the Portuguese." The map also records places sighted by Bartolomeu Dias on his African voyage of 1487–88. Some scholars have suggested that Martin Behaim based the geography of his 1492 globe on Martellus's map. The two are similar, but no direct connection is known; it is possible that they both used the same sources and came up independently with similar maps.

SEE ALSO

Behaim, Martin

Mason-Dixon line

THE MASON-DIXON LINE is one of the best-known boundary surveys in the world. It is the border between the northern and southern parts of the United States. The Mason-Dixon line does not stretch across the whole country, however; indeed, it never extended west of Pennsylvania. It grew out of a dispute between Pennsylvania and Maryland over where the border between them should properly be located.

The Penns of Pennsylvania and the Calverts of Baltimore had each received their colony in a British royal charter, but the charters overlapped. The two families agreed to have the border drawn by impartial surveyors, and the Astronomer Royal in Greenwich sent them

Charles Mason and Jeremiah Dixon. The surveyors arrived in Philadelphia in 1763 and began a series of careful astronomical observations and ground surveys. Not until the spring of 1765 were they ready to mark out the disputed east-west line. They worked until the fall of 1767, pushing the line westward through winter snow and summer heat, working their way along the Susquehanna River and up and down the Blue Ridge mountains, preceded by axmen who hacked a path through the wilderness. They stopped just short of what is now the western border of Pennsylvania when Indians ordered them to go no farther. The border had been established, and in later years it was symbolically extended west along the 40th parallel to mark the differences that developed between the northern and the southern states.

Today, the term refers to the symbolic border between the northern and southern United States, as in "things are different south of the Mason-Dixon line."

Mauro, Fra

ITALIAN MONK AND CARTOGRAPHER

• *Active: 15th century, Venice, Italy*

FRANCIS MAURO—called Fra Mauro (which is Italian for Brother Mauro)—was a monk in the Camaldolese monastery on Murano, an island near Venice, Italy. In 1457 he was hired by King Alfonso V of Portugal to make a map of the world. The map took two years to produce; it was sent to Portugal and later lost. A copy of Fra Mauro's map was preserved in Venice, however, and today it is considered a masterpiece of cartography. More than six feet across, the map is round, in the tradition of the medieval *mappae mundi.* It contains a wealth of up-to-date information about Africa and Asia, gleaned from the reports of travelers and explorers in the previous several centuries. Asia, for example, is dotted with details from the *Travels of Marco Polo,* and the west coast of Africa reflects the explorations there of Portuguese navigators.

In one important respect, Fra Mauro's map was ahead of its time. Unlike Ptolemy, who believed that the Indian Ocean was a landlocked sea closed off by the joined southern coasts of Africa and Asia, Mauro accurately shows a sea route around Africa into the Indian Ocean—three decades before Bartolomeu Dias was to pioneer that route. (This idea was not original with Mauro; ancient writers, including Herodotus and Solinus, had written of a navigable route around Africa, although Dias was the first to verify its existence.)

Mauro was not entirely free of the cartographic errors of his time, however. He placed the mythical kingdom of Prester John on his map, writing "Here Prester John makes his principal

Russia as it appears on Fra Mauro's world map. Southwest is at the top; the Volga River flows into the Caspian Sea at the upper left. The entire map is filled with notes and descriptions.

residence" in the part of Africa that is now known as Ethiopia.

SEE ALSO

Bianco, Andrea; Mappa mundi; Prester John; Ptolemy

Maury, Matthew Fontaine

AMERICAN OCEANOGRAPHER

- *Born: 1806, Spotsylvania County, Virginia*
- *Died: Feb. 1, 1873, Lexington, Virginia*

MATTHEW FONTAINE MAURY was one of the founders of oceanography, the scientific study of the world's oceans. A naval officer assigned to the Naval Depot and Observatory in Washington, D.C., he created a system of collecting data from ships' logs to build up a detailed picture of ocean winds, currents, and temperatures. He created sea charts and cartograms, using ingenious symbols to communicate pages' worth of information at a glance. He also used depth soundings—measurements of the sea's depth made by lowering a weighted rope over the side of a ship—to make maps of the seafloor; his bathymetric chart of the North Atlantic Ocean hints at the existence of the immense undersea mountain range that later explorers were to discover in the middle of the ocean. Maury's book *The Physical Geography of the Sea* (1855) is considered to mark the beginning of modern oceanography.

SEE ALSO

Undersea mapping

FURTHER READING

Goetzmann, William H. *New Lands, New Men: America and the Second Great Age of Discovery.* New York: Viking Penguin, 1986.

Lewis, Charles Lee. *Matthew Fontaine Maury.* 1927. Reprint. New York: Arno Press, 1980.

Stefoff, Rebecca. *Scientific Explorers: Travels in Search of Knowledge.* New York: Oxford University Press, 1992.

Williams, Frances Leigh. *Matthew Fontaine Maury, Scientist of the Sea.* New Brunswick, N.J.: Rutgers University Press, 1963.

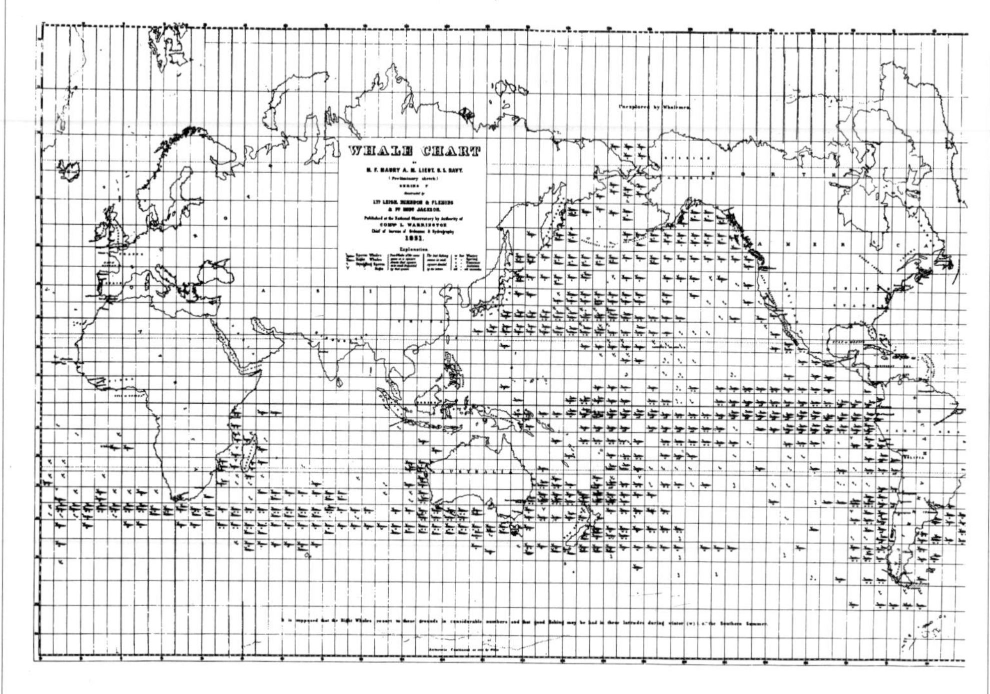

Maury repaid whalers for geographic information by making this chart based on sightings of two species of whale. The chart improved whalers' chances of finding prey.

Medieval mapmakers

SEE Ancient and medieval mapmakers

Mendaña de Neira, Álvaro de

SPANISH EXPLORER OF THE PACIFIC

- *Born: about 1542, Galicia, Spain*
- *Died: Oct. 18, 1595, in the South Pacific*

WHEN ÁLVARO de Mendaña de Neira was 25 years old, he went to South America with his uncle, who was later to become the governor of Peru. Mendaña learned that the Incas, the Native American people of Peru, had many legends about rich islands in the Pacific Ocean. In 1567, he led an expedition in two ships to find these wondrous isles.

Mendaña did find a cluster of islands far across the Pacific from Peru. He named them the Solomon Islands, because he hoped that they contained the legendary land of the biblical king Solomon. Before the expedition returned to Peru, Hernando Gallego, Mendaña's pilot, made a detailed chart of the islands. Unfortunately, however, he lacked a reliable means of determining their longitude, and as a result he could not pinpoint their exact location. So when Mendaña left Peru in 1595 with nearly 400 settlers to establish a colony in the Solomons, he was unable to find the islands again. Mendaña died, and the expedition fell apart in the face of a series of disasters: shipwreck, starvation, and murder. Only 100 of the would-be colonists made their way safely to a Spanish port in the Philippine Islands.

SEE ALSO

Pacific Ocean, mapping of; Quiros, Pedro Fernándes de

Mental maps

DURING THE 20th century, a handful of geographers began studying "mental maps," which are defined by geographers Peter Gould and Rodney White as the "'invisible landscapes' that people carry in their heads." Mental maps are the individual notions about the world around us that each of us has created out of geographic knowledge and geographic ignorance, images from books and movies, stereotypes and prejudices and fantasies. They shape our thinking in ways of which we may not even be aware.

Mental maps can be amusing, as in the novelty map by artist Saul Steinberg that shows the United States from a New Yorker's point of view. Printed on the March 29, 1976, cover of the *New Yorker,* the cartoon shows a hugely exaggerated New York (Brooklyn alone being the size of Texas), a big Florida, a state called Hollywood on the West Coast, and a compressed, wildly misconceived Midwest.

But mental maps can also be signs of social problems. When a researcher asked a number of people living in Los Angeles about their city, he found that people from affluent, mostly white neighborhoods knew a lot about the layout of Los Angeles, its museums and art galleries, and the area around the city. People from low-income, mostly black neighborhoods had more limited mental maps of the city, organized around bus routes rather than freeways. And the mental maps of immigrants in a Spanish-speaking community contained only the local neighborhood, City Hall, and the bus station.

Geographers and psychologists are now studying mental maps to discover more about how people build these internal images. They believe that mental maps have a significant effect on education, on economic trends such as people's decisions to move from one place to another or companies' decisions about where to open plants and offices, and on many other aspects of life.

SEE ALSO
Sketch maps

FURTHER READING

Gould, Peter, and Rodney White. *Mental Maps.* 2nd ed. Boston: Allen & Unwin, 1986.

Mercator family

THE MERCATOR dynasty of map publishers was founded by Gerardus Mercator, the best-known mapmaker in the history of cartography and perhaps the only one whose name is familiar to the general public. Mercator dominated cartography for nearly 60 years.

Born near Antwerp on March 5, 1512, Gerhard Kremer followed the common practice of Latinizing his name and became Gerardus Mercator. He went into business in Louvain as a map and globe maker. In 1544 he became a victim of the religious persecution that convulsed Europe in the 16th century: he was thrown into prison as a heretic. Thanks to the influence of powerful friends, he was released after a few months. He left Louvain for the calmer town of Duisburg, where he lived out his life under the protection of a local duke.

Mercator's single most significant work was a world map he issued in 1569 with the note that he had used "a new proportion and a new arrangement of the meridians with reference to the parallels." Mercator had invented a new projection, which now bears his name. It became the most widely used map projection in the world and is still used on many maps, although in the late 20th century it is being replaced by other projections.

The Mercator projection has several distinctive features. It is rectangular, with the lines of latitude and longitude crossing each other at right angles. It greatly exaggerates the size of land masses, especially near the poles, so that countries such as Greenland, in high latitudes, appear far larger than they really are. But—and this is what made the projection useful—because the map is rectangular and the lines of latitude and longitude are straight, not curved as in some other projections, mariners could use the Mercator map to plot a straight-line course that would carry them across long distances without a change in direction. In other words, a navigator could lay a ruler across the map from the Canary Islands to Rio de Janeiro, determine the compass reading along that line, and

Gerardus Mercator escaped religious persecution to revolutionize cartography with the Mercator projection, designed to aid navigators at sea.

A world map published by Rumold Mercator in 1595 (but not drawn on the Mercator projection). The hypothetical continent of Terra Australis takes up much of the Southern Hemisphere.

sail all the way across the Atlantic Ocean by following that single compass direction. On other maps, by contrast, navigators had to use curved lines for long courses, which meant that they frequently had to change the direction in which they were sailing. Navigators were slow to adopt the Mercator projection, however. The projection did not come into wide use among seamen until the 17th century. Although Mercator's projection had been designed specifically for use by seamen, it was the basis for maps that eventually hung on the wall of nearly every schoolroom.

Mercator's other great achievement was his atlas of 1585–95. Although it was not the first such collection of maps to be published, it was the first to be called an atlas.

Mercator died on December 1, 1594, the year before the third volume of his atlas was issued. His son Arnold, also a mapmaker, had died in 1587, so the Mercator business was inherited by Mercator's younger son, Rumold, who outlived his father by only six years. Gerard, Michael, and John Mercator, sons of Arnold, also made maps, but in 1606 they sold the Mercator map-publishing business—consisting of the copper plates from which the maps were printed—to the Hondius family, which continued to issue Mercator atlases for many years.

SEE ALSO

Atlas; Hondius family; Projections

Mercator projection

SEE Mercator family; Projections

Meridian

SEE Longitude

Meteorological maps

SEE Weather maps

Mid-Ocean Ridge

SEE Undersea mapping

Military maps

MAPS COMPILED for military use are generally of two sorts: maps of enemy territory, to be used in attack, and maps of one's own territory, to be used in defense. Such maps have always been guarded carefully, for when conflict occurs, each side would like nothing better than to get hold of the other side's maps.

Specialized military maps show such things as the positions of forts, gun emplacements, naval bases, docks and canals, military headquarters, armories, and troops; roads and bridges along which armies could travel; railway lines that could be used to move troops and supplies; power stations and other crucial resources; and the topography of the land on which battles may be fought.

Ordinary maps often have military uses. In World War II (1939–45), British pilots who faced the danger of being shot down over German territory carried travel maps of Germany to help them escape. These maps were printed on thin cloth and carried in hollow boot heels or secret pockets. At the same time, Germany prepared maps of England that would have been very useful to German officers and soldiers if their planned invasion of England had taken place.

War often gives cartography a boost. The mapping of the American colonies in the 18th century was due, at least in part, to the French and Indian War and the American Revolution, which created a demand for maps. War has also rescued old maps from obscurity or given new importance to routine geographic maps. When World War II was raging across the atolls of the Pacific, the governments of the Allied nations ransacked their countries' libraries and universities for information about places such as New Caledonia and the Marshall Islands; maps a century old were pressed into use.

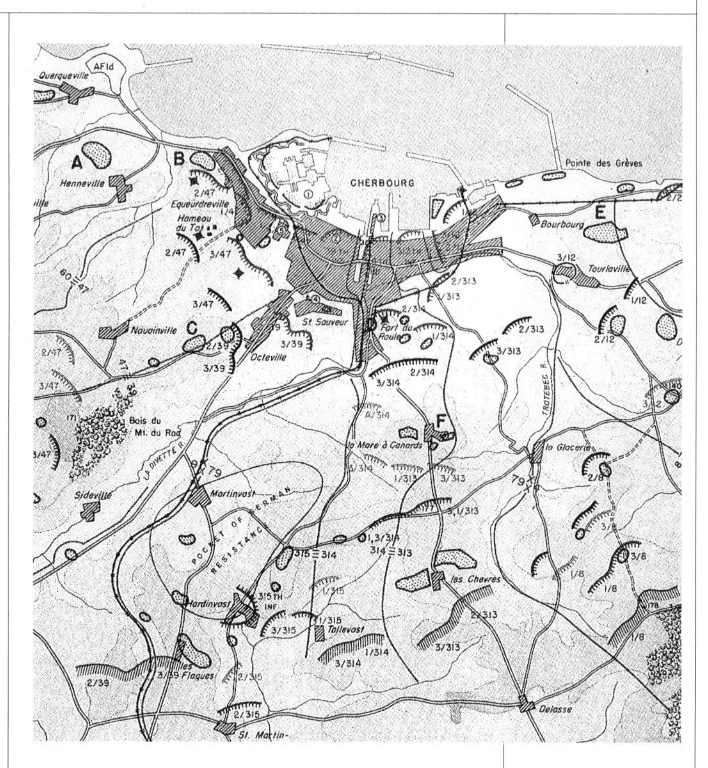

A U.S. map from World War II shows the positions of troops and guns as Allied forces prepare to seize the French port of Cherbourg from the Germans. A "pocket of German resistance" is identified just left of the lower center of the map.

Governments also take pains to keep potentially useful geographic information from falling into enemy hands. Just before World War II broke out, the American Arctic explorer Louise Arner Boyd, who had led a number of expeditions to the northeast coast of Greenland, was asked to delay publication of her most recent research for fear that the Germans might use her maps to build secret bases in Greenland; instead, she turned over her geographic data to the U.S. War Department.

Aerial maps have been used in warfare since the Civil War, when balloonists spied out enemy positions and marked them on maps for the use of commanders in the field. The airplane made it even easier to survey terrain from above; pilots in World War I

(1914–18) were routinely sent aloft to gather information about enemy troop movements, and this information was often plotted on maps. But in the second half of the 20th century, satellites changed the nature of military mapping. With satellite cameras able to distinguish individual buildings and truck convoys from high in space, absolute military security has become harder to achieve, although the overnight invasion of Kuwait by Iraq in 1990 demonstrated that technology alone does not provide military security.

The term *military map* is often used to refer to historical maps that deal with military subjects or past wars. Maps of battlefields or military campaigns fall into this category, as do reproductions of old military maps that no longer serve any strategic purpose. These maps have been collected in military atlases, such as atlases of the world wars.

SEE ALSO
Historical maps

Münster, Sebastian

GERMAN COSMOGRAPHER AND MAPMAKER

- *Born: 1448, Ingelheim, Germany*
- *Died: 1552, Basel, Switzerland*

SEBASTIAN MÜNSTER was a scholar who taught theology and Hebrew at the University of Heidelberg and then taught mathematics at Basel; he also published the first translation of the Bible from Hebrew into German. Like most learned people of his age, Münster was interested in many fields of knowledge, including geography and mapmaking. He produced two major cartographic works. One was his 1540 edition of Ptolemy's *Geography,* with 48 woodcut maps. Münster cleverly set the place names in removable blocks of type so they could be changed, dropped, or added without the entire map having to be recut.

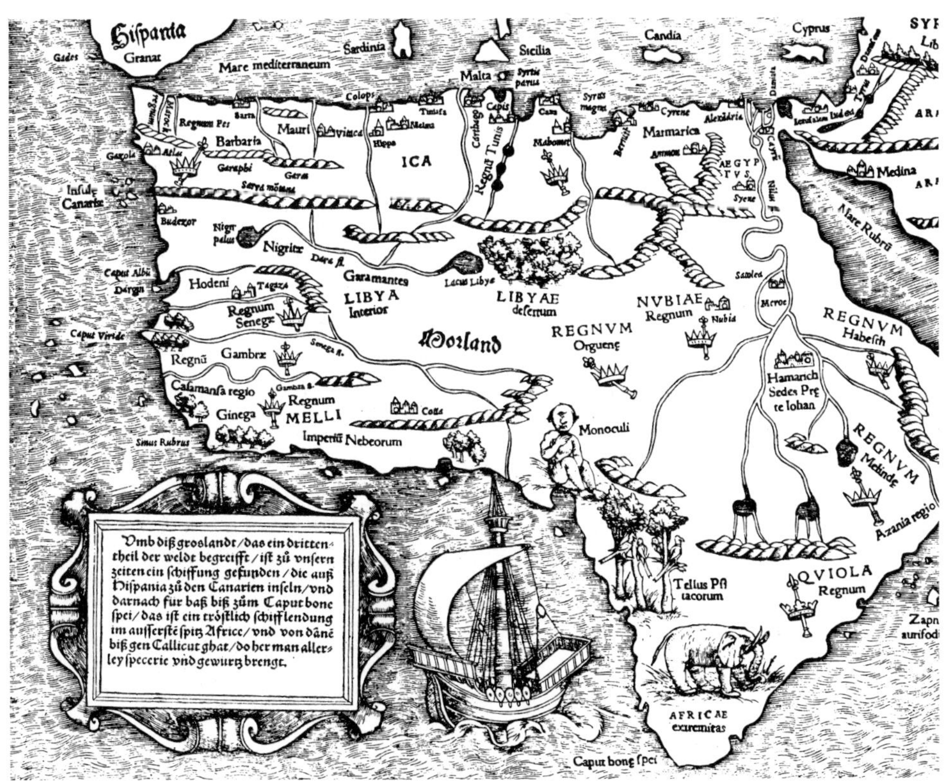

Münster's Africa, complete with elephant and parrots—and a fanciful race of one-eyed beings.

The first edition of Münster's important book *Cosmography* appeared in 1544, with some maps from the earlier book and some new ones. Münster was the first mapmaker to map each continent separately. His map of the Americas is the first to show the Western Hemisphere apart from the rest of the world, and also the first to show North and South America connected by land. Münster was also one of the first mapmakers to list the sources he used when compiling his maps—not just the other maps he adapted but also the books and explorers' reports he had read.

Münster's maps were enormously influential. They were reprinted many times and copied by other cartographers. One of their more endearing—and enduring—features is the images with which they are decorated. Münster filled the blank spaces on his maps with pictures of elephants, shipwrecks, plants and animals, kings, and the strange (and fictional) races of humans described earlier by the Roman historian Solinus: one-eyed people in Africa and one-legged ones in India. These images were copied on many later maps.

SEE ALSO

Solinus, Gaius Julius

National Geographic Society

THE NATIONAL Geographic Society (NGS) was founded in 1888 to publish a magazine about travel and geography; money raised through subscriptions to the magazine would support expeditions of exploration and research. Prominent people who were founding members or early officials of the society included Alexander Graham Bell, the inventor of the telephone, and John Wesley Powell, the explorer of the Colorado River. Since its founding, the NGS has operated from offices in Washington, D.C., where it now has a museum called the Explorers Hall. Membership has grown to include millions of people in all parts of the world, who each month receive an issue of *National Geographic,* with its familiar yellow-bordered cover. The magazine is noted for the high quality of its photographs and for its numerous excellent maps; they are prepared by the NGS's own cartographers. The NGS continues to sponsor expeditions and research projects. For example, in the 1980s the society was a partner in an ambitious international project that used aerial photography and computerized mapping to create a new, more accurate map of the Himalaya Mountains. The NGS also produces television documentaries, books, and a variety of map-related materials and programs (such as the annual Geography Bee) to help schools improve geographic education.

FURTHER READING

Bryan, C. D. *The National Geographic Society: 100 Years of Adventure and Discovery.* New York: Abrams, 1987.

Lutz, Catherine, and Jane Collins. *Reading National Geographic.* Chicago: University of Chicago Press, 1993.

National Geographic. Centennial Issue, January 1988.

Native American mapmakers

THE INDIGENOUS peoples of the Americas did not make maps as they are thought of in the European tradition. They did, however, communicate geographic information. From central Mexico to the Arctic, they traveled and collected information from other travel-

Made in 1569, this map of the Hacienda de Santa Iñes, Mexico, displays some Aztec mapmaking techniques, such as the use of footprints to show roads and pictures of plants and a sheep to indicate crops and livestock.

ers, and they recorded this information in a variety of forms. The ancient civilizations of Mexico, for example, had writing systems that used pictures instead of letters, and these picture alphabets included some symbols for geographic information. Among Indian peoples farther north, geographic lore was part of an oral tradition, in which all kinds of knowledge were passed from generation to generation through the spoken word, in stories, songs, and histories. This oral tradition included a high degree of geographic knowledge, which was part of each tribe's cultural legacy and which helped to shape the Indians' view of the world. The Indians made maps only when they needed to illustrate some part of this information.

Native American maps are concerned with local or regional territory—with the area that might be traversed by a particular individual or tribe. The American Indian cartographers did not try to map the entire continent or the world; rather, they focused on the parts with which they were familiar. Yet their maps were not merely representations of geography. Native American maps express the Indians' way of looking at the world, in which history, tradition, and mythology are as important as geography. As a result, many maps include scenes or symbols from an individual's past or a tribe's history; they may also contain images of spirits, mythological events, or sacred places. Although the Indian cartographers, like European mapmakers, used standardized symbols to indicate rivers, villages, and other features of the landscape, most Indian maps resemble pictures rather than maps in the European sense. But because they show landmarks, boundaries, travel and trade routes, and topographic features, these pictures also function as maps.

Although some original Native American maps have been preserved in museums and libraries, much of what is known about Indian mapmaking comes from copies of Indian maps that were drawn by white explorers, settlers, and historians. Many maps by Europeans have notations explaining that they are based on information given by Indians; some maps were made by Indians and

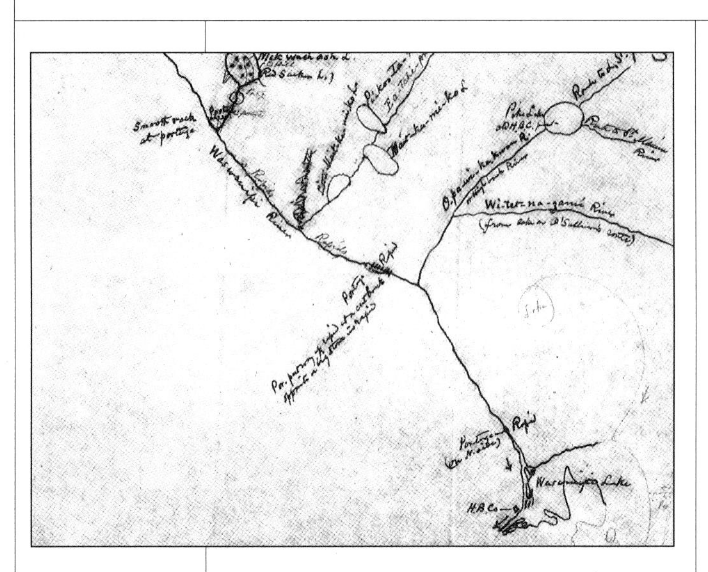

Canadian Indians drew this map in 1896; notes on the map were made by the government surveyors who had asked the Indians for infomation. The map shows part of central Quebec, with a Hudson's Bay Company post at lower right.

whites working together. From the beginning of European settlement in the Americas in the early 16th century, the whites eagerly sought geographic information from the Native Americans. They often relied on Native American guides or maps to help them get around. For example, when the French explorer Samuel de Champlain landed at Cape Ann, Massachusetts, in 1605 and sketched the coast for the local Indians, one Indian responded by drawing a map that showed nearby dwellings and rivers—information that Champlain incorporated into his own 1607 map of the region. The map of Virginia published in 1612 by John Smith, leader of the English colony there, was divided into two parts: the area that Smith had actually explored and an area beyond it that had been described to him by the Algonquian tribes who lived around the Chesapeake Bay.

The Maya civilization that flourished in southern Mexico and Central America from 1500 B.C. to A.D. 1500 may have included mapmaking; scholars are still studying Maya inscriptions to determine their meaning. We know, however, that the Aztecs, who ruled central Mexico at the time of the Spanish invasion in the early 16th century, were practiced cartographers. They drew maps on cloth and on paper made from the fibers of the amatl plant. One type of Aztec map was symbolic or historic, intended to record information about migrations, trade routes, and wars of conquest. The *Mapa Sigüenza,* copied from a 16th-century Aztec map, is an example. It shows the history of the Aztec people, with footprints indicating the route of their migration through Mexico to the site of their capital, Tenochtitlán.

The Aztecs also made maps for route finding, as well as surveys and town plans. These were used—and in some cases copied and reproduced—by the conquering Spanish. Hernán Cortés, the conqueror of Mexico, included a map of Tenochtitlán (now Mexico City), in his 1524 report to the crown. Although drawn by a European, the map contains many Aztec symbols and may have been copied from an Aztec map. The influence of Native American mapmaking traditions can also be seen in two maps that were made in 1569 to settle a dispute over water rights on a farm called the Hacienda de Santa Iñes, located in central Mexico. The identity of the mapmaker is not known, but the maps display a mixture of Spanish and Aztec styles. Roads and paths are shown in the traditional Indian style as rows of footprints; swirling lines, as used by Aztec artists, indicate water.

The Indians of Mexico also made maps that the Spanish called *lienzos* (which means "linen"). These showed the boundaries of communities and also gave information about the history and genealogy of local rulers. More than 50 *lienzos* survive; they are still recognized as important legal documents in their villages. One of these, made in 1829 as a copy of a 16th-century original, shows the community of Miacatlan, near present-day Morelos, Mexico. Spanish elements, including a church in the center of town, are blended with native elements such as the footprints that mark the location of paths.

The Indians who lived farther north, in the region that is now the United States and Canada, had different mapmaking traditions. Their societies were based on oral traditions. These Native Americans seldom made maps as permanent documents, although they clearly understood the concept and purposes of cartography. Often, they drew maps in dirt, sand, or ash—these maps were created for immediate use and vanished as soon as they were no longer needed. However, they sometimes made more permanent maps on a variety of materials: birch bark in the forests of northeastern North America, animal skins across most of the continent, and stone cave or cliff walls in parts of the West. Some of these were route maps, left by traveling Indians to tell companions how to follow them. A birch-bark map was found by a British army officer in 1841, pinned to a tree near the Ottawa River in Canada. The unknown mapmaker used a dotted line to show the route of travel and a triangle representing a tepee as the symbol for a camp.

Some Native American peoples created petroglyphs, images drawn on or carved into stone, usually cliff walls. Although the petroglyphs depict spiritual beliefs, some of them may also include geographic information and thus may be considered maps. One petroglyph near the Snake River in Idaho is called the Map Rock because it is believed to show the course of the Snake River and the territory of the Shoshone people.

Many Native American maps were made to help the Indians communicate with whites in a form that was familiar to European culture. These maps tell the story of the Indians' relationship with the whites, from the early encounters, treaties, and conflicts to the forced migrations of whole peoples and the Indians' desperate struggle to retain control of their lands. Many maps can be seen as statements of the Indians' place in the land. In 1730, for example, a chief of the Catawba people painted a map on deerskin and gave it to the British governor of South Carolina. It shows the streets of Charleston with a British ship in the harbor; it also shows the settlements of the Catawba people as the central feature of the landscape. A map made in Canada in 1869 reflects the encroachment of the whites into Indian territory around Lake Nipigon, north of Lake Superior. Based on information from as many as 10 Chippewa Indians, the map shows British flags at two Indian settlements, indicating that these Native Americans were allied with—or under the control of—the British. Some places on the map are identified by their Indian names, but English place names have begun to appear. A group of islands, for example, has been named after famous English writers, such as Shakespeare and Chaucer.

As white Americans moved westward across the continent, some explorers and historians recorded geographic information gained from the Native Americans. Amiel Weeks Whipple, an army officer who surveyed for a railroad route through the West in the 1840s and 1850s, obtained several maps from the Paiute Indians. Redrawn by Whipple, the maps appeared in his report to the Senate, which was published in 1854. And in 1907, the North Dakota State Historical Society asked a Mandan Indian named Sitting Rabbit to map the location of the Mandan and Hidatsa settlements that had lined the Missouri River in bygone days. Drawing upon traditional Mandan lore, Sitting Rabbit produced a map that included vanished Indian settlements dating from prehistoric times; the accuracy of his map was later proved by archaeological finds.

The Native American peoples of the far north were mapmakers, too. The Inuit (formerly called Eskimos) are the

On this Inuit map, bits of fur and driftwood stitched to sealskin indicate islands and channels.

only Native Americans who made relief maps—three-dimensional maps showing heights and valleys. These maps were modeled in earth, with bits of wood and stone for islands, mountains, and villages. When they drew on skins or paper, the Inuit shaded their hills to show relief, just as European mapmakers did. The Inuit, whose lives were closely tied to the sea and who were expert hunters in their small boats, also made unique coastal maps of carved wood. The features of the coastline are carved along the edges of pieces of wood; sometimes both edges of a plank are used, with one edge picking up the map where the other leaves off. Rods stick out from the plank to indicate offshore islands. The Inuit also carved maps into walrus ivory; these were especially prized by collectors and curio hunters, and some may have been created specifically for sale to visitors.

European explorers in the Arctic, where a wrong turn can lead to starvation or death by exposure, learned to rely on the Inuit for accurate geographic information. British naval officer William Parry, who spent the years from 1821 to 1823 searching for the Northwest Passage, received a hand-drawn map from an Inuit woman of the Melville Peninsula, on the northeastern tip of the Canadian mainland; he reproduced it in his 1824 book about the expedition. Around the same time, another British explorer of the Arctic, John Ross, invited two Inuit aboard his ship to help him map his course. Ross painted a picture of the Inuit man and woman in his cabin, working on the map; Ross also noted that the man's geographic knowledge was highly accurate. American anthropologist Franz Boas, who studied the Inuit of Baffin Island in the 1880s, observed that the Inuit were skillful navigators and cartographers. "If a man intends to visit a country little known to him," Boas wrote of these northern mapmakers, "he has a map drawn in the snow by someone well acquainted there and these maps are so good that every point can be recognized."

SEE ALSO

Parry, Sir William Edward; Ross, Sir James Clark

FURTHER READING

Belyea, Barbara. "Amerindian Maps: The Explorer as Translator." *Journal of Historical Geography* 18, no. 3 (1992): 267–77.

DeVorsey, Louis. "Amerindian Contributions to the Mapping of North America: A Preliminary View." *Imago Mundi* 30 (1978): 71–78.

Lewis, G. Malcolm. "The Indigenous Maps and Mapping of North American Indians." *Map Collector* 9 (1979): 25–32.

Luebke, Frederick D., Frances W. Kaye, and Gary E. Moulton, eds. *Mapping the North American Plains.* Norman: University of Oklahoma Press, 1987.

Warhus, Mark. "Cartographic Encounters: An Exhibition of Native American Maps from Central Mexico to the Arctic." *Mapline* Special Issue, September 1993.

Nautical maps

IN BROAD terms, a nautical map is any map having to do with a body of water—a sea, a lake, or even a river. In practice, though, maps that show a considerable amount of detailed information about channels, harbors, currents, water depths, and navigational markers—in other words, maps that are designed to help guide boats—are called charts. Ordinary geographic maps of the surfaces of bodies of water, showing islands and ports but not such detailed navigational information, can be called marine or nautical maps.

Navigation

NAVIGATION IS the art and science of getting from one place to another. It includes planning a route, following that route, and checking position from time to time along the route.

On a family trip, the person who reads the road map and tells the driver where to turn is acting as a navigator. In a more specific sense, navigation usually refers to the guiding of a ship or boat, an airplane, or a spacecraft.

Cartography and navigation have always had a two-way relationship. Navigation depends on maps—from the portolan charts of the 13th century to the computerized electronic guidance system that is piloting the *Voyager* space probes into deep space. In turn, many navigators have returned from voyages with information to help cartographers make new maps.

Over the centuries, navigators have used a variety of tools to follow their maps. The most basic of these is the compass, which tells direction, the key element of navigation. Position finding—determining one's latitude and longitude—has been refined from the days of the mariner's astrolabe and cross-staff, used since ancient times to find latitude. John Harrison's invention of an accurate timepiece in the 18th century made it possible for navigators to make reliable readings of longitude.

As instruments improved, sea charts became increasingly complex and detailed, as well as increasingly accurate. Thousands of these maps were produced in Great Britain and elsewhere during the 19th century, the age of empire building and heavy sea traffic.

In 1876, the arts of mapmaking and navigation alike were gently satirized by Lewis Carroll, the creator of *Alice in Wonderland,* in his poem "The Hunting of the Snark." Carroll describes the Bellman, who served as both captain and navigator on a singular voyage:

> He had brought a large map representing the sea,
> Without the least vestige of land:
> And the crew were much pleased when they found it to be
> A map they could all understand.
>
> "What's the good of Mercator's North Poles and Equators,
> Tropics, Zones, and Meridian Lines?"
> So the Bellman would cry: and the crew would reply,
> "They are all just conventional signs!
>
> "Other maps are such shapes, with their islands and capes!
> But we've got our brave Captain to thank"
> (So the crew would protest) "that he's brought *us* the best—
> A perfect and absolute blank!"

An array of new navigational instruments appeared in the 20th century, from echo sounders and radar to computers and satellites. Navigational systems that determine latitude and longitude from satellite beacons are now available for every vessel from the family car to the newest interplanetary probe. With the addition of a computerized map, these systems can guide a spacecraft to Jupiter, a guided missile to an enemy target, or

An illustration from The Light of Navigation, *a 17th-century book of sea charts by Willem Blaeu. At right, two apprentices are learning to use the cross-staff; other navigators are busy with charts, globes, and an astrolabe.*

an automobile through the maze of the Los Angeles freeway system.

SEE ALSO

Aeronautical maps; Astrolabe; Celestial navigation; Chart; Compass; Cross-staff; Dead reckoning; Global Positioning System (GPS); Harrison, John; Latitude; Longitude; Orienteering; Sextant

Northeast Passage

THE NORTHEAST PASSAGE is the name given to the sea route from Europe north and east through the Arctic waters north of Russia to the Pacific Ocean. Although not as well known or as ardently sought as the Northwest Passage, the Northeast Passage inspired many voyages of exploration, beginning in the mid-16th century. With overland trade routes to the east blocked by wars, Europeans were eager to find a navigable water route to Muscovy—as the area around Moscow was then called—and beyond to China and Japan.

Europeans knew that it was possible to sail north and east of Scandinavia, into the White Sea and the Barents Sea, but the rest of the eastward route remained a mystery. Several English and Dutch expeditions attempted the route in the 16th and 17th centuries; two of these were led by Willem Barents, for whom the Barents Sea is named. The farthest point reached was the island of Novaya Zemlya in the Kara Sea. In 1619 the Russian czar, seeking to keep international traders out of his country, closed the route to foreign ships.

Throughout the 17th century Russian hunters, adventurers, and explorers pioneered the overland exploration of Siberia, but not until Vitus Bering crossed

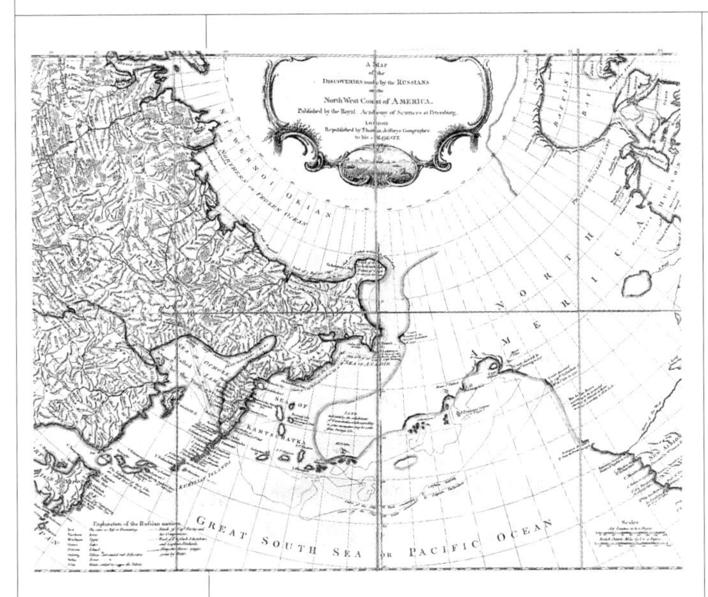

Canada and Alaska were still blank spaces on the map (right) when Russian explorers filled in the map of northern Siberia (left).

the continent and sailed north around the tip of Asia in 1728 did geographers know for certain that Asia and North America were separated by a water channel, now called the Bering Strait: a Northeast Passage did exist.

By the middle of the 18th century the entire north coast of the continent had been mapped. Still no one had sailed the Northeast Passage. That feat was first accomplished in 1878–89 by the Swedish explorer Adolf Nordenskiold. (In addition to being an explorer, Nordenskiold was a collector and scholar of old maps. In 1889 and 1893 he published atlases containing reproductions of many antique maps, thus making these cartographic treasures widely available.)

In the early years of the 20th century, Russian government surveys completed the mapping of the islands and ice fields in the Northeast Passage. The English and Dutch adventurers who sought commercial advantages from the Northeast Passage never did make a penny of profits from the route, but modern Russian vessels routinely ply shipping routes through the passage.

SEE ALSO

Barents, Willem; Bering, Vitus

Northwest Passage

THE NORTHWEST PASSAGE is the name given to the sea route between the Atlantic and Pacific oceans through the Arctic north of North America. It is called the Northwest Passage because European mariners hoped to find the passage by sailing northwest from their home ports. If found, the passage would offer a swifter and safer way into the Pacific than that afforded by the Strait of Magellan at the southern tip of South America. The passage was eagerly sought because traders and adventurers hoped that it would give them easy access to the fabled wealth of China and the Indies.

Although Christopher Columbus thought he had found an outpost of Asia, early in the 16th century European geographers and explorers realized that what he had found was not Asia but America—a huge landmass that blocked the way to the Indies. Europeans probed the American coastline year after year, not just exploring the newfound land but impatiently looking for a way through or around it to the Indies.

The British were particularly diligent in the search for the Northwest Passage. Martin Frobisher looked for it in the 1570s, John Davis in the 1580s, and Henry Hudson in 1610. The closest thing to a passage they found was Davis Strait, a north-trending waterway off Greenland's west coast. Within a few years, however, other explorers had found that the northern reaches of the strait were too choked with ice to allow ships to pass.

The search for the Northwest Passage gave way to commercial missions into the northern waters and the area around Hudson Bay: whaling, cod fishing, and hunting for and trading furs. In 1631–32, Thomas James of Great

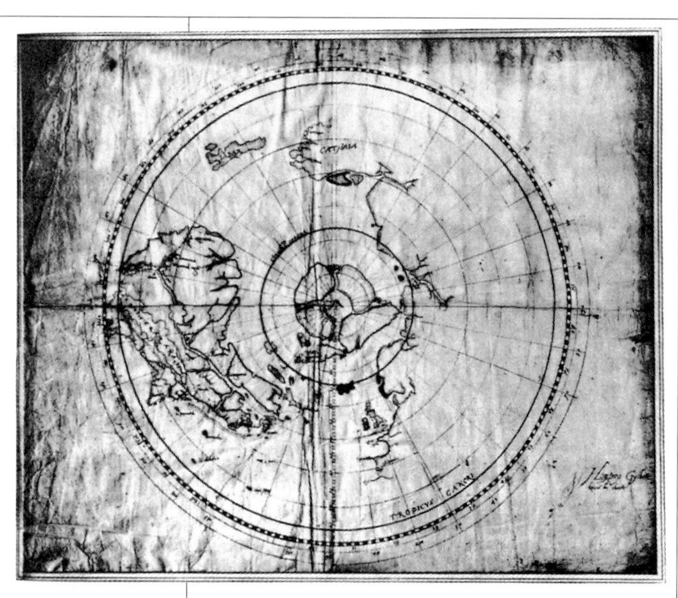

A 16th-century British map of the north pole optimistically shows a wide sea passage north of the Americas (left). It also shows a nonexistent landmass around the pole.

Britain made a final effort to locate an entrance to the Northwest Passage from Hudson Bay. He failed, as no such passage exists, but his account of the hardships he endured on the voyage were later incorporated into Samuel Taylor Coleridge's poem *The Rime of the Ancient Mariner.*

The yearning to find a Northwest Passage had not disappeared, however. Many geographers still believed in the existence of another version of the Northwest Passage called the Strait of Anian. The Strait of Anian first appeared on maps of North America in the 1560s. It seems to have been created by Giacomo Gastaldi from a reference in Marco Polo's *Travels* to a land called Anian on the eastern border of China. By the 18th century, the Strait of Anian was thought to be a river or channel flowing into the North Pacific from somewhere in the interior of North America—perhaps from the Great Lakes. Alexander Mackenzie looked in vain for the strait in the center of Canada. Explorers such as James Cook and George Vancouver looked for its western end in the Pacific Ocean, but they found only the Columbia River and Puget Sound.

By the early 19th century, the British Admiralty was ready to try again to locate a waterway through Arctic Canada to the Pacific. British officials even offered a huge reward for the discovery of the passage. The Admiralty sent expeditions under William Parry, James Ross, and John Franklin to explore the huge realm of ice-choked channels and rugged islands that lies just north of the Canadian mainland and is called the Arctic archipelago. The 1845 expedition of Sir John Franklin proved to be the key to the Northwest Passage—although Franklin and all his men perished miserably without finding the passage.

The search for Franklin went on for years, and by the end of it the Arctic archipelago had been mapped in broad outline. At last geographers knew that the Atlantic and Pacific oceans are indeed connected by several passages that twist and turn amid the ice and rock of the Arctic. But they also knew that the dream of a safe, navigable Northwest Passage was dead. In the end, the shortcut through the Americas that mariners had sought for centuries had to be built, not found. It was the Panama Canal, cut through the narrowest part of Central America and opened to ship traffic in 1914. A decade before the canal opened for business, however, someone finally did sail through the Northwest Passage. In 1903–6, the Norwegian polar explorer Roald Amundsen accomplished the dream of Frobisher and Hudson by sailing from Baffin Bay on the Atlantic side of North America to the Bering Strait on the Pacific side.

SEE ALSO

Amundsen, Roald; Arctic, mapping of; Davis, John; Franklin, John; Frobisher, Martin; Hudson, Henry; Parry, Sir William Edward; Ross, Sir James Clark

FURTHER READING

Berton, Pierre. *The Arctic Grail: The Quest for the North West Passage and the North Pole, 1818–1909.* New York: Penguin, 1988.

Brown, Warren. *The Search for the Northwest Passage.* New York: Chelsea House, 1991.
Day, Alan E. *Search for the Northwest Passage.* New York: Garland, 1986.
Lehane, Brendan, and Time-Life editors. *The Northwest Passage.* Alexandria, Va.: Time-Life Books, 1981.
Morison, Samuel Eliot. *The Great Explorers: The European Discovery of America.* New York: Oxford University Press, 1978.
Scott, J. M. *Icebound: Journey to the Northwest Sea.* London: Gordon & Cremonesi, 1977.

Ogilby, John

BRITISH GEOGRAPHER AND MAPMAKER

- *Born: 1600, Edinburgh, Scotland*
- *Died: 1676, London, England*

JOHN OGILBY started his career as a dancer, but an injury forced him to retire from the stage. After a stint as a dancing teacher and theater manager, in the 1650s he became a publisher, specializing in the Greek classics. He prospered until the London fire of 1666 consumed his business. Approaching 70, Ogilby became a mapmaker and map publisher. Within a few years he had published books about Africa, China, Japan, and Asia; these were translations of Dutch and German books and included maps copied from the original texts. Ogilby also produced his own maps, including a large-scale map of London on 20 sheets.

In 1675 Ogilby produced the first road maps of England and Wales. These were in the form of strips, with five or six strips printed on a page. The route unfolded from strip to strip, and a compass rose on each strip oriented the traveler in the direction of that stretch of road. The same scale was used throughout, so that each strip covered the same distance. This format proved popular and is still used today by the American Automobile Association and other makers of route maps.

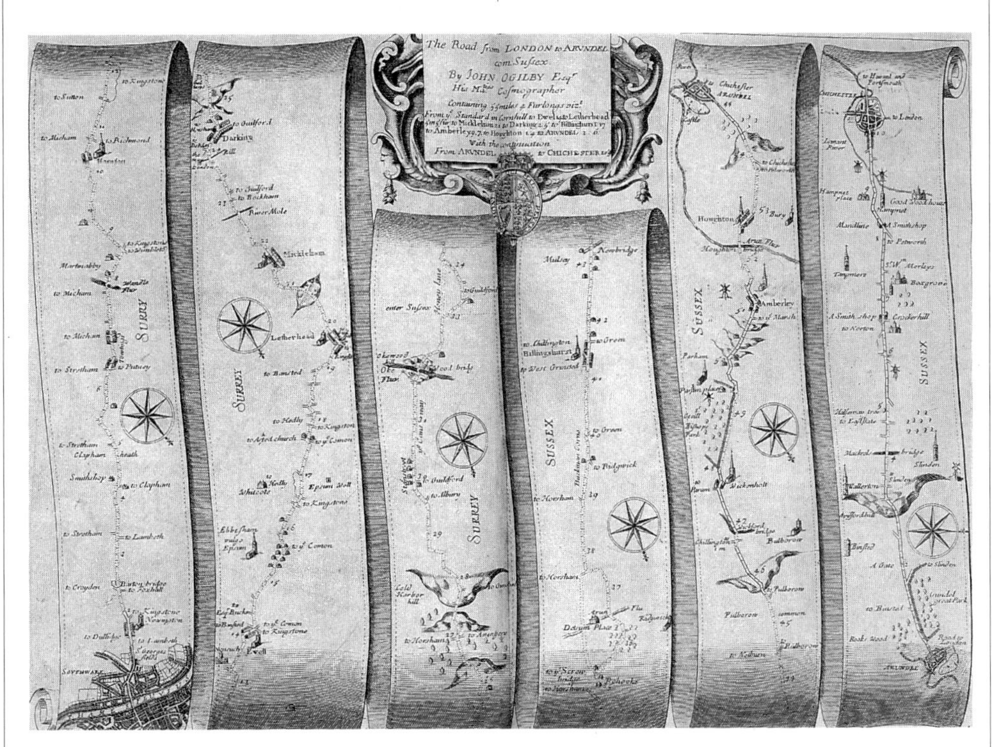

Ogilby's map of the route from London (lower left) to Chicester, by way of Arundel Castle. Each strip is read from bottom to top; villages, crossroads, hills, and estates are shown along the way.

Ordnance Survey

THE ORDNANCE SURVEY is the surveying and mapmaking arm of the British government. It dates from 1787, when the duke of Richmond, master of the king's ordnance (*ordnance* refers to military supplies), provided scaffolding, fireworks, and other equipment to engineers who were making survey measurements across the English Channel. The success of this venture led to a triangulation survey of England under the duke's supervision. In 1791, the survey became the responsibility of the military. Teams of specially trained surveyors and mapmakers began producing detailed, accurate maps of the country; the first of these were engraved and issued in 1801. In 1825, the survey was moved to Ireland, where it remained until 1840. It issued a complete set of 205 maps of Ireland in 1858, and the usefulness of these maps convinced the government to continue the English survey.

The Ordnance Survey remains one of the world's leading map publishers. It produces a wide variety of maps for both government and private use. Among its publications are half a dozen series of maps designed for drivers and hikers. They cover Great Britain and all its counties and cities in varying levels of detail.

SEE ALSO
Triangulation

Orellana, Francisco de

SPANISH CONQUISTADOR AND EXPLORER OF THE AMAZON

- *Born: about 1511, Trujillo, Spain*
- *Died: November 1546, South America*

FRANCISCO DE ORELLANA went to Peru in 1535 to take part in the Spanish conquest of the Incas. In 1541 he joined an expedition led by Gonzalo Pizarro (brother of Francisco Pizarro, who led the war against the Incas) to explore the jungle east of the Andes Mountains, where spices and gold were said to abound. The conquistadors found neither spices nor gold; they encountered only hardship and hunger. Pizarro assigned Orellana to sail down a river in search of food. Orellana and a band of followers did so, following the river to where it entered another, much larger, river. Unwilling or unable to return upstream for Pizarro and the others, Orellana and his men then set sail on the large river and followed it east across South America all the way to the Atlantic Ocean. They were the first Europeans to explore the length of the Amazon River.

Orellana died in 1546 during a second expedition to the river. By that time the river was already appearing on European maps. In 1544, only two years after Orellana reached the Atlantic Ocean, the English navigator and mapmaker Sebastian Cabot made a map that showed a crude version of the river's course; on the map, Cabot credited the discovery of the river to Orellana. For a time the South American waterway was called the Orellana River, but gradually that name fell into disuse and it became known as the Amazon River, the name it bears today. The "Amazons" were women warriors whom Orellana and his comrades reported seeing along the river.

SEE ALSO
Americas, mapping of; Cabot, Sebastian

FURTHER READING

Smith, Anthony. *Explorers of the Amazon.* New York: Viking, 1990.

Orienteering

ORIENTEERING IS the process of finding a route with a map and a compass. A hiker who follows a trail through a wilderness area using a topographic map and a compass is orienteering. The term also has a more specific meaning, referring to a recreational sport in which participants must find their way by using maps and compasses. Usually, these events are competitive races, but sometimes they are more leisurely events in which winning the race is secondary to having fun with a map and compass. In point-to-point orienteering, competitors find their own routes between points announced by the race organizers. In preset-course orienteering, the competitors follow a route marked on a topographic map by the organizers. National orienteering clubs in the United States, Canada, and many other countries organize events such as races and training classes.

SEE ALSO

Compass

FURTHER READING

Kjellstrom, Bjorn. *Be Expert with Map and Compass: The Orienteering Handbook.* New York: Scribners, 1976.

Ortelius, Abraham

BELGIAN CARTOGRAPHER

- *Born: April 4, 1527, Antwerp, Belgium*
- *Died: June 28, 1598, Antwerp, Belgium*

ABRAHAM ORTELIUS is remembered today as the publisher of the first modern atlas. He was also a leading dealer, collector, and scholar of maps. Ortelius began his career as a map colorist, then began dealing in maps. He traveled widely, buying and selling, and lived in London for a time during the 1560s and 1570s. He was a friend of Gerardus Mercator and of many other cartographers and scholars of his time.

Ortelius's first important map was a world map on eight sheets, published in 1564; the only surviving copy is housed at the Basel University Museum. In 1570 Ortelius published the *Theatrum Orbis Terrarum* (*Theater of the World*), a world atlas of 53 map sheets. He may have gotten the idea for his atlas from a local merchant who had hired him to collect a number of single-sheet maps and bind them together into a single volume for ease of use.

The *Theatrum* was unlike any collection of maps that had been made or published before. For one thing, it was the first set of maps in a uniform style and size (Ortelius borrowed maps from many other cartographers but reengraved them all to fit his own format). For another, it was based more on contemporary knowledge than on theory and myth. It contained no maps from the classical era and therefore represented a fairly up-to-date picture of how the world looked to educated people in the 16th century. However, some Ptolemaic misconceptions, such as the belief in a large undiscovered southern continent called the Terra Australis, remained on Ortelius's maps.

The *Theatrum* was extremely popular and did much to make the Low Countries—as Belgium and the Netherlands were called—the new leaders among mapmaking nations. Ortelius's friend and colleague Mercator praised the "care and elegance" with which Ortelius had redrawn the maps, adding that Ortelius had done much "to bring out the geographical truth, which is so corrupted by mapmakers." Ortelius and Mercator enjoyed a cordial relationship,

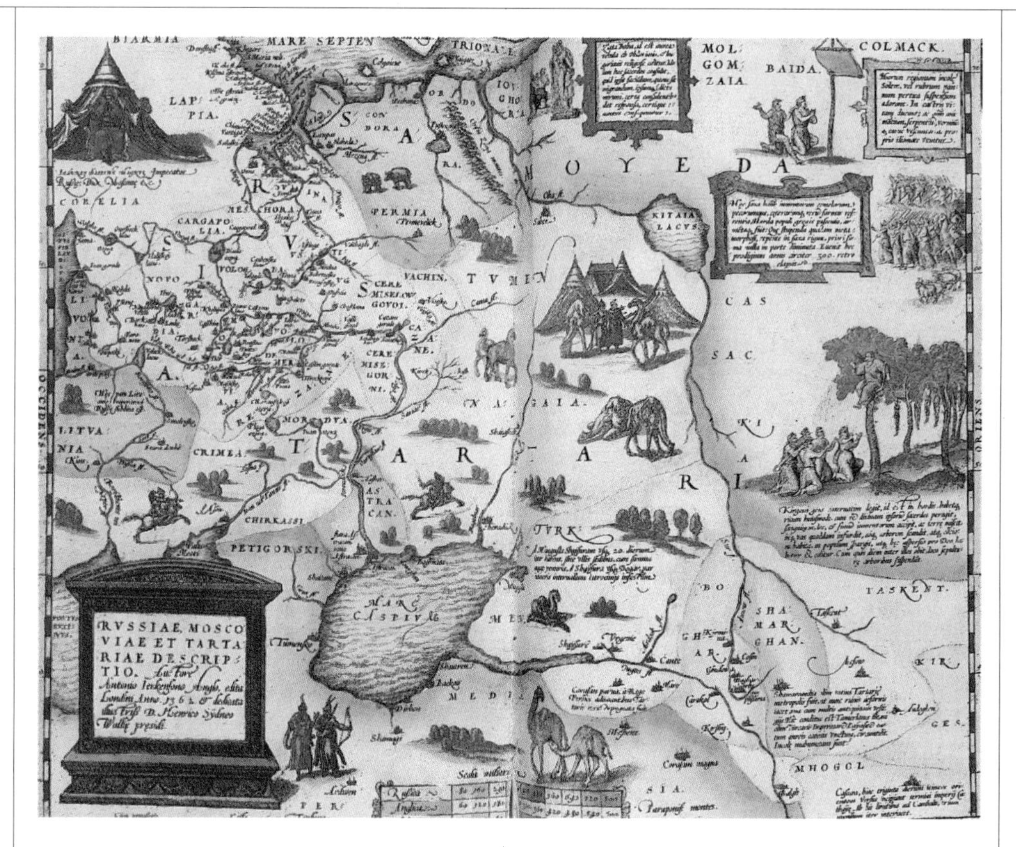

Russia and central Asia from the first atlas, Ortelius's Theater of the World. *"Litvania," on the left, corresponds to modern Lithuania; scenes from travelers' tales adorn the regions farther east.*

untainted by professional jealousy. It has even been suggested that Mercator delayed the publication of his own atlas in order to give his friend's work a chance to succeed.

Ortelius's *Theatrum* was reprinted in many editions during his lifetime and afterward, with new maps added to it all the time. Mapmakers and geographers, eager to be represented in the atlas, sent Ortelius copies of their work to use. There were 119 maps in the atlas at the time of his death and 141 in the final edition, which was produced by another publisher in 1606.

SEE ALSO

Atlas; Dutch mapmakers; Mercator family

Orthographic projection

SEE Projections

Orthomorphic projection

SEE Projections

Orthophoto maps

SEE Aerial mapping

Pacific Islands mapmakers

THE INHABITANTS of the Pacific Ocean islands did not have written languages until Europeans wrote their languages down, beginning in the 18th century. Yet the Pacific Islanders possessed sophisticated geographic knowledge that was transmitted from generation to generation in poems, songs, and artifacts.

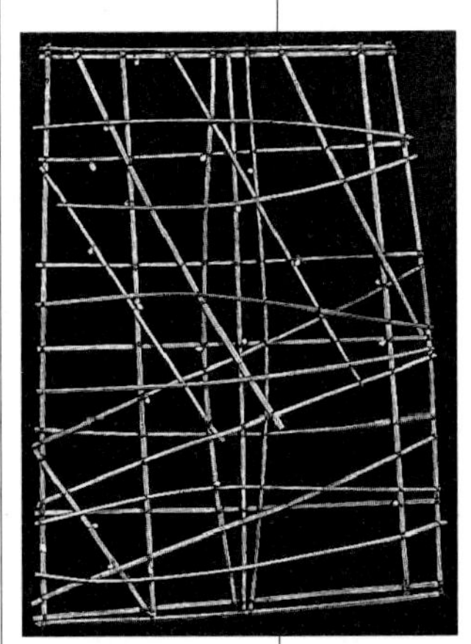

A 19th-century palm-fiber map from the Marshall Islands with wave fronts shown by curved fronds and islands by shells. With these maps and their great skill in navigation, Pacific islanders routinely made long open-sea voyages.

One important feature of island culture was navigation. Over thousands of years, the people of the Pacific had spread from one tiny cluster of islands to another over the largest expanse of water on the face of the earth. We now know that they regularly made epic voyages of trade, war, and colonization, using traditional navigational techniques that relied upon a thorough knowledge of stars, winds, and currents. Many Pacific peoples recorded this geographic knowledge on maps. Almost no examples of indigenous Pacific mapmaking survive today, but European explorers' accounts tell us that the islanders often drew helpful maps for their European visitors. These maps may have been perishable, scratched in the sand or on a large leaf, but they prove that the islanders knew how to make and use maps.

The people of the Marshall Islands, an island group south and west of Hawaii, developed a unique map technology. Their maps consisted of a grid or framework of straight, stiff palm fibers. Other fibers or sticks were lashed onto the grid in curves or angles; these showed how strongly the waves flowed and in what direction. Shells were tied onto the framework to mark the locations of islands. The Marshallese navigated from island to island across wide distances by comparing the waves on the sea's surface to their charts. The secrets of making these maps were handed down from generation to generation and jealously guarded from outsiders, but a number of palm-fiber charts, most made in the 19th century, are preserved in museums.

FURTHER READING

Bellwood, Peter. *Man's Conquest of the Pacific: The Prehistory of Southeast Asia and Oceania.* New York: Oxford University Press, 1978.

Dodd, Edward. *Polynesian Seafaring: A Disquisition on Prehistoric Celestial Navigation and the Nature of Seagoing Double Canoes.* New York: Dodd, Mead, 1972.

Feinberg, Richard. *Polynesian Seafaring and Navigation: Ocean Travel in Anutan Culture and Society.* Kent, Ohio: Kent State University Press, 1988.

Goetzfridt, Nicholas J., comp. *Indigenous Navigation and Voyaging in the Pacific: A Reference Guide.* Westport, Conn.: Greenwood, 1992.

Irwin, Geoffrey. *The Prehistoric Exploration and Colonization of the Pacific.* Cambridge, England: Cambridge University Press, 1992.

Lewis, David. *We, the Navigators: The Ancient Art of Landfinding in the Pacific.* Honolulu: University of Hawaii Press, 1972.

Pacific Ocean, mapping of

The mapping of the Pacific Ocean began with the indigenous peoples of the Pacific Islands, who navigated the world's largest ocean and had colonized all of the major island groups by about A.D. 400. Chinese and Japanese mapmakers, too, may have charted Pacific waters close to their shores, although no such charts have survived. The European exploration of the Pacific began when Magellan

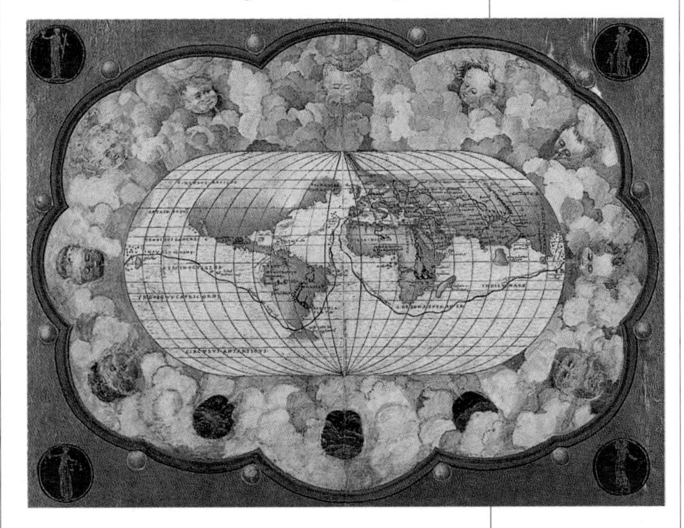

The Victoria, *one of Ferdinand Magellan's fleet, was the first ship to return to Europe after crossing the Pacific. Its track is shown on Battista Agnese's 1544 world map.*

sailed through a strait at the southern tip of South America in 1520 and emerged into the Pacific.

Pacific exploration in the 16th century consisted of two types of voyages. A few mariners, including Francis Drake of Great Britain, circumnavigated the globe as Magellan had done; the routes of Magellan and Drake were shown on many world maps well into the 18th century. Other Pacific journeys were made by Spanish navigators looking for routes to the Indies. These voyages departed not from Spain but from Spain's colonies in Mexico and South America. Mariners such as Álvaro de Mendaña de Neira and Pedro Fernándes de Quiros crossed the ocean in great loops shaped by the ocean's wind patterns. They sailed west at about the latitude of the Tropic of Capricorn and returned east to the Americas near the Tropic of Cancer. These voyagers sighted or landed upon many islands, but their inability to determine the longitude of their landfalls made it impossible to place them on the map with any certainty. Discoveries such as the Solomon Islands pop up in various places on the maps and charts of the 16th and later centuries.

European knowledge at the end of the 16th century is summed up in a 1601 Spanish map by Antonio de Herrera. It shows the southeastern coast of China, Japan, Borneo, and the Philippines, all recognizable though distorted, and also a large stretch of coastline called New Guinea; this coastline trails off indeterminately into blankness.

In the 17th and 18th centuries, explorers of the Pacific searched for the Terra Australis, the hypothetical southern continent that was believed to stretch around the earth south of Africa, Asia, and South America. The Terra Australis appeared on every world map by Gerardus Mercator, Abraham Ortelius, and other mapmakers, even though no one had ever set foot on it, or even seen it. Australia and New Guinea were at first thought to be the northern edges of this great landmass, until explorations by Abel Tasman in the mid-17th century and James Cook in the late 18th century proved them to be islands. (Australia, although it is a continent, is much smaller than the Terra Australis that European explorers had been expecting to find.)

Other mythical lands lurked in the waters of the North Pacific. A 1711 map of China by the British cartographer John Senex shows the "Land of Yedso" and the "Land of Compagnia" north of Japan. Not until the late 18th century did the voyages of Louis-Antoine de Bougainville and especially of James Cook clarify the geography of the Pacific, banishing the Terra Australis from the map but adding much new information. To Cook goes the credit for giving the world its first comprehensive knowledge of Pacific geography, much of which he mapped himself. He charted New Zealand, the east coast of Australia, and the coast of northwestern North America; he also discovered Hawaii

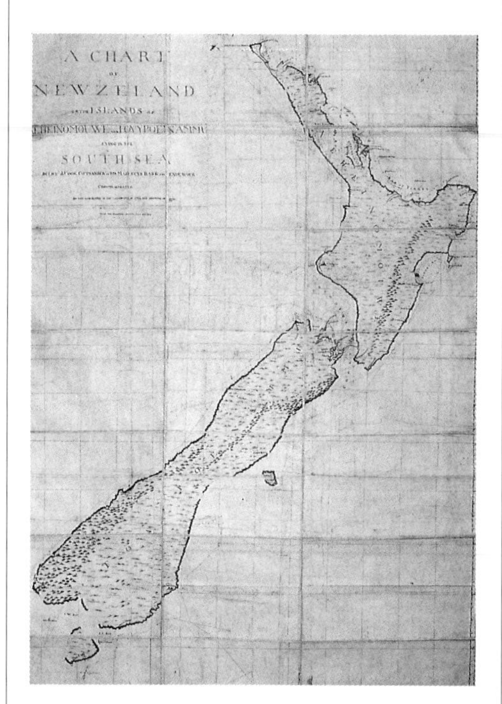

In 1769–70, James Cook spent six months surveying 2,400 miles (3,840 kilometers) of coastline to make this chart of New Zealand.

(which he called the Sandwich Islands) and established precise locations for many of the other islands scattered across the ocean.

Further details of the Pacific map were filled in during the years 1790–93 by an ambitious Spanish expedition under Alejandro Malaspina, who explored the American coast from Chile to Alaska and then charted islands in the Philippines and Melanesia. The discoveries made by Cook, Malaspina, and other navigators were quickly added to the world map. The series of maps made by Aaron Arrowsmith in 1798 is especially useful to historians because it illustrates the progress of Pacific exploration.

By the early 19th century, the map of the Pacific Ocean was fairly complete. Expeditions during that century made territorial claims, carried out scientific studies, and produced detailed charts of islands and coastlines, but they made few significant geographic discoveries.

SEE ALSO

Arrowsmith family; Australia, mapping of; Cook, James; Great U.S. Exploring Expedition; Magellan, Ferdinand; Mendaña de Neira, Álvaro de; Pacific Islands mapmakers; Quiros, Pedro Fernándes de; Terra Australis; Vancouver, George

FURTHER READING

Allen, Oliver, and Time-Life editors. *The Pacific Navigators.* Alexandria, Va.: Time-Life Books, 1980.

Beaglehole, J. C. *The Exploration of the Pacific.* Stanford, Calif.: Stanford University Press, 1966.

Cameron, Ian. *Lost Paradise: The Exploration of the Pacific.* Topsfield, Mass.: Salem House, 1987.

Dunmore, John. *Who's Who in Pacific Navigation.* Honolulu: University of Hawaii Press, 1991.

Gray, William R. *Voyages to Paradise: Exploring in the Wake of Captain Cook.* Washington, D.C.: National Geographic Society, 1981.

Moorehead, Alan. *The Fatal Impact: An Account of the Invasion of the South Pacific, 1767–1840.* New York: Harper & Row, 1966.

Parallel

SEE Latitude

Paris, Matthew

BRITISH HISTORIAN AND MAPMAKER

• *Active: 13th century*

LITTLE IS known of the life of Matthew Paris, who became a Benedictine monk at St. Albans monastery in England in 1217. He wrote several important books about English and European history, and scholars regard him as a leading source of information about events in Europe in the early 13th century.

Paris is also regarded as Great Britain's first serious cartographer. He made a number of maps to illustrate his books. Paris drew the earliest known detailed maps of Britain around 1250. One of these is preserved in the Bodleian Library in Oxford, England. It shows

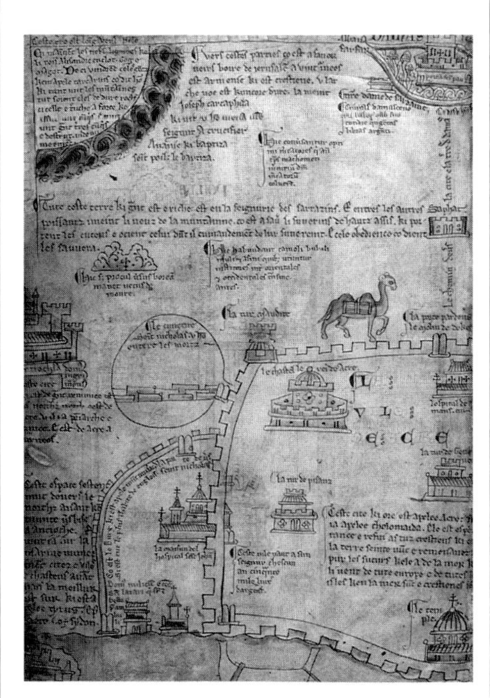

Part of Matthew Paris's guide for pilgrims, showing the city of Acre, a stronghold of 12th-century European crusaders in the Holy Land.

rivers in blue, towns outlined in red, and churches and castles with towers and turrets. Although the map's geography is distorted by modern standards, Britain is recognizable.

Paris also produced an itinerary or road map for pilgrims: a series of strip maps indicating the roads, sea crossings, inns, churches, and mountain passes between London and Jerusalem.

Park, Mungo

SCOTTISH EXPLORER OF AFRICA

- *Born: Sept. 10, 1770, Foulshiels, Scotland*
- *Died: Apr. 1806 on the Niger River, Africa*

MUNGO PARK trained as a doctor in Scotland before going to work as a ship's surgeon in the service of the British East India Company. He traveled to Sumatra in Southeast Asia, where his work in natural history brought him to the attention of Sir Joseph Banks, a leading English botanist. Banks was a member of the African Association, which in 1795 gave Park the job of exploring West Africa. In particular, Park was to trace the course of the Niger River. Europeans knew that this large river flowed through West Africa south of the Sahara Desert, but almost nothing was known about it—not even the direction of its flow.

After many hardships and dangers, Park succeeded in reaching the banks of the Niger in 1796, in the country of Mali. He determined that the river flowed eastward. Upon his return to Great Britain, he published an account of his expedition under the title *Travels in the Interior Districts of Africa* (1799), which aroused great interest in the Niger and the people who lived along it. Park's discovery immediately appeared on the newest maps of Africa. The English mapmaker Robert Wilkinson's "New Map of Africa," published in 1800, showed Park's journey, as did a large map of Africa in four sheets that was published in 1802 by Aaron Arrowsmith, the leading British cartographer of Africa. Arrowsmith summed up the general geographic confusion about Africa's rivers when he wrote, "The general opinion on the interior of Africa is that the Niger and the Nile are one and the same river."

It was partly to clear up this confusion that the British government sent Park back to West Africa in 1805 at the head of a large expedition. One by one, however, the members of the expedition fell victim to disease. Park was finally left with four Europeans and a few African followers. The explorers tried to sail down the Niger but were ambushed by local inhabitants; they either drowned or were killed.

SEE ALSO

Africa, mapping of; African Association; Arrowsmith family

FURTHER READING

Blunt, Peter L. *Black Nile: Mungo Park and the Search for the Niger.* London: Gordon & Cremonesi, 1977.

Lupton, Kenneth. *Mungo Park, the African Traveler.* New York: Oxford University Press, 1979.

Stefoff, Rebecca. *Scientific Explorers: Travels in Search of Knowledge.* New York: Oxford University Press, 1992.

Parry, Sir William Edward

BRITISH EXPLORER OF THE ARCTIC

- *Born: Dec. 19, 1790, Bath, England*
- *Died: July 8, 1855, England*

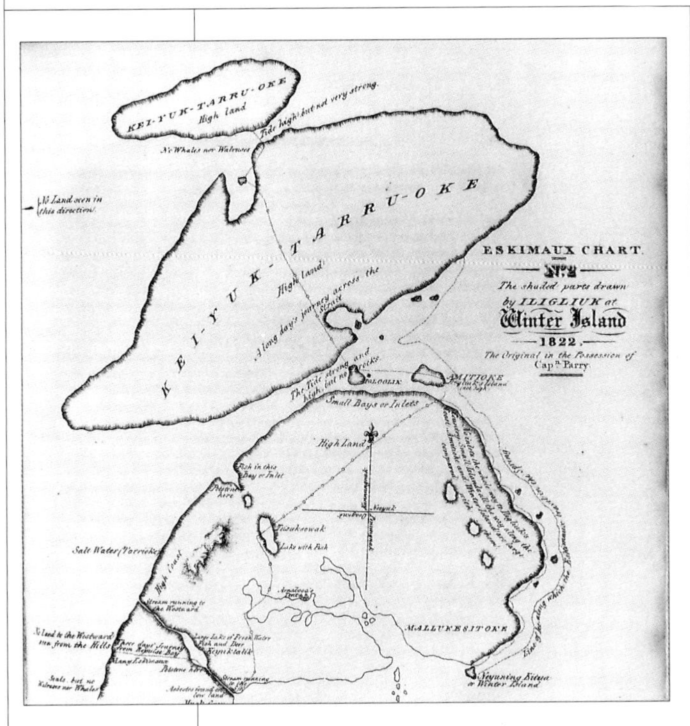

A map of the Melville Peninsula in the Canadian Arctic, drawn for Parry by a local Inuit woman in 1822.

A MEMBER of the Royal Navy from the age of 13, Parry ultimately achieved the rank of admiral. The navy sent him into the Canadian Arctic in 1819 to search for the Northwest Passage. He failed to find this much-desired seaway to the Pacific, but he did manage to force his way into the ice-choked maze of channels that thread Arctic Canada. His observations placed a number of Arctic islands on the map.

In two later expeditions (1821–23 and 1824–25), Parry explored a passage between Baffin Island and the Canadian mainland; he called it the Fury and Hecla Strait after his two ships, the *Fury* and the *Hecla*. Parry's voyages not only shed light on the mysterious geography of the Arctic but also brought to light some new whaling grounds that were later exploited by ships of many nations.

SEE ALSO

Arctic, mapping of; Northwest Passage; Poles

FURTHER READING

Berton, Pierre. *The Arctic Grail: The Quest for the North West Passage and the North Pole, 1818–1909*. New York: Penguin, 1988.

Peary, Robert Edwin

AMERICAN EXPLORER OF THE ARCTIC

- *Born: May 6, 1856, Cresson, Pennsylvania*
- *Died: Feb. 20, 1920, Washington, D.C.*

EDUCATED AT Bowdoin College in Maine, Robert Peary became an engineer for the U.S. Navy, but he yearned to make a name for himself in some grand enterprise. "I must have fame," he once wrote to his mother. Beginning in 1886, he devoted much of his life to Arctic exploration, first in northern Greenland and later in the race for the north pole. He obtained a leave of absence from the navy and explored as a private citizen, raising money from wealthy backers and from sponsors such as the National Geographic Society to pay for his expeditions.

Peary's first major contribution to cartography involved the northern coast of Greenland. Geographers did not yet know how to draw Greenland on the map. Did the huge island extend all the way to the pole? If not, where did its northern coastline lie? In two trips across the Greenland ice cap, Peary took scores of position sightings that helped to fix the northern coast of Greenland on the map.

Another of Peary's contributions to science involved meteorites. For years, explorers had wondered where the Inuit, the Native American people of Greenland, obtained the iron for their knives. Peary, who spent long stretches of time living with the Inuit and gained their confidence, learned that the iron came from three large meteorites, iron-rich stones that had fallen from space years before. He succeeded in locating the stones and shipping them to the

American Museum of Natural History in New York City.

Beginning in 1898, Peary made several expeditions onto the Arctic ice pack from Ellesmere Island, the northernmost island in Arctic Canada. His goal was the pole. Finally, after years of setbacks and disappointments, he reached the pole on April 6, 1909—or so he said. Some researchers have questioned Peary's claim, but most experts believe that Peary either reached the pole or was very close to it. He is generally credited with being the first person to stand at the north pole, thus capturing one of the greatest remaining geographic prizes of the 20th century.

SEE ALSO

Arctic, mapping of; Cook, Frederick; Poles

FURTHER READING

Anderson, Madelyn. *Robert E. Peary and the Fight for the North Pole.* New York: Franklin Watts, 1992.

Berton, Pierre. *The Arctic Grail: The Quest for the North West Passage and the North Pole, 1818–1909.* New York: Penguin, 1988.

Herbert, Wally. *Noose of Laurels: Robert E. Peary and the Race to the North Pole.* New York: Doubleday, 1990.

Maxtone-Graham, John. *Safe Return Doubtful: The Heroic Age of Polar Exploration.* New York: Scribners, 1988.

Rawlins, Dennis. *Peary at the Pole: Fact or Fiction?* Washington, D.C., and New York: Robert Luce, 1973.

Pei Hsiu

CHINESE MAPMAKER

- *Born: 224*
- *Died: 271*

PEI HSIU is considered the founder of Chinese cartography. Although maps had been made in China before Pei Hsiu's time, he wrote a handbook for mapmakers that was published by the imperial government and brought a new level of professionalism and standardization to the craft of cartography. Pei Hsiu also made maps of his own, including a map of China on 18 sheets. None of his works survive; they are known only through references by other writers.

FURTHER READING

Temple, Robert. *The Genius of China: 3000 Years of Science, Discovery, and Invention.* New York: Simon & Schuster, 1986, pp. 30–33.

Wanru, Cao. "Maps 2,000 Years Ago and Ancient Cartographical Rules." In *Ancient China's Technology and Science,* by the Chinese Academy of Sciences. Beijing: Foreign Languages Press, 1983.

Periplus

SEE Portolan

Peters projection

SEE Projections

Peutinger table

THE PEUTINGER TABLE is one of the very few surviving clues to what Roman maps were like. Made in the 12th or 13th century, it is a copy of a Roman map that probably originated in the 1st century A.D., to which new details were

A section of the Peutinger Table showing Rome (far right).

apparently added over the centuries. The Peutinger table takes its name from a German collector who acquired it in the 16th century.

If the Peutinger table is a fair example, Roman maps were above all functional. It is really a road map: a scroll 21 feet long and 1 foot wide (6.4 meters by .3 meters), showing routes between cities and distances along roads. The shapes of land and water masses are distorted almost beyond recognition, but the network of roads that connected places in Roman times is easy to trace.

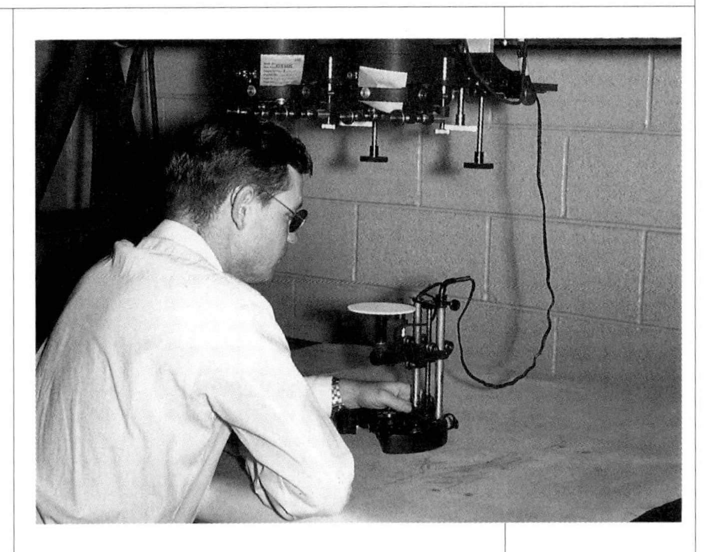

During the 1950s, photogrammetry became the primary technique for making new maps. Using projectors that allowed them to view multiple images at once, technicians converted aerial photographs into accurate maps.

Photogrammetry

PHOTOGRAMMETRY IS the science and craft of obtaining accurate measurements from photographs. Maps made through the use of photogrammetry are not themselves photographs, although aerial photographs can be made into maps (called orthophoto maps). Rather, photogrammetric maps are simply maps based on information from photographs. A special kind of photograph is needed, however, to yield the necessary survey information.

Photogrammetry was invented in 1859 by Aimé Laussedat, a French army officer and photographer who developed the photogrammetric camera, a combination of the camera and the theodolite, an instrument that measures angles and is used in surveying. Using this new device to photograph the same object from two different angles a set distance apart, he could plot the position of that object. The whole process of triangulating distances and measuring location became simpler. In the early 20th century, photogrammetry was widely used by European surveyors; it proved particularly useful in mountainous areas such as the Alps because photographers could perch their cameras on mountainsides and take pictures of very large areas.

In order to cover still larger areas, surveyors wanted to take photogrammetric pictures from the air. Aerial photogrammetry progressed rapidly after World War I (1914–18). Viewing machines were invented by the Brock and Weymouth Company of Philadelphia, Pennsylvania, in the 1920s. Film taken the air was viewed through these machines, which allowed the operator to see two images, taken from slightly different points, at the same time. This revealed even tiny variations in the terrain, which were then double-checked against points on the ground whose location had been verified by ground surveys. By this means, accurate contour maps could be made from vertical aerial photographs.

By the 1950s, virtually all large-scale mapping projects relied on aerial photogrammetry, which continued to improve with the development of new and better cameras and with the introduction of electronic distance-measuring devices for making highly accurate baselines. In the 1960s and 1970s, the National Aeronautics and Space Administration (NASA) made photogrammetric maps of the moon using photos taken by the Lunar Orbiters, a series of camera-carrying

A photomap of the moon reveals a surface covered in craters.

spacecraft that took about 2,000 pictures of the moon's surface. Maps made from these photos helped scientists choose landing places for the manned Apollo missions.

SEE ALSO

Aerial mapping; Surveying; Triangulation

Photomosaic

SEE Aerial mapping; Satellites

Pigafetta, Antonio

ITALIAN TRAVELER

- *Born: before 1491, Vicenza, Italy*
- *Died: after 1525, probably in Italy*

ANTONIO PIGAFETTA chronicled one of the most adventurous journeys of all time. For reasons that are not known, Pigafetta volunteered to accompany Ferdinand Magellan's 1519 expedition. Perhaps he was simply curious, or perhaps he wanted to spy out new Spanish discoveries and trade routes on behalf of his fellow Italians. Pigafetta was one of the few members of the expedition who survived to return to Spain on the *Victoria,* the first ship ever to sail around the world. A keen observer and a gifted writer, Pigafetta published an account of the journey that remains the best source of direct information about Magellan and the first circumnavigation of the globe. Pigafetta's maps of the islands and ports he had seen were copied by many mapmakers in the 16th century and later.

SEE ALSO

Magellan, Ferdinand

FURTHER READING

Cameron, Ian. *Magellan and the First Circumnavigation of the World.* New York: Saturday Review Press, 1973.

Pigafetta, Antonio. *Magellan's Voyage: A Narrative Account of the First Circumnavigation.* 2 vols. Translated and edited by R. S. Skelton. New Haven, Conn.: Yale University Press, 1969.

Piri Reis map

PIRI REIS was a Turkish navigator and chart maker of the 16th century. As a boy serving on Turkish ships, he kept records of harbors and sailing directions. He became a skilled cartographer. He included 125 maps in his 1521 book *Kitab-i Bahriye* (*The Book of the Navy*). But the map for which he is best remem-

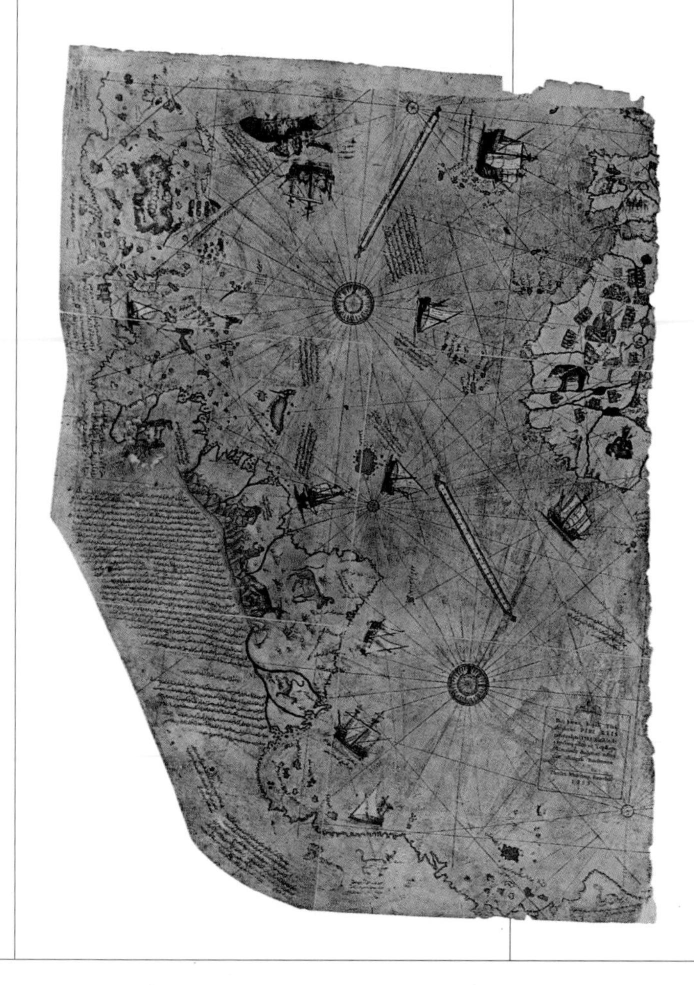

Piri Reis's map of the Caribbean Sea and the coast of Brazil (left) may be based on a chart drawn by Christopher Columbus.

bered is a world map he completed in 1513. Piri Reis spent two years preparing this map and drew upon numerous sources: according to his own note on the map, he used nine Arab maps, "four recent Portuguese maps," and "a map of the western regions drawn by Colombo." Most scholars believe that this last reference is to a chart of the American voyages drawn by Christopher Columbus himself. Piri Reis's map is thus the only surviving map based directly on Columbus's own map.

It seems that the Columbus map belonged to a Spanish sailor who was captured in 1501 by Kemal Reis, Piri Reis's uncle, who is variously described as an admiral in the Turkish fleet and a pirate. The sailor also possessed a feathered headdress. He told Kemal Reis that the headdress came from a land in the western sea that had been discovered by a man named Colombo. He also said that Colombo had drawn a chart of this land, which was in the sailor's possession. Kemal Reis got hold of the chart and passed it on to his nephew, who incorporated it into his own world map.

The map shows 42 place names in the New World; nearly all of them can be identified with places explored and named by Columbus. In a long note in the margin, Piri Reis gives a narrative account of Columbus's first three voyages. The map also shows the coast of Brazil, and a note explains how Portuguese explorer Pedro Álvars Cabral found Brazil while sailing southwest to catch a wind that would carry him around Africa. Piri Reis decorated his map with pictures of several varieties of ships and also of mythical creatures, such as a unicorn roaming the Brazilian forest. In all, the Piri Reis map is not only a masterpiece of cartography from a culture often overlooked by Western historians but it is also a unique record of how Columbus described his own voyages. "No one now living," wrote Piri Reis in the margin of his map, "has seen a map like this."

Piri Reis probably presented his map to his patron, the sultan of the Ottoman (Turkish) Empire. It was stored away in the vast Ottoman archives and was lost from sight, only to be rediscovered in 1929 in the Topkapi Palace Museum in Istanbul, where it is now preserved.

SEE ALSO

Americas, mapping of; Cabral, Pedro Álvars; Columbus, Christopher; Islamic geographers and mapmakers

Pisan chart

THE PISAN CHART is sometimes called by its Italian name, the *carte Pisano*. It is the oldest surviving example of a portolan, a type of navigational chart that originated in the Mediterranean Sea. The chart dates from around 1290 and was probably made in the Italian port of Genoa, although it was later found in Pisa, from which its name is taken. The Pisan chart was based on earlier charts and probably reflected the experiences of many mariners and chart makers, rather than the knowledge of a single mariner. It is kept in the French National Library in Paris.

SEE ALSO

Portolan

Drawn on sheepskin, the Pisan chart shows part of the Mediterranean Sea.

Plancius, Peter

DUTCH MAPMAKER

- *Born: 1552, Holland*
- *Died: 1622*

PETER PLANCIUS was born Petrus Plant but, like many learned people of his time, adopted a Latinized form of his name. He was a minister of the Dutch Reformed Church, but his interest in geography and exploration led him to become an expert in the history of Dutch navigation. He left the clergy and became a cartographer for the Dutch East India Company, the firm that oversaw Dutch exploration and trade in Africa and Asia. His position there gave him access to the most up-to-date geographic knowledge.

Plancius made several world maps in the 1590s. His 1594 map of the world in two hemispheres is especially important because it is the first world map published with very elaborate decorative borders. Around the edges of the two hemispheres Plancius arranged a number of images: female figures in various states of undress representing the continents (Europe and Asia are fully clad, Africa and the Americas nearly naked); armies on the march; the constellations; and creatures such as elephants, rhinoceroses, crocodiles, giraffes, and aardvarks. Many of the illustrations echo Théodore de Bry's accounts of the New World. They ushered in an era of lush decoration on maps because they were

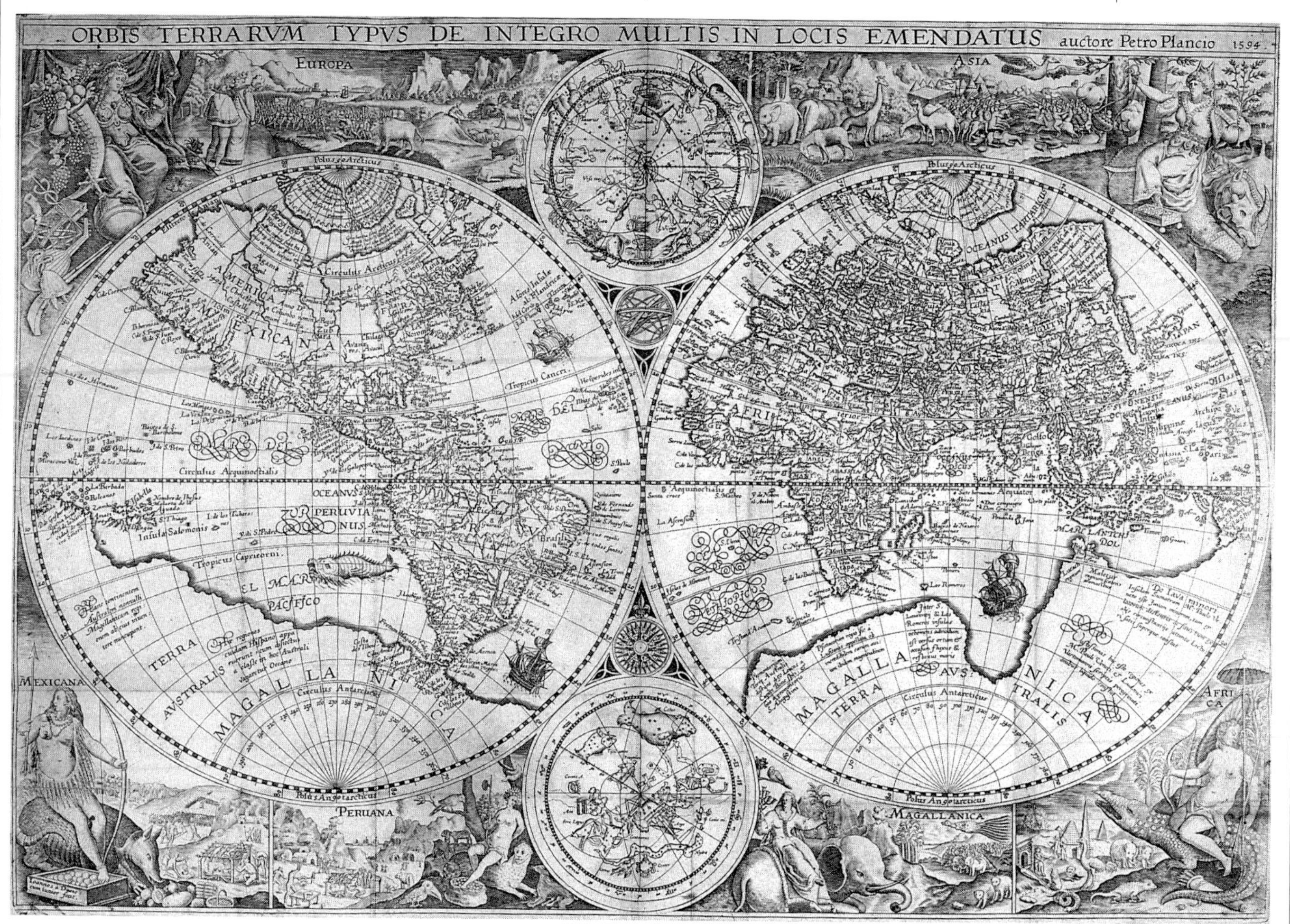

Plancius's decorative 1594 world map. The female figures representing various parts of the world were widely copied and reinforced Europeans' idea that Africa and the Americas were "primitive."

widely imitated by other cartographers, especially among the Dutch.

Other works by Plancius include the maps for Jan Huyghen van Linschoten's 1596 travel book, maps for the Bible (1609), and a set of gores—that is, a map in segments designed to be pasted onto a globe (1615). Plancius is less well known today than Gerardus Mercator or Abraham Ortelius because he did not make an atlas; atlas publishing was the surest road to fame for a 16th- or 17th-century mapmaker. But historians of cartography consider Plancius a major figure because the hundred or so maps that he produced were used by so many later mapmakers.

SEE ALSO

Bry, Théodore de; Decoration; Linschoten, Jan Huyghen van

Planimetric map

A PLANIMETRIC MAP is a map that does not show relief—that is, elevations and depressions in the terrain. The terrain and its features are presented as though they were a flat plane. The term *planimetric* comes from the Latin word *planum* (flat surface or plane) and the Greek work *metron* (measure).

Poles

THE EARTH has two kinds of poles: geographic and magnetic.

There are two geographic poles, a north pole and a south pole, located at 90 degrees of latitude north and south of the equator. These poles are the northernmost and southernmost places on earth. The earth's axis, an imaginary line around which the earth rotates (in the way that a top spins on its axis), runs through the poles, which remain in the same place while the earth spins.

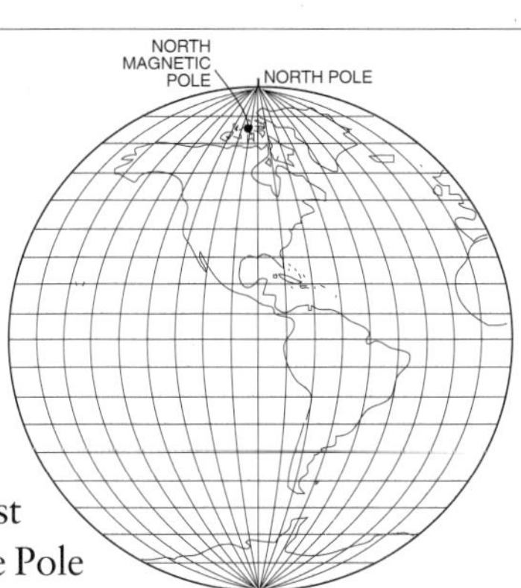

The north pole is almost directly beneath Polaris, the Pole Star (also called the North Star), which is part of the constellation Ursa Minor, more commonly known as the Little Dipper. From ancient times mariners and astronomers used the Pole Star as a guide to navigation. Sailors in the waters off Southeast Asia and India still use the Pole Star from time to time. It is low on the horizon but still visible on clear nights.

The north pole is covered by ice in the middle of the Arctic Ocean. The south pole is covered by ice near the center of Antarctica. Partly because of their inaccessible, inhospitable terrain, and partly because they are literally "the ends of the earth," the poles were the goal of many adventurers and were among the last places on earth to be explored. American explorer Robert Peary was the first to reach the north pole, in 1909; the Swedish explorer Roald Amundsen beat a rival British expedition to the south pole in 1911.

The earth's magnetic pole is the pole to which compass needles point. Everywhere on earth, the compass needle points to a location in the Queen Elizabeth islands off the north coast of Canada, about 1,000 miles (1,609 kilometers) from the geographic pole. The magnetic pole has not always been located in that same spot, however; it moves slowly over time according to variations in the earth's magnetic field. Within recorded history, the magnetic pole has occupied several points near the north geographic pole. All of these points have been in the high northern latitudes, and compass needles have always pointed north.

But the earth's magnetic pole has not always been located in the north. By studying iron particles in lava flows and rocks from different periods in the history of the planet, scientists have found that the earth's magnetic field occasionally reverses itself completely. The reasons for these reversals are not yet understood, but in the past 76 million years the magnetic field has reversed itself more than 170 times. Although in theory a reversal could occur at any time, switching the planet's magnetic pole to the south and causing all the compass needles in the world to spin 180 degrees, the odds are greatly against such an event's happening during any individual lifetime.

SEE ALSO

Amundsen, Roald; Arctic, mapping of; Antarctic, mapping of; Compass; Cook, Frederick; Peary, Robert Edwin; Scott, Robert Falcon

Political map

POLITICAL MAPS show the outlines of the world's political boundaries—the borders of countries, states, and other political creations, usually with their capitals or centers of government shown. They are produced by numerous map publishers. Some maps are purely political, and some combine political outlines with topographic data.

Polo, Marco

ITALIAN TRAVELER TO CHINA

- *Born: about 1254, Venice, Italy*
- *Died: Jan. 9, 1324, Venice, Italy*

MARCO POLO was born into a family of merchant traders in the Italian port of Venice, which in the 13th century was a bustling center of commerce with Asia. In 1271 he set out on a journey across Asia to the court of Khublai Khan, the emperor of the Mongols, who had conquered China. Polo made this remarkable trip in the company of his father and uncle, Niccolò and Maffeo Polo, who had already visited the khan once and had promised to return.

The Polos' route took them along centuries-old caravan trails through Turkey, Armenia, Persia (today called Iran), and Afghanistan. At Kashgar, an ancient trading city in the heart of Central Asia, they joined the route called the Silk Road. From there, after a hazardous journey across the fringes of several large deserts, they reached China. They arrived at the khan's court in 1275.

The Polos stayed in China, working in the khan's service, for 17 years. During this time, Marco Polo traveled widely through China; he also claimed to have visited Tibet, Burma (now Myanmar), India, and other Asian lands on the khan's business. In 1292, the khan allowed the three Venetians to escort a Mongol princess to her wedding in Persia. From Persia they made their way back to Italy, arriving in Venice in 1295 after an absence of nearly 25 years.

Marco Polo's journey was one of the great exploring trips of all time, but his real importance to history and geography grew out of an accident of war. In 1296, while serving with a Venetian force, he was captured by troops from the rival city of Genoa. Imprisoned in Genoa, he met a writer known as Rustichello of Pisa, who listened to his stories about Asia and made them into a book (much embroidered with Rustichello's own imaginative cre-

A scene from the Catalan atlas shows the Polos' caravan crossing Asia. Camels carry goods; the passengers ride on horses.

ations). Marco Polo called the book *A Description of the World;* it later came to be known as the *Travels of Marco Polo.*

The book was published widely throughout Europe and had an extraordinary effect. In it, Polo lovingly described the wealth of the East—jewels piled on docks, warehouses bursting with spices and silks, and limitless gold in Japan (which he had never seen). These descriptions made Europeans more eager than ever to open up direct trade with China and other Asian realms. Princes and merchants who read Marco Polo's book were inspired to seek a sea route to Asia—a search that resulted, several centuries later, in Columbus's first voyage. Columbus owned a copy of the *Travels* and studied it closely for clues about a possible sea route to Cipangu and Cathay, as Polo called Japan and China. Thus, Marco Polo's description of his stay in the Mongol Empire was at least partly responsible for the European encounter with the Americas.

Polo himself made no maps—none, at least, that have survived the centuries. But his book stimulated cartographers and geographers all over Europe to draw new maps of the places he had described. One of the most important of all medieval maps, a collection of charts called the Catalan atlas that was assembled in 1375, shows the Polos' routes across Asia, illustrated with tiny drawings of the explorers themselves, Khublai Khan sitting regally in his tent, and many other scenes from Marco Polo's book. From the 14th century on, Marco Polo's influence was reflected in every new world map or map of Asia.

SEE ALSO

Asia, mapping of; Catalan atlas

FURTHER READING

Lattimore, Owen, and Eleanor Lattimore. *Silks, Spices, and Empire: Asia Seen Through the Eyes of Its Discoverers.* New York: Delacorte, 1968.

Polo, Marco. *The Travels.* Translated and edited by W. Marsden. Annotated by John Masefield. New York: Viking, 1980.

Severin, Timothy. *Tracking Marco Polo.* New York: Peter Bedrick, 1964.

Stefoff, Rebecca. *Marco Polo and the Medieval Explorers.* New York: Chelsea House, 1992.

Ponce de León, Juan

SPANISH CONQUISTADOR AND DISCOVERER OF FLORIDA

- *Born: about 1460, Servas, Spain*
- *Died: June 1521, Cuba*

JUAN PONCE DE LEÓN, one of the Spanish conquistadors who came to the Caribbean in the wake of Columbus's

While searching for the legendary fountain of youth, Ponce de León became the first European to discover Florida.

voyages, served as governor of the island of Puerto Rico in 1510–11. In 1513 he left Puerto Rico in search of an island that was rumored to lie in the waters north of the Bahamas. This island—called Bimini by the native peoples of the Caribbean—was said to contain a magical spring or fountain. Anyone who drank its water would remain young forever. The legendary island even appeared on an important map that was published in 1511 by Peter Martyr. It is called the Illa de Beimeni and is just north of Cuba. The mapmaker, of course, had no real knowledge of the island; he simply followed the custom of placing imaginary or legendary places on the map. But although Peter Martyr seems to have believed that the island existed, he was skeptical about the power of the magical water to restore youth. "I believe," he wrote, "God has reserved this prerogative to himself."

Ponce de León took two ships north through the Bahamas and then sighted an unknown land, which he believed to be an island. He named it Florida. He explored part of Florida's coast, but found no fountain of youth. His pilot made an important discovery, however: a current of water that flowed strongly from south to north along the east coast of Florida. This was the Gulf Stream, and its discovery revolutionized shipping in the Atlantic Ocean. Captains soon found that riding the Gulf Stream was the fastest way to get from the Caribbean back to Europe.

Ponce de León returned to Florida in 1521, still hoping to find the magical fountain. He died of a wound he received in a fight with the Native Americans there. By that time, Florida was taking its place on European maps, although it was most often incorrectly shown as an island. It even appeared on a globe made in 1515 by the great Italian artist and scientist Leonardo da Vinci, who called it La Florida.

SEE ALSO

Americas, mapping of; Atlantic Ocean, mapping of

Portolan

A PORTOLAN (from the Italian word *portolano,* meaning "book of directions") is generally defined as a sea chart for use in navigation. In fact, a true portolan is a written book of sailing instructions. Many such books contained charts, and gradually the charts became more common than the books, so that the term was used for charts by themselves.

Portolans are associated with the Middle Ages and the age of European seaborne discovery, but sailing guides existed in ancient times. A Greek sailor of the 1st century B.C. wrote a *periplus,* or handbook, of sailing directions for the Arabian Sea; no doubt other, similar works existed around the same time but have not survived. These Greek *periploi* did not contain maps, but later, drawings of the coasts and sketches of currents and reefs were inserted into these guides.

The charts had one important advantage over texts: a mariner did not have to know how to read in order to use them, although most charts did contain some writ-

A portolan chart by Petrus Vesconte, showing the coasts of Spain, France, and southern England and Ireland. West is at the top of the map.

ten information. Additionally, charts could convey a lot of information in a hurry, so that a pilot at the tiller of his ship in a storm, or as darkness was falling, did not have to page through a book to get directions to the nearest harbor.

One of the most useful features of the portolans was the geometric pattern of lines that covered them. Compass roses with markers pointing in as many as 32 directions were often placed at the points where these lines intersected. The lines are called rhumb lines, or loxodromes, and are guides to the wind—the force that powered ships in the age of sail. Over the centuries, mariners had learned that certain winds carried ships to certain points. The rhumb lines showed lines of direction so that mariners would know which winds to choose. They worked this way: Suppose you wanted to sail from Marseilles to Sardinia. You would spread out your portolan with its web of rhumb lines, and you would lay a ruler or other straight object on the chart between Marseilles and Sardinia. Then you would look for the rhumb line that was closest to your ruler (most portolans had a great many rhumb lines, so you would be fairly sure of finding one near your desired route). You would compare that rhumb line with the lines coming out of the compass rose and discover that it was a northwest bearing. Then you would know that you needed winds from the northwest—which you could identify in relation to the sun—to carry you to Sardinia. To make the rhumb lines easier to use, some chart makers followed a color system: black or gold lines for the 8 principal directions, green for the 8 half-winds (located halfway between each two principal winds on the compass rose) and red for the 16 quarter-winds (midway between each wind and half-wind).

Because the information they held was so important to trade and defense, portolans were closely guarded and passed from generation to generation as prized possessions. The oldest portolan that survives today, the Pisan chart, was made around 1290. The oldest dated portolan that has been preserved was made by Petrus Vesconte in 1311. In 1320

Vesconte produced a book or set of 10 maps: a map of the world, six portolan charts of the Mediterranean and Black seas, and three other regional or city maps. The Laurentian portolan, a collection of eight maps made by an unknown mapmaker in 1351, contains a remarkably accurate early chart of the west coast of Africa, which suggests that the mapmaker may have sailed to the Gulf of Guinea. A portolan made by German cartographer Henricus Martellus in 1492 is the first known chart to show Bartolomeu Dias's voyage around the southern tip of Africa.

Portolans were hand-drawn until 1569, when the first printed charts were made. In 1584–85, the Dutch cartographer Lucas Janszoon Waghenaer published a two-volume set of printed portolan charts. This was published in English in 1588 under the title *The Mariners Mirrour*. It became immensely popular with English pilots—so much so that they dropped their usual term for sea charts, "rutters" (from the French word *routier*, or "route finder"). Instead, the English began calling portolan atlases "waggoners," a form of Waghenaer's name.

At first, the portolans dealt only with strips of coastline and did not include geographic information about the lands beyond the coasts. As time went on, however, chart makers began filling in the continents with a mix of geographic information and decorative illustrations. The Catalan atlas, compiled in 1375, is part portolan and part world map, reflecting the way portolans merged with other cartographic traditions. They gradually became maps of the landmasses as well as the seas, not just sea charts. The portolans are, nonetheless, the ancestors of today's highly technical marine charts.

SEE ALSO
Catalan atlas; Chart; Compass rose; Navigation; Pisan chart; Waghenaer, Lucas Janszoon

Poseidonis

SEE Ancient and medieval mapmakers

Powell, John Wesley

AMERICAN GEOLOGIST AND SURVEYOR

- *Born: Mar. 24, 1834, Mount Morris, New York*
- *Died: Sept. 23, 1902, Haven, Maine*

AFTER FIGHTING in the Civil War (he lost an arm at the Battle of Shiloh), Powell began exploring Colorado and other western regions. In 1869, after raising funds from a variety of private individuals, businesses, and scientific organizations, he led the first expedition through the Grand Canyon of the Colorado River. Powell then received funding from the federal government for an official survey of the canyon and led a second expedition through it in 1872, this time making careful scientific observations.

Powell was made director of the Geographical and Geological Survey of the Rocky Mountains, which would merge with the U.S. Geological Survey (USGS) in 1879. He became the second director of the USGS.

With his brother-in-law, Almon Harris Thompson, Powell pioneered the development of topographic and geologic maps that contained detailed information about soils, springs, and other resources. He believed that careful classification of the various types of land was essential if the land and its resources were to be wisely used. His own maps are landmarks of western mapping and set a high standard for later USGS maps.

FURTHER READING

Dellenbaugh, Frederick S. *A Canyon Voyage: The Narrative of the Second Powell Expedition down the Green-Colorado River from Wyoming, and the Explorations on Land, in the Years 1871 and 1872.* 1908. Reprint. Tucson: University of Arizona Press, 1984.

Gaines, Ann. *John Wesley Powell and the Great Surveys of the American West.* New York: Chelsea House, 1991.

Goetzmann, William H. *Exploration and Empire: The Explorer and the Scientist in the Winning of the American West.* New York: Norton, 1978.

Stegner, Wallace. *Beyond the Hundredth Meridian: John Wesley Powell and the Second Opening of the West.* Boston: Houghton Mifflin, 1954.

Prester John

PRESTER JOHN (the title is short for presbyter, or priest) was a mythical king who haunted the European imagination for many years, beginning in the 12th century. Prester John was rumored to be a powerful Christian priest-king who ruled an empire somewhere beyond the borders of the known world. Prester John's kingdom was said to be rich—so rich that the pebbles in the streams were diamonds and rubies. In addition, the kingdom contained many wondrous things, including the Tower of Babel that was mentioned in the Bible, a magic mirror that let the emperor see what was happening in any part of his kingdom at any time, and a fountain of youth.

At first, people placed Prester John's kingdom in India, the home of colonies of Christians, who might have inspired the legend. Then, when India proved not to contain any mysterious, magical, wealthy Christian emperors, people decided that the kingdom probably lay somewhere in the huge expanse of Central Asia. Early travelers to the Mongol Empire in China looked eagerly for Prester John but failed to find him.

The mythical Christian king Prester John eluded explorers.

By the late 14th century, Prester John had failed to turn up in Asia. Mapmakers—reluctant to discard such a well-loved legend—began situating Prester John's kingdom in Africa, possibly confusing it with the real Christian kingdom of Abyssinia (now called Ethiopia). Many journeys of exploration were made with the goal of reaching Prester John's kingdom, which continued to appear on maps until the beginning of the 17th century. In 1573, for example, the Dutch cartographer Abraham Ortelius published a map of Africa on which Ethiopia was labeled "the empire of Prester John."

FURTHER READING

Stefoff, Rebecca. *Accidental Explorers: Surprises and Side Trips in the History of Discovery.* New York: Oxford University Press, 1992.

Prime meridian

SEE Greenwich, England; Longitude

Printing

THE INTRODUCTION of printed maps—as opposed to hand-drawn, or manuscript, maps—marks a turning point in cartography. Printed maps could be reproduced many times and were the same each time. Printing made it possible for maps to circulate more widely and be purchased by more people than had ever been possible with laboriously hand-drawn maps. Printing also introduced a higher standard of accuracy by eliminating the accidental errors that crept in during hand copying of individual maps.

The Chinese invented printing before the 8th century A.D. They carved pages of text onto blocks of stone, wood, or metal so that the part to be printed was raised above the rest of the block; this method is known as relief printing. The printing block was then pressed to an inked surface and then onto a blank sheet of paper or cloth, transferring the inked characters onto the sheet. Maps were also printed from blocks in both China and Japan long before printed maps existed in Europe.

Movable type for printing text, using a separate small block for each letter of the alphabet so that the same letters could be used again and again in countless combinations, was invented in Germany in the mid-15th century. By that time Europeans were already printing simple pictures, mostly of religious scenes, from carved wooden blocks and engraved copper plates. Once the technology of movable type was applied to making books, booksellers discovered that people liked their books to be illustrated, and maps helped fill the need for illustrations. So maps began to be printed from wood blocks and copper plates to accompany the books issued by the printing presses of Europe.

In the early 16th century, booksellers began making single-sheet printed maps. By about the middle of that century, copperplate engravings had almost entirely replaced woodcuts, and engraving remained the standard way of making printed maps until the 1820s, when a new printing process called lithography was introduced.

SEE ALSO
Engraving; Lithography; Woodcut

Projections

IF YOU were able to remove the skin of an orange in a single piece, you would find it impossible to flatten that skin onto a tabletop without some buckling, tearing, or trimming. Like the orange, the earth is a sphere, and its "skin"—the surface that makes up the map of the world—cannot be flattened onto a plane surface like a sheet of paper without some distortion. The only kind of map that has no distortion, the only map that is a true representation of the earth's surface, is a globe.

Though every flat map has some distortion, maps of very small areas have very little distortion; they can be almost as true as if they were traced from the globe. But distortions creep in when maps cover more than a few square miles or kilometers. The larger the area, the greater the distortion. World maps are the worst. No map projection offers a true picture of the whole world; on every one, area, distance, direction, or shape must be distorted.

For centuries, mapmakers have experimented with ways of portraying the spherical earth on a flat map. They have come up with hundreds of solutions called projections, of which about a hundred are in use today. No projection can be true for shape, area, distance, and direction; each projection offers its own strengths and weaknesses. One of the cartographer's most crucial responsibilities is to choose the best projection.

Conformal projections

Conformal projections received their name because they conform to, or match, the shape of the land. They are also called orthomorphic projections, from the Greek words *orthos* ("correct") and *morphe* ("form"). Conformal projections preserve the true shapes of small areas, although shapes and angles are distorted in large areas such as continents. Engineers, ship pilots, and military strategists like conformal maps because these maps reveal the shape of local terrain. Many topographic maps are made on conformal projections. Among these projections are the Mercator and

Mercator Projection (Conformal)

Central meridian (selected by mapmaker)
Great distortion in high latitudes
Examples of rhumb lines (direction true between any two points)
Equator touches cylinder if cylinder is tangent
Reasonably true shapes and distances within 15° of Equator

Albers Equal-Area Conic Projection

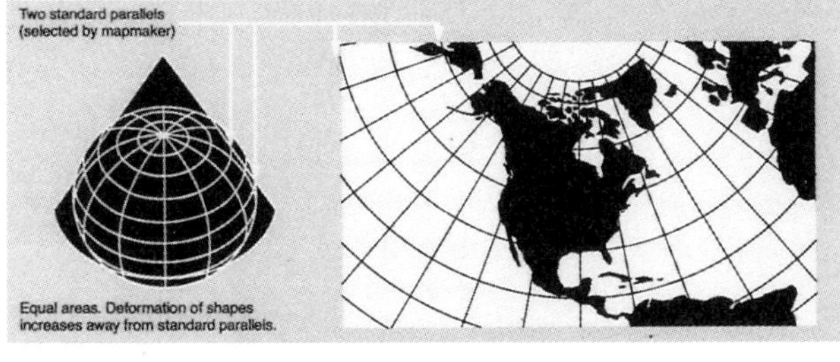

Miller Cylindrical Projection

Central meridian (selected by mapmaker)
Change in spacing of parallels is less than that on Mercator projection
Equator always touches cylinder

Lambert Conformal Conic Projection

various specialized versions of it (such as the Space Oblique Mercator, which is used for satellite mapping because it allows a band of no distortion along the satellite's flight path), the Stereographic, the Lambert Conformal Conic, and the Bipolar Oblique Conic Conformal projections.

Equal-area projections

Equal-area projections show the true area of places in relation to one another. In other words, the scale remains constant on every part of an equal-area map; a square inch or centimeter of Antarctica on the map represents the same amount of territory as does a square inch or centimeter of New Guinea. Equal-area projections are also called equivalent projections (because the area remains equivalent, or equal), proportional projections (because areas on the map remain in true proportion to one another), or homolographic projections (from the Greek *homos,* "same," and *graphein,* "write," meaning "drawn the same way"). Geographers, scientists, and statisticians like equal-area maps because such maps permit fair comparisons between areas. For example, a map showing the size of countries around the world should be on an equal-area projection. If the map is not an equal-area map, areal distortion will cause some countries to appear much larger than they really are. The Albers Equal-Area Conic, the Sinusoidal, and the Lambert Azimuthal Equal-Area projections all have the property of true area.

Methods of projection

Map projections fall into several categories, according to the mathematical method by which the globe is projected onto the flat surface.

Cylindrical projections are made by projecting the globe onto a cylinder, as though a sheet of paper were rolled into a tube with a globe fitting snugly inside it. If the pattern of the globe were projected outward onto the paper and the paper were unrolled, the result would be a world map on a cylindrical projection. The Mercator projection is cylindrical, with the equator being the line at which the cylinder touches the globe. Variations on the Mercator projection are made by tilting the globe within the cylinder so the cylinder touches the globe along some other line.

Conic projections use the image of a cone rather than a cylinder. Picture the cone dropped over the globe like a dunce cap over a bowling ball. If the pattern of the globe were projected outward onto the cone and then the cone were unrolled and spread flat, you would have a conic projection. Such projections are very old: The curving, fan-shaped world maps of Ptolemy and other early mapmakers are conic projections. Today conic projections are not used for mapping the whole world, but they are useful for mapping areas in the middle latitudes that have a broad east-to-west extent, such as the continental United States. The Lambert Conformal Conic, the Albers Equal-Area Conic, the Simple Conic, the Bipolar Oblique Conic Conformal, and the Polyconic are the most commonly used conic projections.

Azimuthal projections are projections from a point onto a plane. All azimuthal maps have a center called an azimuth, the point from which the projection originates. Directions are true only from that point. The word *azimuth* is adapted from the Arabic *as-summut,* which is the plural form of an Arabic term meaning "the way."

Azimuthal projections are often used for maps of the polar regions; the map is a circle centered upon the pole. Azimuthal projections can also be used to map continents, oceans, and hemispheres. The best-known azimuthal projection is the Orthographic, which is used for perspective views of the entire earth. The Orthographic projection resembles a picture of the earth as it would appear from far out in space. The earth can be seen from above one of the poles, the equator, or any other point. The Orthographic projection is often used for maps of the moon and planets. Similar to the Orthographic is the Stereographic projection, which projects the map from a point on the earth's surface directly opposite the center of the projection. Other azimuthal projections include the Azimuthal Equidistant, the Lambert Azimuthal Equal-Area, and the Gnomonic.

Several important projections are neither cylindrical, conic, nor azimuthal. These require the cartographer to adjust the point of projection and the math-

Orthographic Projection (Azimuthal)

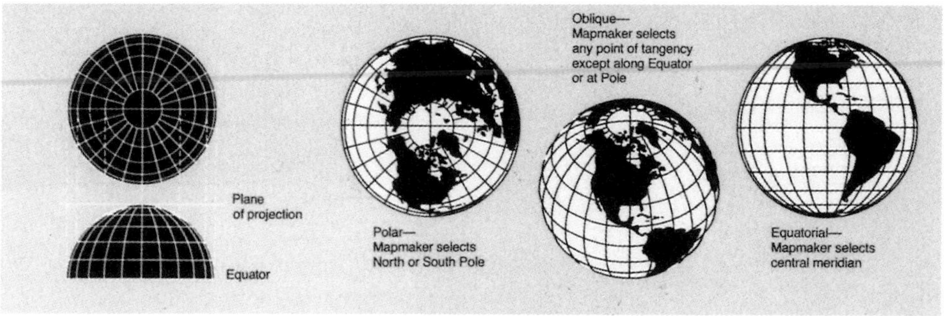

Plane of projection
Equator
Polar—Mapmaker selects North or South Pole
Oblique—Mapmaker selects any point of tangency except along Equator or at Pole
Equatorial—Mapmaker selects central meridian

Stereographic Projection (Azimuthal)

Plane of projection
Equator
Point of projection
Polar—Mapmaker selects North or South Pole
Oblique—Mapmaker selects any point of tangency except along Equator or at Pole
Equatorial—Mapmaker selects central meridian

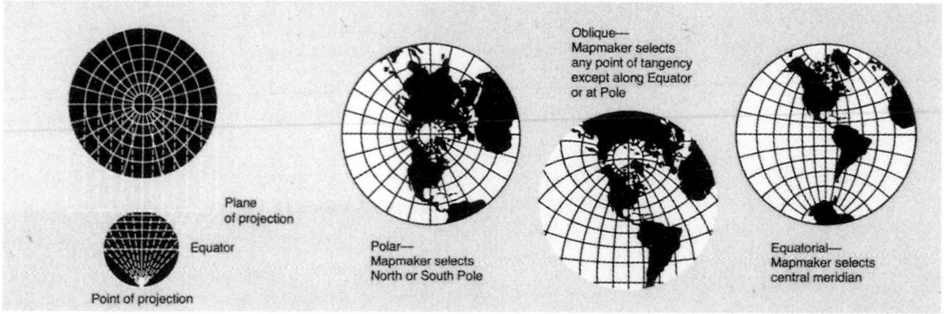

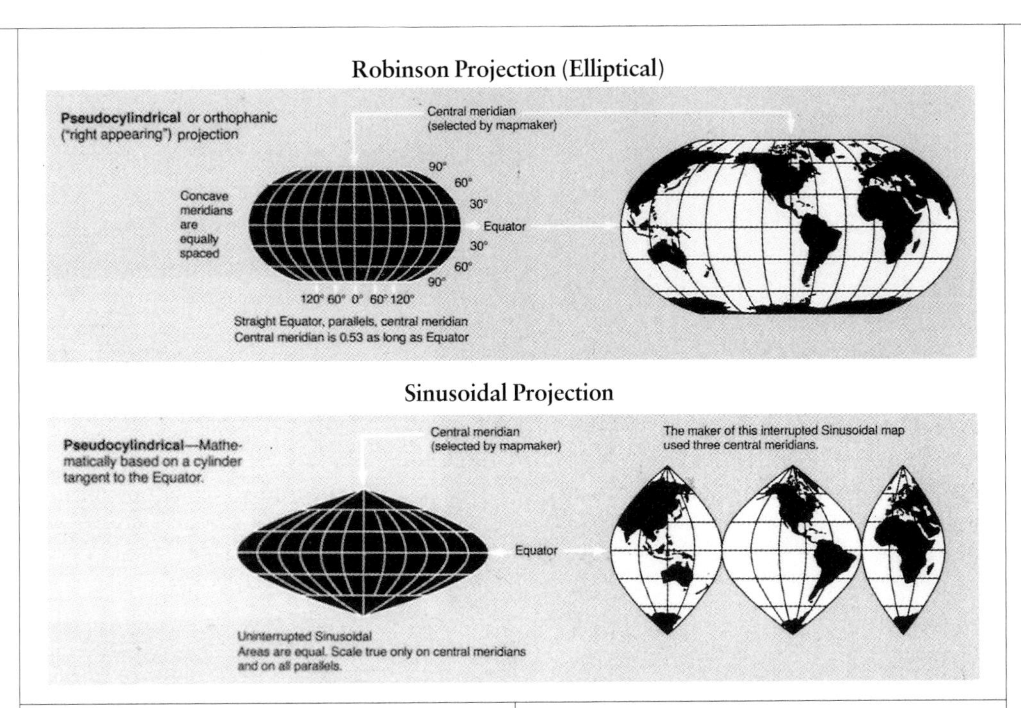

ematical formulas to achieve a compromise map that minimizes distortion and gives a realistic appearance. Elliptical and interrupted projections made in this fashion are increasingly used for world maps.

Elliptical projections, such as the Mollweide, Hammer, and Robinson projections, show the entire world within an ellipse, or flattened oval. These are used to show the whole world on one map.

Interrupted projections are those in which the cartographer has interrupted, or cut, the projection along lines of longitude to "spread" it out more effectively. A set of globe gores, tapered strips ready to be pasted onto a globe, is an interrupted projection. It is highly accurate but would be very hard to use because it is in at least 12 pieces. The Sinusoidal Equal-Area and Goode's Homolosine Equal-Area projections offer useful compromises between accuracy and readability by splitting the world map into only three or four sections, or lobes. They are used for statistical and educational maps.

Importance of projection

Just as every map represents the world, every map also misrepresents the world because every map projection introduces some distortion in the shape, area, distance, or direction of the world's lands and seas. Maps can have a powerful effect on how people think about the world, as the American Cartographic Association (ACA) acknowledged in 1991: "A poorly chosen map projection can actually be harmful. We tend to believe what we see, and when fundamental geographical relationships, such as shapes, sizes, directions, and so on, are badly distorted, we are inclined to accept them as fact when we see them on maps." The ACA meant that, for example, people who see only maps made with projections that are centered on points in the Northern Hemisphere would come to think of the world as dominated by the Northern Hemisphere; this would seem natural, because it would be reinforced by their mental image of the world, created by the maps they had seen.

In the 1970s and 1980s, debate centered on the widespread use of the Mercator projection in wall maps for classrooms. Arno Peters, a German historian, claimed that the Mercator projection glorified Europe and North America at the expense of Third World nations by making landmasses in the Northern Hemisphere look larger than they really are, thus diminishing the size of tropical countries and encouraging the Northern Hemisphere to feel superior to the southern.

In 1974, Peters presented a new world map that tried to correct the alleged injustices of the Mercator projection by giving each country its true area. Cartographers hastened to point out that the Peters projection had already been invented by James Gall, a Scottish clergyman, in 1855, and that it had long since been superseded by other, better equal-area projections. The Peters controversy subsided, but it did bring about some welcome changes in the use of projections. The Mercator projection, originally designed as an aid to navigators rather than a geographically accurate map, is being replaced in schools by more plausible projections.

Chief among these is the Robinson projection, invented by the American cartographer Arthur H. Robinson in 1963. Although area is distorted in the Robinson projection, the distortion is much less than in the Mercator projection: Mercator exaggerates the area of Greenland by several hundred percent, Robinson by only 60 percent. The National Geographic Society, a major producer of maps for everyday use, adopted the Robinson projection in 1988, and the following year the American Cartographic Association and five other geographic societies endorsed it as a worthy replacement for the Mercator projection. A growing number of geographers and educators are working to educate map users about the pitfalls and benefits of various map projections. They claim that a map user's first question upon seeing a map, especially a world map, should be: "What's the projection?"

SEE ALSO
Globe

FURTHER READING

American Cartographic Association. *Which Map Is Best? Projections for World Maps.* Falls Church, Va.: American Congress on Surveying and Mapping, n.d.

Snyder, John. *Map Projections: A Working Manual.* Washington, D.C.: U.S. Government Printing Office, 1987.

Propaganda

PROPAGANDA IS information or rumors deliberately spread to help a cause or hurt an opposing cause. Maps have often been used in works of propaganda. Mark Monmonier, a leading American geographer and scholar of cartography, has compiled many examples of what he calls "cartopropaganda": the use of maps to mislead or to encourage a particular point of view.

Maps are powerful visual images that can often communicate quickly and directly. Furthermore, maps have an air of scientific respectability, so that many people think that if they see something on a map, it must be "true." This makes maps useful tools for the propagandist, who can tilt the message of the map in many ways. For example, during World War II (1939–45), Nazi propaganda articles often included maps that were designed to make the reader feel sympathy for Germany. After Germany had invaded Poland and Great Britain had declared war on Germany, one propaganda piece contrasted a map of Germany with a whole page of maps of the various countries of the British Empire. The message was clear: Germany could not possibly be the aggressor, when Great Britain was so much bigger and more powerful.

Another form of propaganda is the disinformation map. Starting in the 1930s, the Soviet Union deliberately altered geography on many official maps, moving towns, railroads, boundaries, and even rivers from their true locations in order to foil enemy spies or invading armies. Such practices are rare now that satellites can make accurate maps from above the earth.

Another way maps serve propagandists is by making political and territorial claims look official. During the Falklands War of the 1980s, when Great Britain and Argentina fought over the ownership of the Falkland Islands in the South Atlantic, Argentina asserted its claim to the islands by issuing its own maps of them, calling them the Malvinas.

Because maps do not represent reality in all its complexity but only those features that the mapmaker has selected, they offer many opportunities for distortion or slanting. Every choice of projection, scale, or color affects how the viewer will respond to the map. Users of maps should be thoughtful consumers, aware of maps' potential not only for deliberate propaganda but also for fallibility and accidental misrepresentation.

SEE ALSO
Projections

FURTHER READING

Lobeck, Armin. *Things Maps Don't Tell Us: An Adventure in Map Interpretation.* Chicago: University of Chicago Press, 1993.

Monmonier, Mark. *How to Lie with Maps.* Chicago: University of Chicago Press, 1991.

Wood, Denis. *The Power of Maps.* New York and London: Guilford, 1992.

Proportional projection

SEE Projections

Ptolemy

GREEK GEOGRAPHER AND ASTRONOMER

• *Active: 2nd century* A.D.

ALTHOUGH CLAUDIUS PTOLEMAEUS, called Ptolemy, is one of the towering figures of geography and mapmaking, his life is a mystery. We know only that he lived and worked in

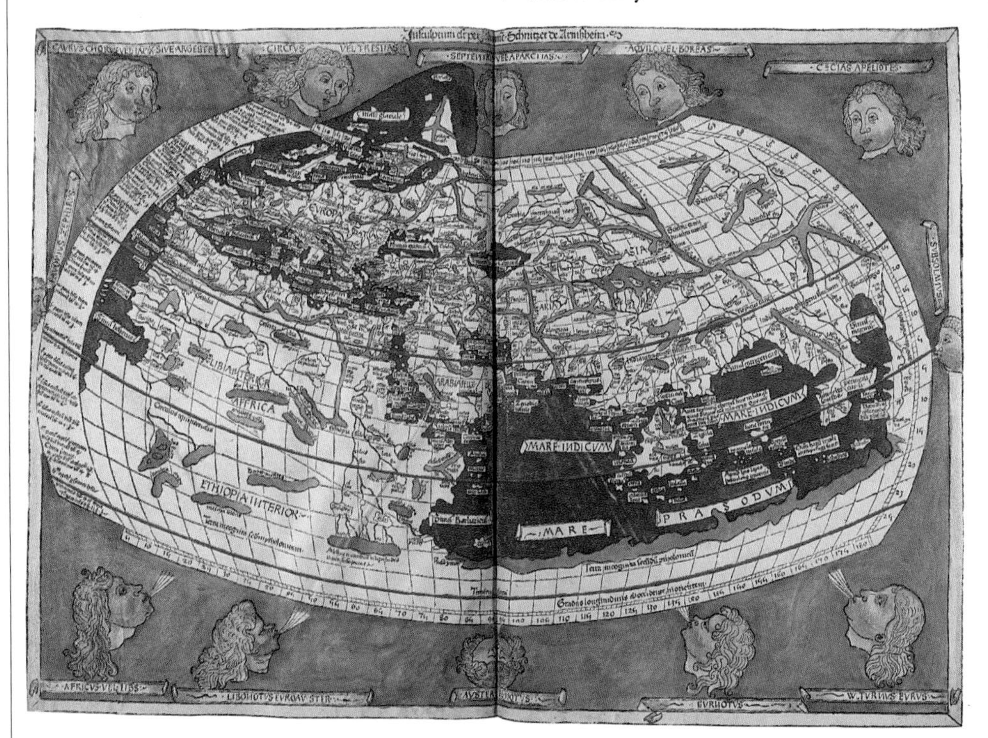

The world according to Ptolemy, from the Ulm edition of the Geography, *published in Germany in 1482.*

Alexandria, the Greek and Roman administrative capital of Egypt, between 127 and 145 A.D. Some early sources give the years of his birth and death as 87 and 151, but these are not necessarily reliable dates.

One of Ptolemy's achievements was to sum up Greek astronomy, together with his own theories and observations about the stars and planets, in an encyclopedic work that came to be known as the *Almagest* (from the Arab *al-Majusti,* which meant "great book"). Some of the basic principles set forth in the *Almagest* were incorrect—for example, Ptolemy rejected the idea that the earth moves and claimed that the stationary earth was the center of the universe. Yet the power of the work lay in its orderly organization, its huge store of facts and figures (many culled from earlier works), and its insistence on science and logic. These same qualities appear in Ptolemy's great work on geography and mapmaking, the *Geography,* which, like the *Almagest,* influenced other philosophers and scientists for nearly 1,500 years. It is both a distillation of the work of other geographers and an orderly, thoughtful set of principles to guide later scholars.

"Geography is a representation in picture of the whole known world together with the parts that are contained therein," wrote Ptolemy at the beginning of the *Geography*—a definition that has never been improved upon. He discussed the difference between mapping the world and mapping local features such as harbors and farms, and he then explained the importance of mathematics and astronomy to geography. Ptolemy's most important contribution was his insistence upon the use of latitude and longitude to fix the locations of places. He pointed out that latitude and longitude form a reference system based on fixed, unchanging features—the sun and the stars. Only by using such a system could mapmakers ensure that a given map was accurate and could be reproduced. Once the latitude and longitude of a place were known, that place could be correctly positioned on any map. (Unfortunately, Ptolemy and his contemporaries were unable to practice this sound theory with much accuracy, for they lacked a reliable method of determining longitude precisely. Such a method was not developed until the 18th century. The importance of Ptolemy's contribution lies in the idea of a grid of latitude and longitude lines, not in his imperfect practice.)

Ptolemy introduced other ideas that formed the basis of cartography in later centuries. He gave instruction on two methods of projection—that is, of representing a spherical world on a flat sheet of paper. He also experimented with the use of scale, breaking the world map up into regional maps so that some areas could be shown in larger scale and greater detail than others. Ptolemy also began the practice of putting north at the top of the map.

In addition to the mapmaking principles, the *Geography* contained a list of about 8,000 place names, each with a latitude and longitude. It is not known whether Ptolemy made maps to go with the *Geography;* no such maps survive. But scholars and publishers found it easy to draw maps using his coordinates. Such maps are called Ptolemaic maps because they reflect Ptolemy's view of the world.

The earliest known copy of the *Geography* is in Arabic and dates from the 12th century. Like other scientific works from the Greek and Roman eras, the *Geography* was unknown to Europeans until the end of the Middle Ages. However, it was preserved and studied by Islamic scholars, who later introduced it to Europe. The *Geography* was translated from Arabic into Greek in the 13th century and into Latin in the 15th century, although these translations were not widely read at first. Hand-copied versions of the *Geography,* usually with maps attached, began to circulate through Europe in the 15th century. The first printed edition appeared in 1475, and from then on new editions poured off the presses of Europe. The Ulm edition, printed in 1482 in Germany with woodcut maps, was the first to add modern discoveries such as Greenland to the Ptolemaic world. Following his "rediscovery" by Europeans, Ptolemy was revered as the founder of geography. For several centuries, Ptolemaic maps were copied and recopied, becoming the cartographic standard of excellence.

One of the strengths of Ptolemy's geography was his insistence on carefully measured, verified data as the basis for all mapmaking. Yet his own data were not very sound. Many of the latitudes and longitudes he cited were known only from hearsay or guesswork. He also had a tendency to fill the edges of his world with theory or legend rather than observation. For example, he believed that a large unknown continent must exist in the Southern Hemisphere to "balance" the weight of Europe and Asia in the north. He also believed that the Indian Ocean was a landlocked sea, enclosed by the joined coasts of Africa and Asia—a feature that continued to appear on Ptolemaic maps centuries later, even after Portuguese navigators had disproved it by sailing around Africa.

Ptolemy's most significant error, though, was his estimation of the earth's size. He underestimated the circumference of the earth by about 30 percent. He also overestimated the width of Asia. The combination of these two errors suggested that the known world covered a much greater portion of the globe than was really the case, and that the distance between the east coast of Asia and the west coast of Europe was much smaller than it really is. Christopher Columbus, who studied Ptolemy's work closely, found this notion encouraging. Who knows if he would have been so eager to venture westward in search of Asia if he—and Ptolemy—had known just how far away Asia really was. Ptolemy's 2nd-century view of the world thus contributed to the world-changing encounter between Europe and the Americas in the 15th century.

SEE ALSO

Ancient and medieval mapmakers; Latitude; Longitude; Terra Australis

FURTHER READING

Brown, Lloyd. *The Story of Maps.* 1949. Reprint. New York: Dover, 1979.

Campbell, Tony. *The Earliest Printed Maps, 1472-1500.* Berkeley: University of California Press, 1987.

———. *Early Maps.* New York: Abbeville, 1981.

Dilke, O. A. *Greek and Roman Maps.* Ithaca, N.Y.: Cornell University Press, 1985.

Harley, J. B., and David Woodward, eds. *The History of Cartography.* Vol. 1, *Cartography in Prehistoric, Ancient, and Medieval Europe and the Mediterranean.* Chicago: University of Chicago Press, 1987.

Shirley, Rodney W. *The Mapping of the World: Early Printed World Maps, 1472–1700.* London: Holland Press Cartographica, 1983.

Wilford, John Noble. *The Mapmakers: The Story of the Great Pioneers in Cartography from Antiquity to the Space Age.* New York: Knopf, 1981.

Pythagoras

SEE Ancient and medieval mapmakers

Pytheas

GREEK MARINER AND EXPLORER

• *Active: 4th century* B.C.

PYTHEAS WAS a native of the Greek colony of Massalia, located where the city of Marseilles now stands on the Mediterranean coast of France. A skilled astronomer and navigator, Pytheas made the first recorded voyage by a Greek into northern seas. He sailed west into the Atlantic Ocean and then north to the British Isles. He made landfall in Great Britain, and it is possible that he circumnavigated the British Isles; his own account of the trip has not survived.

Leaving Britain, Pytheas sailed on to a place he called Thule. He described it as cold and dark for part of the year; he also hinted that he had seen pack ice a short distance north of Thule. These details suggest that he may have visited Iceland, although it might have been Norway or Scotland. At any rate, he added the word *Thule* to the vocabulary of geographers. It was used to refer to any region in the remote north, and the phrase Ultima Thule, or Farthest Thule, came to mean any very distant place at the edge of the known world.

Quadrangle maps

QUADRANGLE MAPS (called quads) are maps of rectangular areas whose boundaries are lines of latitude and longitude. The topographic maps issued by the U.S. Geological Survey are quadrangle maps. Depending upon the scale of the maps, they cover areas from 7.5 minutes square (a minute is 1/60th of a degree of latitude or longitude) to 4 degrees by 6 degrees. Because their shape is consistent, quadrangle maps on the same scale can be joined edge to edge to form a larger map. Depending upon the maps' scale, for example, a given territory can be covered by 1, 4, 32, 64, or 128 USGS quadrangle maps.

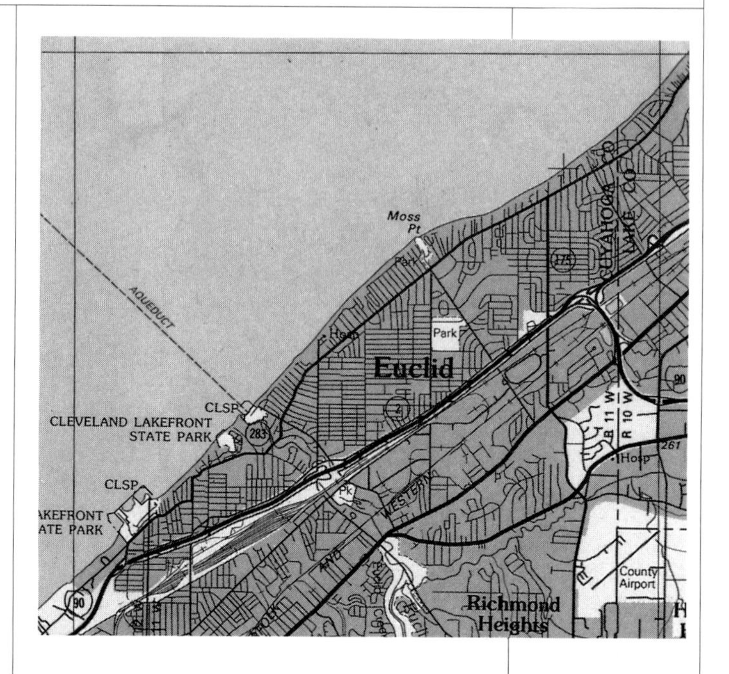

A large-scale U.S. Geological Survey quadrangle map of part of Cleveland, Ohio. Each map in this USGS series is exactly the same size, shape, and scale.

Quadrant

A QUADRANT is an instrument once used to measure the altitude of the sun or a star in astronomy and navigation. It consisted of a 90-degree arc marked off in degrees and a movable arm with a sighting hole. The navigator would line up the sun through the sighting hole, and the angle of the sun's altitude above the horizon would be indicated on the arc. The instrument was called a quadrant because *quad* means "four" in Latin and the 90-degree arc represented one-fourth of a circle.

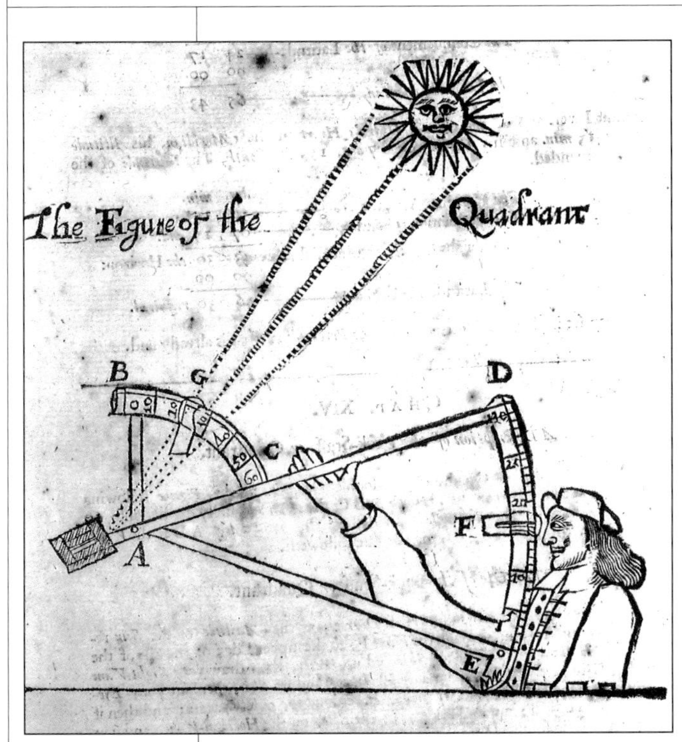

A 16th-century English quadrant, also called a backstaff. The user lines up sight F with the horizon, then adjusts the movable, curved scales until the sun is reflected in mirror A. The figures on the scales will give the angle of the sun above the horizon.

SEE ALSO

Cross-staff; Navigation

Quiros, Pedro Fernándes de

SPANISH NAVIGATOR AND EXPLORER OF THE PACIFIC OCEAN

- *Born: about 1565, Spain*
- *Died: 1615, Peru*

PEDRO FERNÁNDES DE QUIROS became a clerk in Spain's Pacific trade, serving aboard the galleons that made yearly voyages between two Spanish colonies: Manila, in the Philippine Islands, and Acapulco, on the west coast of Mexico. In this way he learned about navigating the Pacific trade routes, although the greater part of the vast Pacific was still unexplored.

In 1595, Quiros served as a pilot aboard one of the ships in Álvaro de Mendaña's expedition in search of the Solomon Islands. After Mendaña's death, Quiros piloted the survivors of that disastrous expedition to the Philippines. Like many people of his time, Quiros believed that a large unknown continent, the Terra Australis (Latin for "southern land"), awaited discovery in the Southern Hemisphere. Quiros, who fancied himself a second Columbus, decided to discover the Terra Australis. He journeyed to Europe to get the support of King Philip III of Spain and Pope Clement VIII.

In 1605, Quiros's little fleet of discovery set sail from Peru. He landed in an island group that he called Espíritu Santo (Holy Spirit); later these islands were called the New Hebrides, and today they form the independent nation of Vanuatu. Quiros convinced himself that these islands were part of the mysterious Terra Australis, but he gave up his attempt to colonize them when his people complained that the native inhabitants were "devils with poisoned arrows."

Quiros returned to Spain in 1607 to seek support for another voyage to what he still believed was the Terra Australis. Finally, in 1614, he received permission to sail the Pacific again. He hastened to Peru but died there, exhausted and dispirited, before he could organize another expedition. The legacy of his 1605 voyage, however, lived on to perplex mapmakers for many years. Mapmakers trying to place the nonexistent Terra Australis on the map confused Quiros's discovery with Australia, the coast of which was just beginning to be explored. As late as 1756, the French cartographer Gilles Robert de Vaugondy placed Espíritu Santo on the east coast of Australia.

SEE ALSO

Mendaña de Neira, Álvaro de; Pacific Ocean, mapping of; Terra Australis

FURTHER READING

Dunmore, John. *Who's Who in Pacific Navigation.* Honolulu: University of Hawaii Press, 1991.

Radar

RADAR (the name stands for radio detection and ranging) is a form of remote sensing in which a radio signal is beamed outward, bounces off an object or surface, and is recorded back at its source. First used during World War II (1939–45) for military purposes, radar proved useful to surveyors after the war. Using airborne radar, technicians could get an accurate picture of landscapes that were difficult to survey on the ground by using standard aerial photography. Radar offered a revolutionary new way to survey tropical areas, where the ground is obscured by both cloud cover and dense vegetation.

The most ambitious radar mapping program was carried out in Brazil, where the government hired U.S. aerial surveyors to map the Amazon River basin in the 1970s. Before that time, the region was largely unmapped, except for the river and its tributaries. The results of the radar survey were published in 18 volumes and included the first comprehensive topographic maps of the vast Amazon rain forest. Scientists used these as the basis for geologic, agricultural, and land-use maps.

This cartographic triumph did not come without a price, however: Brazil had commissioned the survey to speed the economic development of the Amazon, and the radar maps produced in the early 1970s helped make possible the surge of road building, forest burning, strip mining, and other activities that threaten the survival of the rain forest.

SEE ALSO

Americas, mapping of; Remote sensing

Raleigh, Sir Walter

ENGLISH EXPLORER OF SOUTH AMERICA

- *Born: about 1553, Budleigh Salterton, England*
- *Died: Oct. 29, 1618, London, England*

WALTER RALEIGH'S services as a soldier in the English army won him the favor of Queen Elizabeth I. With her permission, he launched the settlement of a colony at Roanoke, Virginia, in 1584. By 1589, the colony had failed, but one of the colonists, a surveyor named John White, drew a map of Roanoke that was later engraved by several European cartographers, including Théodore de Bry and Jodocus Hondius (White's original sketch map is preserved in the British Library).

Raleigh's next venture was the search for El Dorado, a legendary city of gold that was rumored to exist at the headwaters of the Orinoco River in South America. In 1595 Raleigh led an expedition into Guyana to look for El Dorado but failed to find the fabled city. More than 20 years later, Raleigh tried once more to find gold in Guyana. Again he failed. He was executed by King James I, having been convicted—on

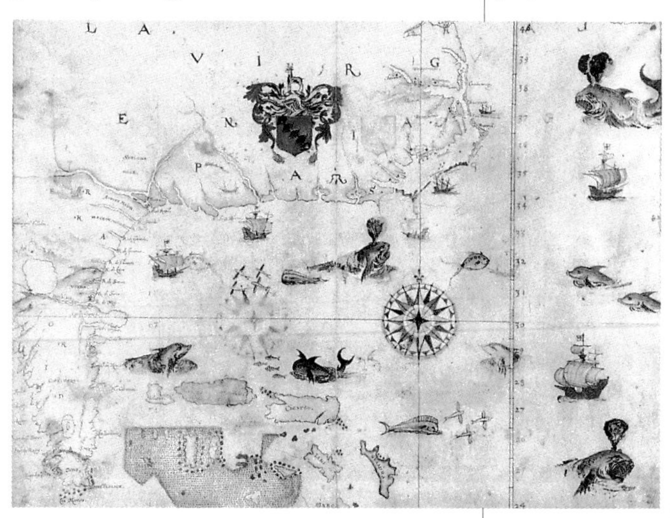

John White's map of the Virginia colony, with Raleigh's coat of arms prominently displayed.

flimsy evidence—of taking part in conspiracies against the government. But Théodore de Bry had produced a map based on Raleigh's theory that a golden city stood on the shores of a great inland sea somewhere in Guyana or Venezuela. That map continued to spread the legend of El Dorado long after Raleigh's head had rolled in the Tower of London.

FURTHER READING

Quinn, David. *Set Fair for Roanoke: Voyages and Colonies, 1584–1606*. Durham: University of North Carolina Press, 1985.

Ross, Williamson H. *Sir Walter Raleigh*. 1951. Reprint. Westport, Conn.: Greenwood, 1978.

Relief

RELIEF REFERS to the elevations and depressions—the highs and lows—of terrain. A map that shows these features is called a relief map. A relief map or globe may actually be three-dimensional, showing hills and valleys as bumps and grooves on the map surface. Or it may simply appear three-dimensional, using line and color on a flat surface to create the illusion of heights and depths.

Cartographers have developed many ways of showing relief on flat maps. Early mapmakers used what is sometimes called the "molehill" system, in which mountains were drawn as rows of little bumps, sometimes shaded. In this type of relief drawing, all mountains looked fairly similar, and the line of "molehills" strung across a map's surface did not say much about the actual shape or extent of the mountain range in question, only that there were mountains of some kind in that general vicinity.

The demand for greater accuracy and detail produced new methods of illustrating relief. Over the years, these methods have included hachures or hill shading, contour lines, symbols, and hypsometric tinting, or the use of different colors to show differences in altitude. The most realistic-looking appearance of relief is created by hill shading, in which the terrain is drawn as though lit by

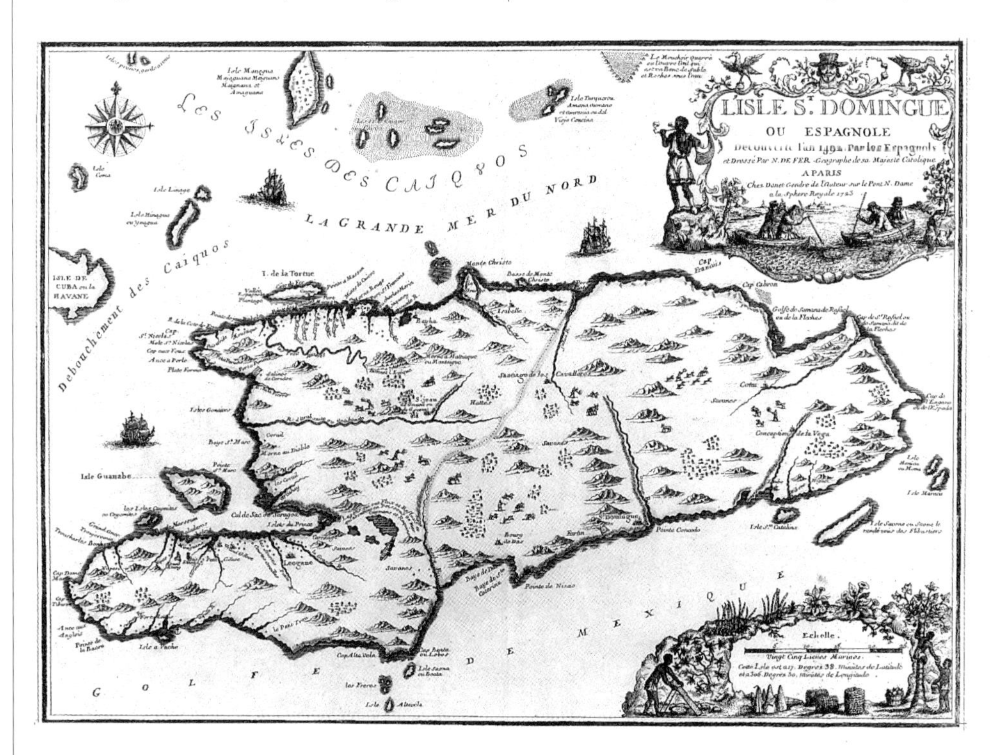

On this 18th-century French map of Hispaniola island, a sprinkling of separate hills indicates a generally mountainous terrain.

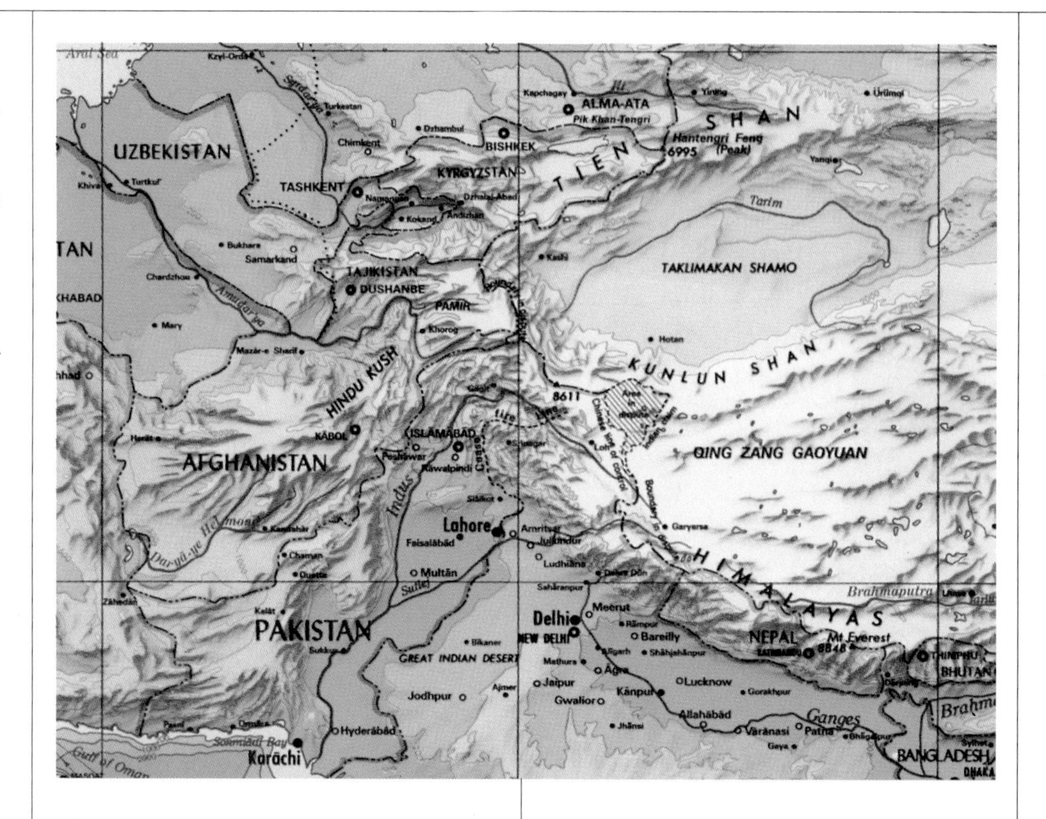

Hill shading, contour lines, and color show relief on a map of southwest Asia. Lowlands are green, highlands are tan, and the highest regions are shown in white.

slanting sunlight; the details of mountains and valleys are revealed by the contrast of light and shadow. Most two-dimensional relief maps today employ either hill shading or contour lines, often in combination with color.

SEE ALSO
Contour lines; Hachure; Hypsometric shading; Topographic maps

Remote sensing

REMOTE SENSING is the process of examining something without touching it or approaching it directly. The British officials in 19th-century India who sent specially trained Indians into regions that were off-limits to foreigners to carry out secret surveying missions were engaging in a form of remote sensing; the native experts, called pandits (the source of the modern word *pundit*) were their sensing devices. Today the term applies to the gathering of data by probes and satellites from places where humans cannot easily go, such as the depths of the sea and outer space.

In terms of mapmaking, remote sensing generally refers either to the study of the earth and other planets from artificial satellites or to the gathering of data from which new maps can be made. For example, the *Mariner* and *Viking* probes that orbited Mars carried remote sensing devices—cameras, thermometers, and other instruments—that transmitted images and information about Mars back to earthbound scientists who used the data to produce maps of the Martian surface. Similarly, remote-sensing devices using sound waves have allowed scientists to map the landscape that lies beneath the Antarctic ice and at the bottom of the seas.

SEE ALSO
Satellites

FURTHER READING

Baker, D. James. *Planet Earth: The View from Space.* Cambridge: Harvard University Press, 1990.

National Air and Space Museum and Smithsonian Institution. *Looking at Earth.* Washington, D.C.: Smithsonian Books, 1994.

Rennell, James

BRITISH MAPMAKER

- *Born: 1742, England*
- *Died: 1830*

JAMES RENNELL went to India as a midshipman on a Royal Navy ship at the age of 18. He then went to work as a chart maker for the British East India Company. When he was only 21, he was named surveyor-general of Bengal, a large province in northeastern India; he wrote to a friend, "I was never so surprised in my life."

Rennell spent 13 years surveying Bengal and then returned to London to work on his maps. His *Bengal Atlas* was issued in 1779 and was the first reliable atlas of any part of India. Rennell also prepared the first relatively accurate map of the whole Indian subcontinent. It was published in 1782 and remained the standard map of India until the mid-19th century, when it was replaced by the maps produced by the Great Trigonometrical Survey of India. Throughout his life Rennell collected maps and information about India. He also served as an adviser to the African Association.

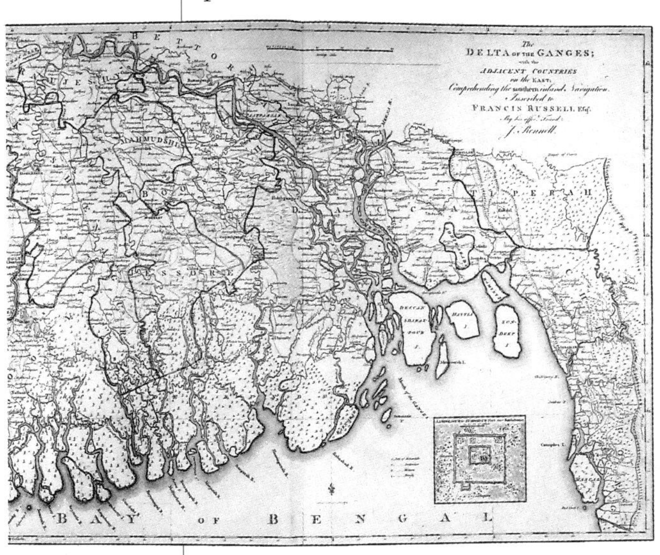

Rennell's Bengal Atlas *included this map of the delta of the Ganges River in what is today the nation of Bangladesh.*

SEE ALSO

African Association; Asia, mapping of; Great Trigonometrical Survey of India

Rhumb line

SEE Portolan

Ricci, Matteo

ITALIAN MISSIONARY AND MAPMAKER

- *Born: Oct. 6, 1552, Macerata, Italy*
- *Died: May 11, 1610, Beijing, China*

MATTEO RICCI (or Father Ricci, as he is sometimes called) joined the Society of Jesus, the Jesuit order of Roman Catholic scholars and missionaries, at the age of 19. When he was 30, he was sent to the Jesuit mission at Macao, a port on the south coast of China—one of the very few places where Europeans were allowed into China. Determined and diplomatic, Ricci gained permission both from his Jesuit superiors and from the Chinese officials to open a mission in the interior of China. He kept moving north, into the heart of the country, until by 1601 he was established in the imperial capital of Beijing, where he remained until his death.

Ricci earned the admiration of the Chinese by showing respect for their customs; for example, he scandalized other missionaries by dressing in the traditional robe and cap of a Chinese scholar. He also learned to speak, read, and write

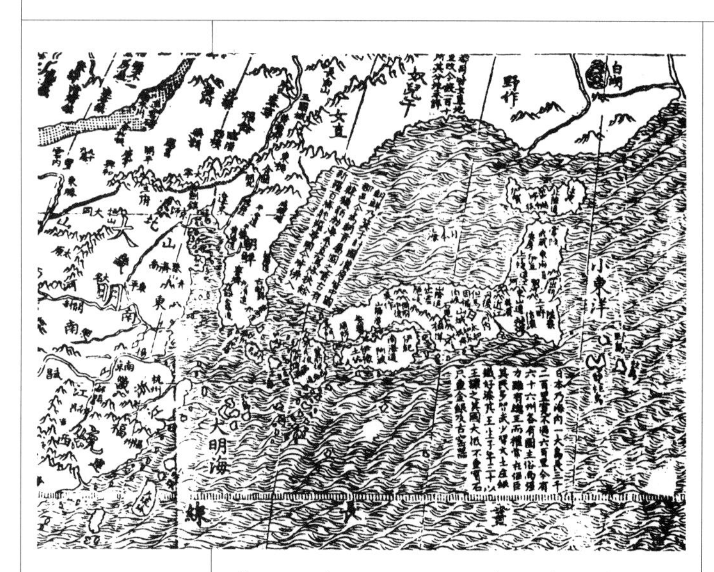

Northeast China, Japan, and Korea, as shown on the world map that Ricci prepared for his Chinese hosts.

fluent Chinese. Ricci is thought to have converted about 2,000 high-ranking Chinese officials to Christianity. But religion was not his only concern during his years in China. Ricci was a skilled mathematician and was interested in history and geography. He played a key role in educating the Europeans about China and the Chinese about the rest of the world—mostly through maps.

The medieval Chinese regarded their country as the center of the world and the people of other countries as uncultured barbarians. To prove that Europeans were people of culture and learning, Ricci showed the Chinese clocks, paintings, and astronomical instruments. He also showed them maps of the world outside China, and he drew maps that showed how China was related to the rest of the world. One world map that he drew offended his hosts, who were displeased that China appeared so small. Ricci tactfully redrew the map to place the Middle Kingdom, as China was called, closer to the center of the map. Named "The Great Map of Ten Thousand Countries" in the flowery language of the Chinese court, this map was printed from wood blocks onto strips of paper that were then glued to silk. The finished map measured 6 feet high by 12 feet wide (1.8 by 3.7 meters). Ricci filled it with detailed notes in Chinese and presented it to the emperor; it is still preserved in China. But the wealth of geographic knowledge contained in this map was not confined to the emperor's library. Ricci also sent copies of it to Europe.

Ricci used Chinese maps of Asia as sources for making his world map, which reflected what he had learned from the Chinese cartographers, including the fact that Korea was a peninsula, not an island, and the proper location of Japan. When Ricci's maps reached Europe, this new geographic information gradually began making its way onto the world maps made by Europe's cartographers. Ricci's geographic scholarship had given Europeans their most accurate picture yet of Asia.

SEE ALSO

Asia, mapping of

FURTHER READING

Spence, Jonathan. *The Memory Palace of Matteo Ricci.* New York: Viking, 1984.

Road maps

ROAD MAPS are one of the most common kinds of maps. Nearly everyone has used a road map at one time or another. In many a car, the glove compartment or side pocket is full of maps.

Road maps have a long history. Nearly 2,000 years ago, the Romans were making maps of their elaborate system of highways. In the Middle Ages, itineraries, or route maps, were produced to guide pilgrims to holy sites. Some medieval itineraries are long and narrow, similar to the strip road maps of today. The British mapmaker John Ogilby perfected the format of the strip map when he made route maps for travelers in Britain in the late 17th century.

Modern road maps date from the early 20th century, when the automobile came into use. Many road maps have been published and distributed by businesses connected with the automobile industry, such as oil companies and tire makers. In the United States, road maps are also distributed by state government agencies, as well as by the American Automobile Association (AAA). A similar service is provided by the Automobile Association (AA) in Great Britain, the Canadian Automobile Association (CAA) in Canada, the Australian Automobile Association (AAA) in Australia, the Automobile Association of Upper India (AAUI) and the Automobile Association of Southern India (AASI) in India, the New Zealand Automobile Association (NZAA) in New Zealand, and similar organizations in many other countries.

Road maps—which require frequent revision to keep up with the openings and closings of roads—are often gathered into atlases, with a page or two for each state and separate road and street maps for major cities. Today's road maps are made for drivers. They indicate whether a road is a toll road, a highway, a state road, a county road, or an ordinary street. They often contain other information of interest to travelers, such as the location of tourist attractions, national and state parks, and campgrounds. Symbols may be used to show features such as hospitals and airports, and the map may include charts or mileage figures to help motorists determine the distances between points.

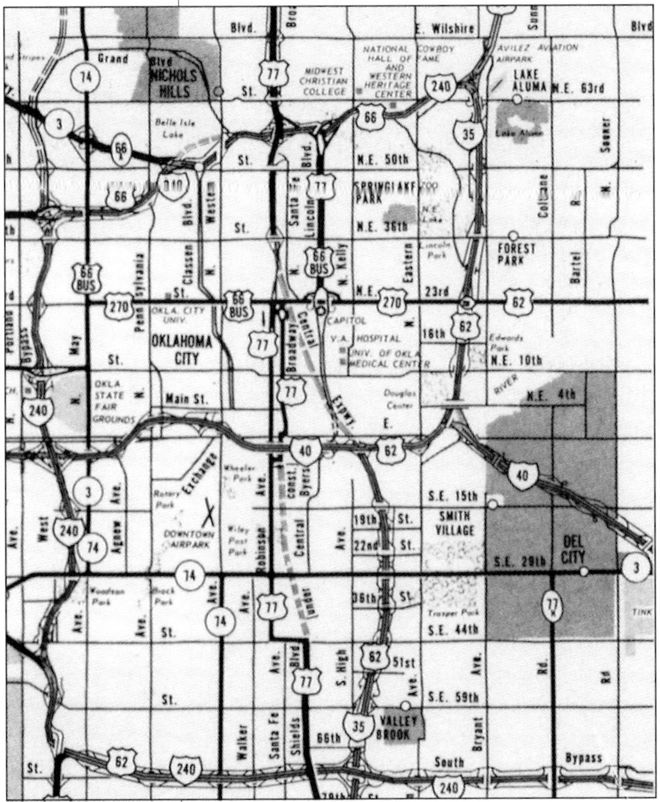

Junctions between major highways and smaller streets and roads are a key feature of modern road maps, such as this map of Oklahoma City.

SEE ALSO

Ogilby, John

Robinson projection

SEE Projections

Rogers, Woodes

ENGLISH PRIVATEER

- *Born: about 1679*
- *Died: July 16, 1732, Nassau, Bahamas*

WOODES ROGERS was a privateer—a pirate who operated with the permission of the English crown and preyed upon the ships of countries that were at war with England. A group of merchants in Bristol, England, who had lost ships to foreign pirates and privateers hired Rogers to lead a privateering expedition on their behalf, and in the years 1708–11 he took his band of privateers around the world, looting Spanish settlements and Spanish ships.

One incident during this cruise resulted in a literary classic. In 1709

Rogers stopped at the tiny island of Mas à Tierra, in the Pacific Ocean off the coast of Chile. There he rescued a Scottish sailor named Alexander Selkirk, who had been marooned on the island by his shipmates five years earlier. Selkirk's story was one of the inspirations for the 1719 book *Robinson Crusoe,* a classic tale of shipwreck and survival by the English novelist Daniel Defoe.

Another incident that occurred during Rogers's round-the-world journey is an interesting indication of the value of maps. Among the treasures that he captured were some Spanish maps and charts. At the time, each country—and sometimes each sea captain—guarded sea charts very carefully. A country's wealth and power might well depend on its secret trade routes and its knowledge of ports and harbors. Ships' logs and chartbooks were often weighted or kept in metal boxes so that they could be thrown overboard, sacrificed to keep them out of enemy hands, in an emergency. But Rogers acquired some of these priceless charts, which were promptly engraved and published by the English mapmaker John Senex.

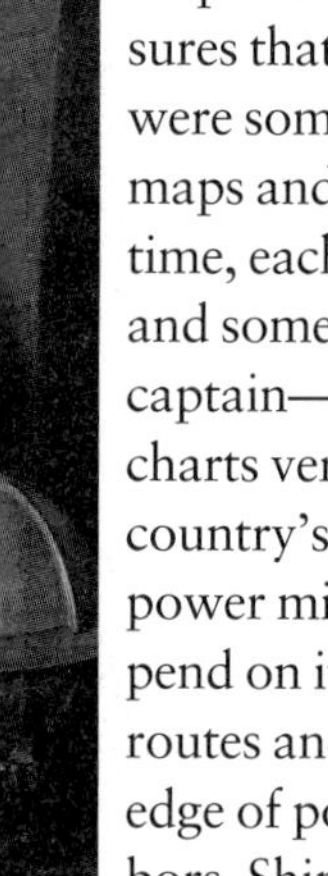

The stately governor of the Bahamas, several decades after he circled the globe as a pirate.

Later in life, Rogers became a lawman. In 1717 he was named governor of the Bahamas, which was then a haunt of pirates. He arrived in Nassau, the capital of the islands, to find that the town was the headquarters of more than 2,000 pirates, many of whom he forced to surrender.

FURTHER READING

Botting, Douglas, and Time-Life editors. *The Pirates.* Alexandria, Va.: Time-Life Books, 1978.

Roman mapmakers

SEE Ancient and medieval mapmakers

Ross, Sir James Clark

BRITISH POLAR EXPLORER

- *Born: Apr. 15, 1800, London, England*
- *Died: Apr. 3, 1862, Aylesbury, England*

JAMES CLARK ROSS entered the Royal Navy at the age of 12; by 18 he was serving as a junior officer on an expedition in search of the Northwest Passage. In 1819 he made a second voyage into the Canadian Arctic under the command of William Parry. Frozen in the ice with the onset of the bitter Arctic winter, Parry's two ships were unable to set sail for England until the autumn of 1821.

Throughout the 1820s, Ross continued to sail with Parry on naval expeditions in search of the passage. Each voyage probed a bit farther into the confusing maze of Arctic Canada. Then, in 1829, Ross served on a private expedition led by his uncle, John Ross. They took a steamship and a sailing vessel into the Arctic and became icebound for three years near the Boothia Peninsula, which juts north from the mainland of Canada.

During this time, James Clark Ross made an important geographic discovery. Scientists knew that the earth's magnetic pole, the point to which compass needles point, was not located precisely on the geographic north pole, but no one had yet pinpointed the location of the magnetic pole. Ross tracked the magnetic pole to its location on the west coast of the Boothia Peninsula (it has

Ikmalik and Apelagliu, an Inuit man and woman (center), added their local knowledge to one of Ross's Arctic maps.

shifted since Ross's day). Ross compared finding the pole to seeing "an object as conspicuous and mysterious as the fabled mountain of Sindbad . . . or a magnet as large as Mont Blanc." A member of his crew drew a lighthearted sketch of the great discovery, complete with the British flag proudly flying under the northern lights and a band of Inuit capering joyously, harpoons in hand. The Rosses and the other expedition members made their way home by abandoning their ships and hitching a ride with a passing whaler.

James Clark Ross's achievement in locating the magnetic pole was recognized by scientists and geographers. He was assigned to make a magnetic survey of Great Britain, measuring the intensity of the magnetic field in many locations around the country; such surveys often reveal variations in local magnetism, which may indicate useful deposits of mineral-bearing ores. Then, in 1839, he was given command of an expedition to the southern polar region. Antarctica was as yet unknown. Earlier navigators had sighted islands and bits of coastline in the Antarctic region, but no one was certain whether there was a continent at the south pole, or only empty sea. Ross was told to find out.

Learning of Britain's plans, both France and the United States also organized expeditions to the Antarctic. The Americans and the French, in fact, made the first recorded sightings of the Antarctic landmass. But Ross managed to sail farther south than the other ships. He pushed through a screen of icebergs and found himself in the large gulf that today is called the Ross Sea. Geographers were astonished by his description of the volcanoes along its coastline and the immense slab of floating ice now known as the Ross Ice Shelf. Ross went farther south than any explorer had gone up to that time, setting a record that would not be broken for 60 years. His book *A Voyage of Discovery and Research in the Southern and Antarctic Regions* (1847) contained the most detailed maps of Antarctica that had yet appeared.

SEE ALSO

Antarctica, mapping of; Arctic, mapping of; Northwest Passage; Parry, Sir William Edward; Poles

FURTHER READING

Berton, Pierre. *The Arctic Grail: The Quest for the North West Passage and the North Pole, 1818–1909.* New York: Penguin, 1988.

Maxtone-Graham, John. *Safe Return Doubtful: The Heroic Age of Polar Exploration.* New York: Scribners, 1988.

Royal Geographical Society

THE ROYAL Geographical Society (RGS) was founded in 1830 under the name Geographical Society of London. In 1831 the new society absorbed the African Association, which had been formed in 1788 to promote exploration. The RGS took its present name in 1859.

The RGS was founded to increase geographic knowledge, and it has done this in several ways. Since its early days, it has given money and other support to explorers, including David Livingstone, Richard Francis Burton, John Hanning Speke, John Franklin, Ernest Shackleton, and Robert Falcon Scott; it has also helped to publicize their discoveries. The society makes geographic knowledge available to the public through lectures and publications and by encouraging the teaching of geography in universities. Its library and map collection in London are open to the public.

SEE ALSO

African Association

FURTHER READING

Cameron, Ian. *To the Farthest Ends of the Earth: 150 Years of World Exploration by the Royal Geographical Society.* New York: Dutton, 1980.

Rutter

SEE Portolan

Sanson family

17TH-CENTURY FRENCH GEOGRAPHERS AND CARTOGRAPHERS

THE SANSON dynasty of geographers and mapmakers began with Nicolas Sanson, who was born in 1600 in Abbeville, France. Sanson studied the writings of ancient geographers and historians. He began drawing maps to illustrate his studies, then publishing the maps to earn an income to support his wife and children. In 1627, plagued with debts, he moved to Paris. His only possession was a map he had drawn of ancient Gaul, as France had been called in the days of the Roman Empire. This map happened to catch the attention of Cardinal Richelieu, an adviser to King Louis XIV, who got Sanson a job tutoring the king in geography and gave him a handsome salary. Sanson continued to make maps as a hobby, and by the time of his death he had published maps of the world and of Asia, Africa, and the Americas.

Sanson's three sons, trained by him, also became geographers and mapmakers. Nicolas (who died in 1648) was the eldest; he produced a map of the Russian czar's estates. Guillaume (who died in 1703) and Adrian (who died in 1708) continued the Sanson map business after their father's death, and Guillaume inherited his father's title of royal geographer.

Nicolas Sanson the elder is considered the founder of the French school of geography and mapmaking, not only because of his own influence and that of his sons but because he taught geography and cartography to others, including the prominent L'Isle family.

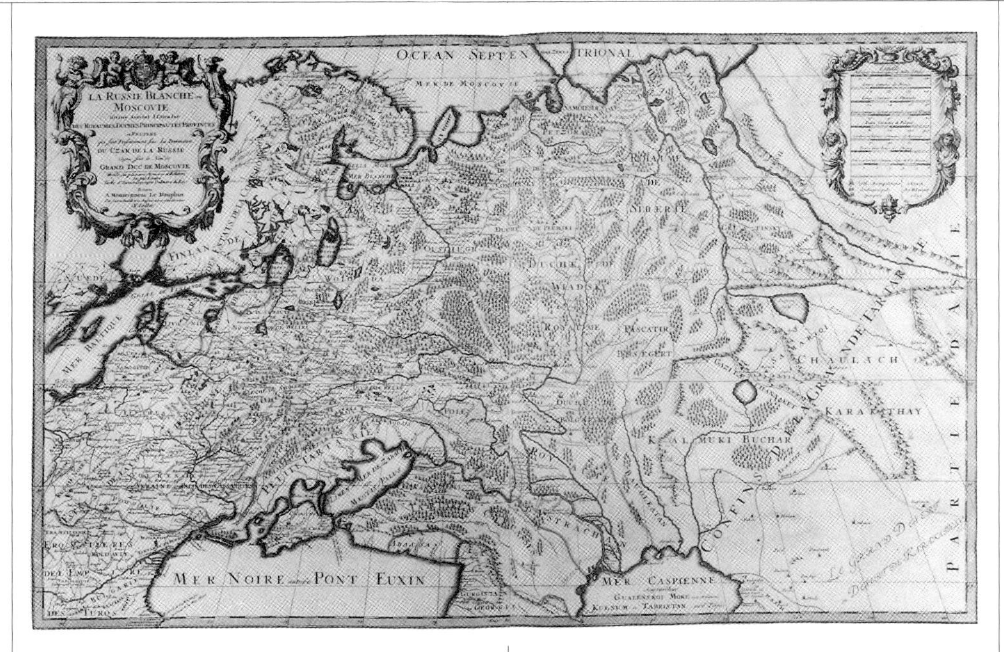

Nicolas Sanson the younger's map of Russia. The Baltic Sea is at the left, the Black and Caspian seas at the bottom. The map identifies the estates of the Russian royal family and nobility.

SEE ALSO
French mapmakers; L'Isle family

Satellites

SATELLITES ARE objects placed in orbit around the earth that are designed to collect and transmit information back to earth. They play an enormous role in modern mapmaking. The Soviet Union launched the first satellite in 1957, and since that time satellites have been used as vantage points from which to study the earth. Military purposes were paramount at first, but scientists soon realized that satellites offered an unparalleled view of the world for mapmaking purposes.

The most important contribution of satellites has not been mapping in the traditional sense: outlining unknown territory. Most of the earth's outlines were fairly well known when the space age began, although satellite data have corrected many errors and made traditional maps more accurate. But the real benefit of satellites is that they can provide up-to-the-minute, constantly changing maps of what is happening on earth: the locations of fires, storm systems, and the like. Satellites carry cameras and other remote-sensing equipment that sees what is happening and then sends the information back to earth in the form of radio signals. Satellites can also serve as fixed points and location beacons for surveying and navigating.

The first weather satellite was launched in 1960. Then, when U.S. astronaut Gordon Cooper said after a 1963 space flight that he had seen individual vehicles and buildings from space (a claim that was first dismissed as a hallucination but later verified), scientists realized that they could survey the use—and abuse—of resources from space. In 1972, the first of five U.S. satellites called *Landsat* went aloft to monitor geologic formations, land use, and vegetation patterns. The American *Seasat* (1978) and *Geosat* (1985) satellites performed similar functions for the sea. Other nations have also sent remote-sensing devices into space. The most powerful satellite in orbit in the early 1990s was *SPOT,* owned by a French company: it can detect objects only 33 feet (10 meters)

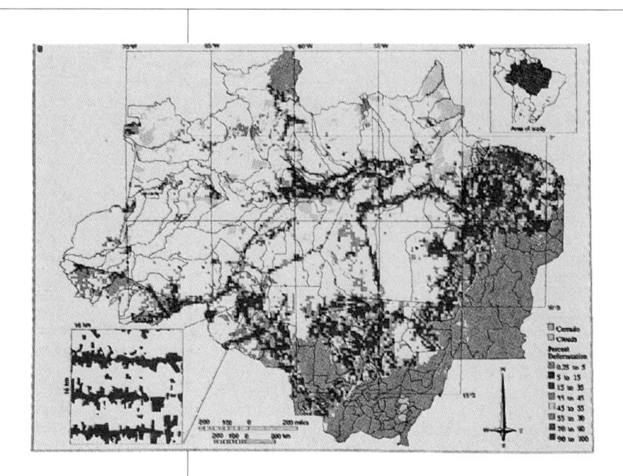

This image, produced in 1988 from data acquired by the Landsat *satellites shows deforestation in the Brazilian Amazon. Deforestation has been most extensive in the bright red areas.*

wide. The United States is planning to launch a satellite called the Earth Observing System in the late 1990s. Each month it will beam down so much data that, if it were printed in book form, it would fill millions of volumes.

Early remote-sensing devices were ordinary cameras taken into space. Conventional photographs are still taken from space, but now much satellite data comes from scanners that "read" radiation and transfer their readings into bits of data that a computer can arrange into a visual format. The result looks like a photograph but is computer generated. The term *satellite imagery* has been coined to describe these pictures.

Data from satellites can be shown on two kinds of maps. One type is the line map or traditional drafted map (although today the cartographer is just as likely to draw at a computer terminal as at a drafting table). These maps may be based on satellite imagery, but they are not themselves images. The other kind of map is the photomap, a photo or satellite image with map information—boundaries, coordinates, and the like—superimposed on it.

Satellite data are used in a number of ways. Information about the earth's magnetic field has enabled geodesists to map the earth's shape and contours with great precision. Nearly all new topographic maps are based on photogrammetric data from space. Mining, oil, and timber companies use *Landsat* maps to locate resources. Low-altitude earth resources satellites provide information about subjects as varied as the spread of plant diseases and the flow of water pollution. Higher-altitude environmental satellites monitor the earth's atmosphere for weather reports and scientific studies. Maps made from these studies are used to explain global issues such as desertification, deforestation, and ozone deterioration. Many experts believe that the combination of satellites and maps has given us a priceless tool in the crucial struggle to preserve and protect the earth's environment.

SEE ALSO

Photogrammetry; Remote sensing; Space mapping; Undersea mapping

FURTHER READING

Hall, Stephen. *Mapping the Next Millennium: The Discovery of New Geographies.* New York: Random House, 1992.

Images of the World: An Atlas of Satellite Imagery and Maps. Chicago: Rand McNally, 1983.

Stewart, Doug. "Eyes in Orbit Keep Tabs on the World in Unexpected Ways." *Smithsonian,* December 1988, pp. 70–81.

Wilford, John Noble. *The Mapmakers.* New York: Knopf, 1981.

Saxton, Christopher

ENGLISH MAPMAKER

- *Born: about 1542*
- *Died: 1606*

CHRISTOPHER SAXTON, a surveyor and draftsman, was the first person to make a systematic survey of England and Wales. He published maps of the individual counties from 1574 to 1579; a collection of all 35 surveys was published in 1579, with a title page showing Queen Elizabeth I flanked by Strabo and

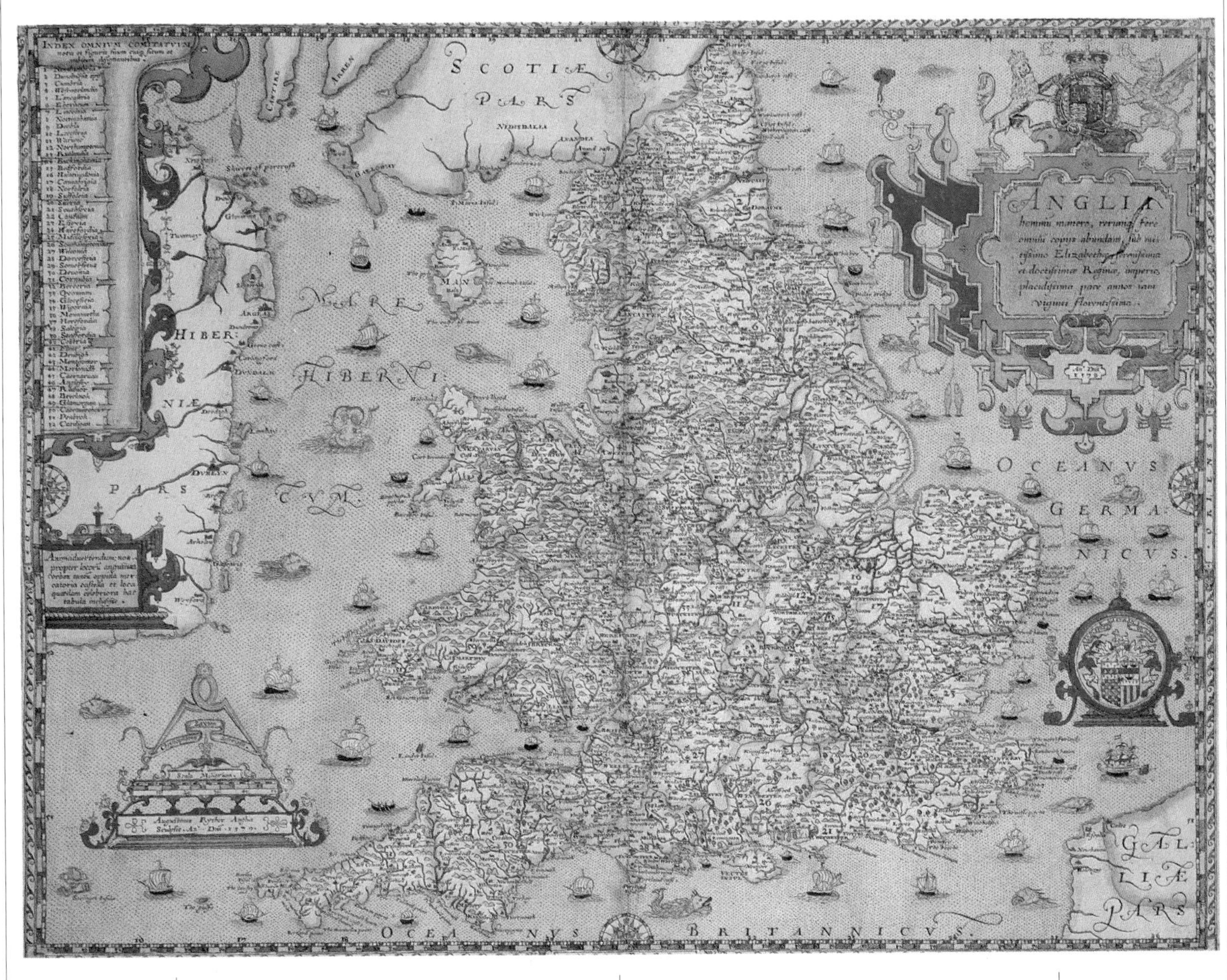

Some historians of cartography regard Saxton's map of England and Wales as the finest British map of the 16th century. It was copied by many later mapmakers.

Ptolemy, the leading geographers of the ancient world.

Elizabeth deserved this honor, for it was her patronage, along with that of her councillors, that had allowed Saxton to complete the enormous task of surveying England and Wales in only five years. The queen's council issued a special order on Saxton's behalf, requiring mayors and other local officials throughout the land "to see him conducted unto any towre, castle, high place or hill to view that country . . . and that at his departure from any place that he hath taken the view of, the said towne do set forth a horseman that can speak both Welsh and English to safe-conduct him to the market-towne."

In 1583 Saxton produced a single map of England and Wales. This, as well as his county surveys, served as the basis for many later maps and atlases of Great Britain.

SEE ALSO
British mapmakers

Scale

A MAP'S scale is the relationship between a unit of distance on the map and a unit of distance on the earth's surface. Scale may

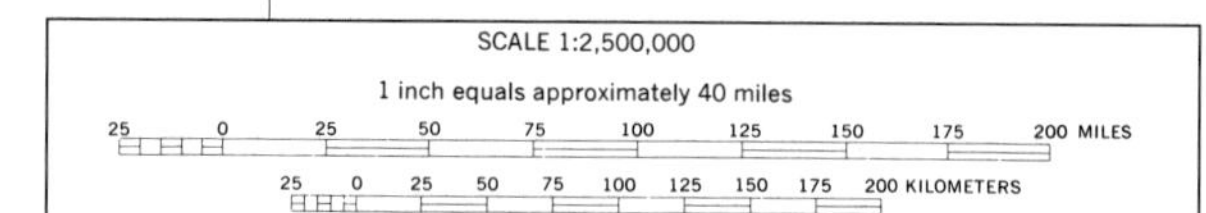

be expressed in the form of a representative fraction, as in 1/100,000th, more commonly written as 1:100,000. This means that one inch on the map equals 100,000 inches (1.578 miles) in the real world; 1 centimeter on the map equals 100,000 centimeters (1 kilometer).

A map's scale may also be shown graphically, using a bar like a small ruler printed somewhere on the map. The bar is marked into segments, and the corresponding distance in miles or kilometers (or both) is printed at various points on the bar. Using all or part of the bar as a guide, you can then measure distance on the map. Certain maps, including Mercator projections, have "sliding scales"—scales that vary from one part of the map to the next—because the ratio between distance on the map and distance in the world varies at different latitudes on these projections.

A third way of presenting a map's scale is in text form. Written scales such as "One inch equals 1,200 miles" appear on many maps.

A map can be drawn to any scale, although a 1:1 map would be rather impractical because it would be the same size as the terrain it showed. Cartographers must select the right scale for each map's purpose, and users of maps must be aware of the drawbacks and virtues of different scales. The larger the scale, the larger the detail but the smaller the area shown in a given space. For example, a map on a scale of 1:50,000 will offer a much more detailed look at the terrain than a 1:1,000,000 map, but the latter—the smaller-scaled map—can show a much larger area. A world map on an atlas page may have a scale of 1:75,000,000, whereas a close-up map of Istanbul later in that same atlas might have a scale of 1:300,000.

FURTHER READING

Greenhood, David. *Mapping*. Chicago: University of Chicago Press, 1964.

Schematic maps

TECHNICAL DIAGRAMS are sometimes called schematic maps. The word *schematics* refers to certain kinds of technical information, such as the layout of an electrical circuit. Such schematics are a kind of simplified map, showing what wire goes where and how it relates to the other wires. If you have ever followed the diagram in an instruction booklet to hook up the wiring of a stereo or videocassette recorder, you have used a schematic map.

Scott, Robert Falcon

BRITISH EXPLORER OF ANTARCTICA

- *Born: June 6, 1868, Devonport, England*
- *Died: on or about March 27, 1912, Antarctica*

ROBERT FALCON SCOTT, an officer in the Royal Navy, was chosen by Great Britain's Royal Geographical Society to lead an expedition to Antarctica in 1901; Ernest Shackleton, who later led his own expedition to Antarctica, was one of his companions.

Scott's expedition gathered a great deal of scientific and geographic information about the southern polar continent. Among other achievements, Scott made the first balloon ascent in Antarctica and reached a new farthest-south point. Upon returning to England in 1903, Scott found himself a popular hero. He became fired with the desire to be the first person to reach the south pole. Shackleton tried and failed to reach the pole, and then in 1910 Scott set out for Antarctica to make his own attempt. At the same time, the Norwegian explorer Roald Amundsen was also headed for the pole.

Although the valor and determination of Scott and his men are unquestioned, neither their ship nor their equipment was as efficient as Amundsen's. When Scott and his exhausted crew finally reached the pole on January 17, 1912, after a long struggle over the ice, they found that the Norwegians had beaten them to this geographic goal. Scott and his companions died of starvation and exhaustion on the way back to their base.

SEE ALSO

Amundsen, Roald; Antarctica, mapping of; Shackleton, Ernest Henry

FURTHER READING

Huntford, Roland. *Scott and Amundsen: The Race to the South Pole.* New York: Putnam, 1980.

Sea charts

SEE Chart; Portolan

Sextant

A SEXTANT is an instrument used for measuring the angle of the sun or a star above the horizon; it is mainly used for navigation at sea. The altitude of the sun or star is the basis for a calculation of latitude.

Invented in 1757, the sextant is a variation of the quadrant, a navigational instrument that in the early 18th century replaced the cross-staff as the mariner's chief means of determining latitude. The quadrant used a curved measuring scale that covered 90 degrees, or one-quarter of a complete circle. The sextant was a bit smaller and therefore easier to use; it had a scale that covered 60 degrees, or one-sixth of a circle, and so could measure any angle up to 60 degrees.

SEE ALSO

Cross-staff; Quadrant

Shackleton, Ernest Henry

BRITISH EXPLORER OF THE ANTARCTIC

- *Born: Feb. 15, 1874, Kilkee, Ireland*
- *Died: Jan. 5, 1922, South Georgia Island*

ERNEST SHACKLETON served as a junior officer with Robert Falcon Scott's first Antarctic expedition in 1901–3. In 1907–9 he led his own expedition and discovered several important geographic features in Antarctica, including the Beardmore Glacier. Shackleton was the first explorer to climb off the coastal ice shelf and up onto the 10,000-foot-high (3,048 meters) Antarctic Plateau, significantly extending the world's knowledge of Antarctica's topography. He got within 97 miles (156 kilometers) of the south pole before a shortage of food made him turn back.

In 1914–17, Shackleton returned to Antarctica with an ambitious plan: to make the first crossing of the southern continent. But when his ship, the *Endurance,* was crushed by ice, he was forced instead to make a heroic sea journey. He and his crew drifted for five months on an iceberg before reaching an island.

From there Shackleton and five others sailed 800 miles (1,287 kilometers) in a small open boat to South Georgia Island, where he was able to get help for his stranded crew. This epic voyage is considered one of the world's most remarkable stories of survival at sea. It was the first known passage in those waters.

Shackleton planned a third expedition, but on his way to Antarctica he died suddenly of a heart attack in South Georgia Island.

SEE ALSO

Antarctica, mapping of; Scott, Robert Falcon

FURTHER READING

Maxtone-Graham, John. *Safe Return Doubtful: The Heroic Age of Polar Exploration.* New York: Scribners, 1988.

Worsley, F. A. *Shackleton's Boat Journey.* New York: Norton, 1977.

Skalholt map

SEE Viking explorers

Sketch maps

A SKETCH MAP is drawn, or sketched, by hand. Many of the basic features of ordinary maps are simplified or even omitted on sketch maps, which are meant to serve specific, limited purposes. The informal maps we make and use in everyday life are sketch maps. For example, when you draw a quick map to guide a friend to your house, you are probably making a sketch map. Unless you are a skilled and conscientious cartographer, you probably do not bother with such details as exact compass bearings, longitude and latitude, and scale—nor are such things necessary. You can simply show your friend what road to follow, where to turn, and what landmarks to look out for. Your sketch map sacrifices strict geographic accuracy in the interest of being easy to use for a particular purpose.

Since the 19th century, sketch maps have been used for things other than routes and directions. The German explorer and scientist Alexander von Humboldt made sketch maps of subjects as varied as the patterns of vegetation on the sides of South American volcanoes and the layout of Aztec ruins in Mexico. His hand-drawn sketch maps were later made into engravings to illustrate his books, inspiring many later writers to use sketch maps to convey similar information.

Smith, Jedediah Strong

AMERICAN FUR TRADER AND EXPLORER

- *Born: Jan. 6, 1799, Bainbridge, New York*
- *Died: May 27, 1831, on the Santa Fe Trail*

JEDEDIAH STRONG SMITH was born in New York State, but by the age of 23 he had made his way westward to the American frontier. He joined the fur business, first as a trapper and guide, and later as part owner of a fur trading company. He is best remembered, however, for a journey he undertook in 1826–27, partly in search of new beaver trapping territories and partly out of geographic curiosity. He led a party of fellow trappers south from the Great Salt Lake of Utah through the Mojave Desert and into southern California; they were the first people besides Native Americans to reach California overland from the east. Smith then traversed the Sierra Nevada range and made the first known crossing of the Great Basin in present-day Nevada.

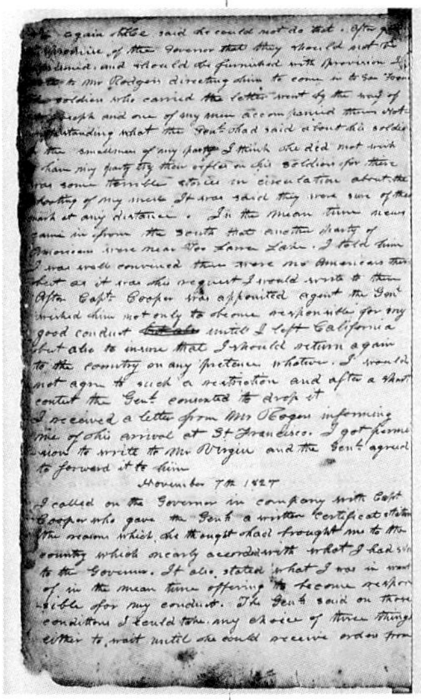

Smith was one of the foremost trailblazers of the American West. At the time of his death, he knew as much about the geography of the region as anyone who had ever lived. Many of his papers and all of his maps were scattered and lost when he died, but enough of his knowledge had been communicated—in the form of letters he wrote and copies of his maps made by his fellow fur traders and others—to bring about a great improvement in the accuracy of western maps.

Smith's California journal describes his grueling trip across the scorched, barren Mojave Desert in southern California.

SEE ALSO

Americas, mapping of

FURTHER READING

Allen, John L. *Jedediah Smith and the Mountain Men of the American West.* New York: Chelsea House, 1991.

Stefoff, Rebecca. *Accidental Explorers: Surprises and Side Trips in the History of Discovery.* New York: Oxford University Press, 1992.

Solinus, Gaius Julius

ROMAN GRAMMARIAN

• *Active: 3rd century* A.D.

ON CERTAIN old maps and in geography books, capering about on the fringes of the known world are all manner of remarkable people and animals: men with ears so enormously long that they can wrap those ears around themselves like clothes, or with a single leg ending in a foot big enough to serve as an umbrella when they are lying on their backs, or with a single eye in the middle of their foreheads; man-eating birds called griffons, lions with the tails of snakes, and ants as big as dogs. These and other curious cartographic fantasies are the legacy of Gaius Julius Solinus.

Solinus was a Roman of the 3rd century whose specialty was the study of grammar. Today, however, he is remembered for a book of tall tales called *Collection of Memorable Things.* It was a compilation of material borrowed from other writers, gleaned from travelers' stories, and in some cases simply made up for the fun of it. In short, the book was a treasure trove of geographic myth and misinformation, with a few nuggets of truth carelessly scattered around in it. But it was immensely popular, especially after it was revised and republished under the title *Polyhistor* (meaning "many stories") in the 6th century.

In later centuries, Solinus's wild tales took on an aura of geographic reality when mapmakers such as Sebastian Münster and Théodore de Bry, looking for something lively to fill up the blank spaces on their maps, began drawing the dog-headed people, giant ants, and other curiosities that Solinus had described. These creations appeared on many maps of the Middle Ages, and they lingered on in the outlying suburbs of cartographic respectability until the 18th century.

Two of Solinus's monstrous races.

For many people, part of the charm and interest of old maps lies in these pictures of remarkable monsters and bizarre beings. Yet some 20th-century historians, including J. B. Harley, a leading scholar of cartography who taught geography at the University of Wisconsin, have pointed out that by picturing the *terra incognita* (the "unknown world" beyond the borders of Europe) as peopled by monsters, mapmakers contributed to Europeans' ideas of superiority.

Such notions encouraged European explorers and colonists to view the native peoples they encountered as savages—with tragic results for the native peoples. Stephen S. Hall was referring to Solinus's creations when he wrote in *Mapping the Next Millennium: The Discovery of New Geographies,* "For centuries the relentless, vivid, and visually powerful lesson was that terra incognita is a place inhabited without exception by freaks and monsters—a powerful, thousand-year lesson taken to heart by the time of the great discoveries in the fifteenth and sixteenth centuries."

Sonar

SONAR (the name stands for **so**und **na**vigation **r**anging) is a method of using sonic and supersonic vibrations to measure distances underwater. Also called echo sounding, sonar was developed in 1914 by the U.S. inventor Richard A. Fessenden; French scientists were also working on a method of echo sounding at around the same time. Sonar equipment consists of an oscillator, which is an electric instrument that broadcasts a sound wave that travels through water, and a hydrophone, or underwater microphone. The sound wave sent out by the oscillator strikes an object under the sea and emits a sound that is picked up by the hydrophone. By timing the period that elapses between the sending of the signal and the reception of the echo, the operator can tell how far away the underwater object is.

Sonar was immediately put to use to protect ships from obstacles such as hidden rock reefs and icebergs; it was used during World War I (1914–18) and World War II (1939–45) to hunt for enemy submarines. But it was also applied to the study and mapping of the seafloor. French, German, American, and British research ships began using sonar to survey first the Atlantic and then the Pacific Ocean. Sonar technology was refined; pens were attached to the equipment so that sonar readings were recorded in visual form, as graphs. Wide-beam sonar was developed, allowing scientists to scan a strip of seafloor several miles wide at one time. Sonar contributed greatly to the picture of the undersea world that began to emerge in the 1950s, and it is still one of the principal tools in the exploration and mapping of the ocean floors.

SEE ALSO
Undersea mapping

Space mapping

PEOPLE HAVE been studying the heavens since ancient times and mapping them for thousands of years. Many early cultures, including the Babylonians and the Mayas, knew the night skies well.

The heavens were filled with mythological figures in Albrecht Dürer's star charts, the first such charts to be printed.

The 2nd-century Greek astronomer Ptolemy cataloged more than a thousand stars in his *Almagest,* and later astronomers added to his list.

The first printed star charts were made by the German printmaker Albrecht Dürer in 1515; the first atlas of star maps was published in 1603. Most of the great 16th- and 17th-century atlases, such as those of Willem Janszoon Blaeu, included one or more star maps. These were usually ornamented with the mythological figures for whom the constellations are named. Mapmakers also produced celestial globes; Blaeu and many other cartographers made pairs of globes, in which one globe was terrestrial and the other celestial. The high point of decorative star maps was reached in an atlas called *Harmonica Macrocosmica* that was issued in 1660 by Andreas Cellarius. Sumptuously painted and highlighted with gold, Cellarius's charts showed the stars and the figures of the constellations. By the 18th century, star maps were becoming less mythological and more astronomical, and by the 19th century most atlases included at least two star charts, one for the Northern Hemisphere and one for the Southern Hemisphere.

Astronomers' star maps today list hundreds of thousands of stars, most of them far too faint to be seen with the naked eye or even the amateur telescope. Many maps and charts are available, however, that show the night sky as it appears to the eye or to the hobbyist with a small telescope or a pair of binoculars. Some of these are planispheres, wheels that can be turned to show the star patterns for any time of night and any day of the year.

Lunar mapping When the Italian scientist Galileo Galilei first turned his telescope on the moon in 1609, he began a process of lunar mapping that continues today. The moon's features were named by Giovanni Riccioli and appeared on his 1651 map of the moon, but the principal early mapper of the lunar surface was Johannes Hevelius, who published the first lunar atlas around the same time. Other maps followed, each adding a bit more detail, and in the 19th century astronomers measured craters to serve as control points for a grid of latitude and longitude lines on the moon's surface.

In the late 1950s, the U.S. Air Force and the National Aeronautics and Space Administration (NASA) began producing topographic maps of the moon. Photogrammetric mapping, using images obtained by the lunar orbiter satellites between 1966 and 1977, produced even better maps of both sides of the moon; photos from the first satellite were processed into maps that were used during the first manned mission to the moon in 1969. The manned *Apollo* missions produced the best photographs

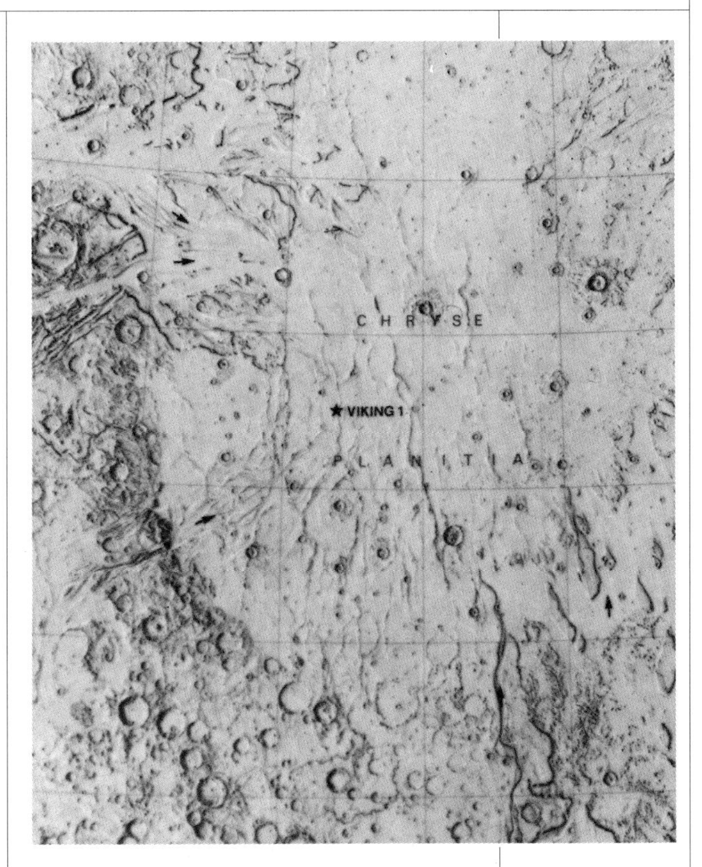

The Viking 1 *landing area on the Chryse Plain of Mars. The landing site is in the center, marked by a star. This photomap shows several channels that were carved by immense martian floods ages ago.*

ever, as well as data on the moon's chemistry, geology, and magnetic field; these were later made into topographic maps and a variety of specialized maps.

Planetary mapping Other parts of the solar system are being mapped from data sent back to earth by planetary landers and satellites. Photographs, geomagnetic readings, and radar scans have resulted in maps of Venus, Mars, and the outer planets and their moons. Mars is the best known of the planets; the *Mariner* and *Viking* probes of the 1960s and early 1970s sent back images that banished forever the century-old fantasy of canals and dying civilizations on Mars but introduced a new landscape of giant mountains, plunging crevasses, and broad plains covered with dried lava.

Cosmic mapping The mapping of the universe beyond our solar system is the work of astrophysicists as well as astronomers. From radio emissions and other forms of radiation, they are piecing together a map of the farthest reaches of space. They are mapping time as well as space, for the radiation they perceive on earth left its sources thousands, or millions, or even billions of years ago. In the mid-1980s, physicist Margaret Geller and astronomer John Huchra began an ambitious project: mapping the universe. Working at the Harvard University–Smithsonian Institution Center for Astrophysics, they have discovered that galaxies are distributed through the universe in vast, complex patterns. "This business of mapping the universe is part of our attempt to write our address on a very large scale," Geller said in 1994. "By studying this ancient light, we are reaching for an understanding of how the universe is made."

SEE ALSO

Satellites

FURTHER READING

Lewis, H. A. G., ed. *The Times Atlas of the Moon*. London: Times Publishing, 1969.

Wilford, John Noble. *The Mapmakers*. New York: Knopf, 1981.

Speed, John

BRITISH MAPMAKER

- *Born: 1552, Farndon, England*
- *Died: 1629*

JOHN SPEED was a historian who in 1603 began making maps of the counties of England and Wales to illustrate a history book he was writing. The maps, however, proved more popular and enduring than the book. Some of these maps were published separately, but in 1611 he issued a collection of them under the title *Theatre of Great Britain*. The atlas was reprinted many times until 1676. During Speed's lifetime, he corrected each edition, changing the spelling of place names and adding new information.

Speed's county maps were based in part on earlier county maps, such as those by Christopher Saxton. But Speed introduced features such as the coats of arms of local families and descriptions of the countryside on the back of each map. He was also the first to produce separate printed maps of the Isle of Wight, the Isle of Man, and the counties of Ireland.

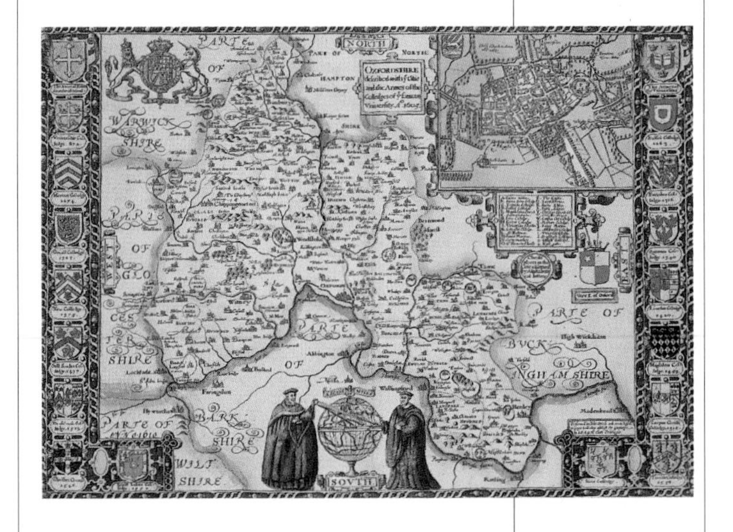

Speed's map of Oxfordshire, decorated with the shields of the colleges of Oxford University.

SEE ALSO

British mapmakers; Saxton, Christopher

Speke, John Hanning

BRITISH EXPLORER OF AFRICA

- *Born: May 3, 1827, Bideford, England*
- *Died: Sept. 15, 1864, Bath, England*

JOHN HANNING SPEKE accompanied Richard Francis Burton on two expeditions into East Africa before leading his own. In 1858, he and Burton were in the vicinity of Lake Tanganyika, searching for the source of the White Nile, the longer of the Nile River's two main branches. Speke, exploring northward while Burton was ill, sighted another large body of water, which he named Lake Victoria. Speke claimed that this lake was the source of the Nile, while Burton insisted that the river flowed out of Lake Tanganyika. Neither man could muster solid geographic evidence to prove his case, and they feuded publicly over the matter.

Speke (left) and Grant sought to answer age-old questions about the source of the Nile River.

The source of the Nile was one of the great geographic mysteries of the 19th century and aroused widespread public curiosity. In 1860, the Royal Geographical Society sent Speke back to Africa to settle the issue. With him was James Augustus Grant, a young army officer. The two men made a long, wearisome journey through a series of kingdoms in what is now Uganda before returning to England by way of Egypt. Speke was now certain that the Nile flowed out of Lake Victoria, but Burton still argued that he did not have sufficient proof. The two explorers were supposed to debate the question, but on the afternoon before the debate Speke died of a gunshot wound—probably an accident but just possibly suicide. In fact, however, Speke was perfectly right about the Nile source, although his evidence *was* incomplete. Today he is credited with having discovered the source of the Nile. His sketch map of the lakes and rivers of east central Africa was engraved by Keith Johnston, a Scottish cartographer, and published in Speke's 1863 book about his trip. It set a new standard for completeness and detail in maps of the African interior.

SEE ALSO

Africa, mapping of; Burton, Sir Richard Francis

FURTHER READING

Moorehead, Alan. *The White Nile.* New York: Harper & Row, 1960.

Stanley, Sir Henry Morton

BRITISH-AMERICAN JOURNALIST AND EXPLORER OF AFRICA

- *Born: Jan. 28, 1841, Denbigh, Wales*
- *Died: May 10, 1904, London, England*

RAISED IN an orphanage, Henry Morton Stanley fled his humble surroundings and eventually became a U.S. citizen, although he ended his days in England. In the 1860s he became a newspaper reporter, and in 1871 his employer, James Gordon Bennett of the *New York Herald,* sent him to Africa to look for David Livingstone, a Scottish missionary explorer who had not been heard from for a long time and was believed to be lost in Africa—perhaps even dead. Stanley located Livingstone on the shores of Lake Tanganyika in East Africa, and his account of their meeting, when Stanley uttered the greeting "Dr. Livingstone, I presume," made him famous all over the world.

Stanley's contributions to geography and mapmaking came after Livingstone's death in 1873. Returning to Africa at the head of a large expedition, Stanley explored Lake Victoria. Then he worked his way west across the continent, making the first descent by a European of the Congo River. The trip involved tremendous suffering and loss of life, but it placed the entire length of the Congo on the map for the first time.

Stanley's explorations filled in many of the blank spaces in the map of Africa's interior. They also opened the door to a century of change and confusion in the political map of Africa, as the nations of Europe scrambled to stake territorial claims and establish colonies in the regions he had traversed.

SEE ALSO

Africa, mapping of; Livingstone, David

FURTHER READING

Bierman, John. *Dark Safari: The Life Behind the Legacy of Henry Morton Stanley*. New York: Knopf, 1990.

Stefoff, Rebecca. *Accidental Explorers: Surprises and Side Trips in the History of Discovery*. New York: Oxford University Press, 1992.

Statistical maps

STATISTICAL MAPS are maps that arrange a particular set of data on a geographic framework. Population distribution maps, which show how people are spread across a landscape, are statistical maps; so are maps that show average family income in different places, or the proportion of rural to urban residents, or any other unit of information that can be gathered. Many statistical maps are compiled from censuses, or surveys of the population. In the United States, the Census Bureau produces maps that show the results of the census.

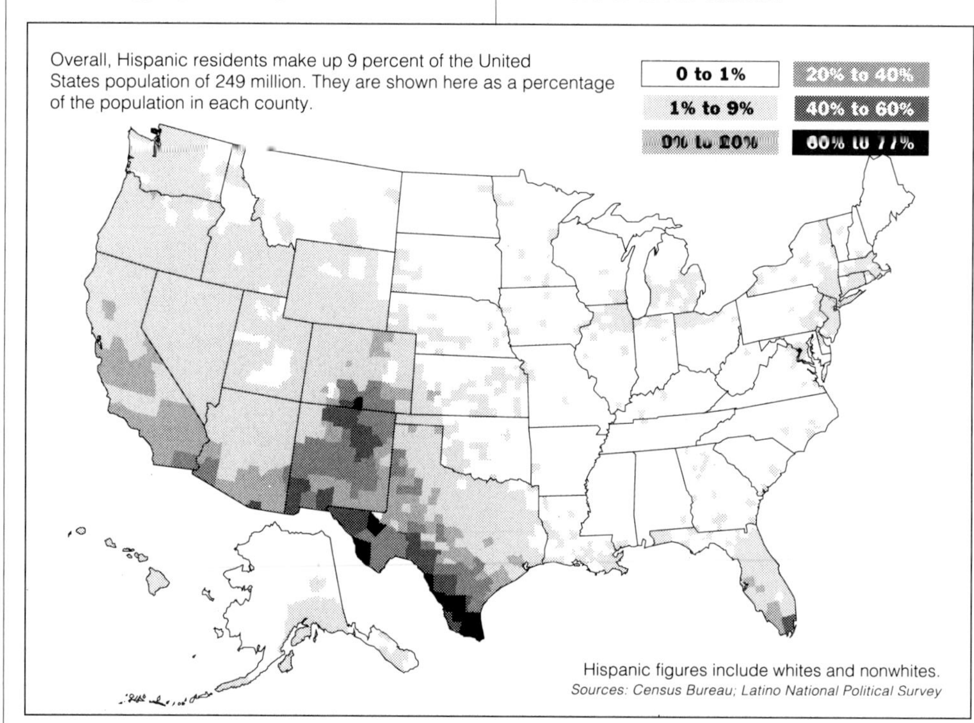

A 1992 map shows the distribution of people of Hispanic descent in the U.S. population.

Strabo

GREEK GEOGRAPHER

- *Active: 1st century* B.C.

STRABO WAS probably born around 63 B.C. in Pontus, a Greek colony near the Black Sea. He traveled widely, boasting that he had gone as far south as Ethiopia, as far east as Armenia, and as far west as Sardinia. At various times, he lived in Rome and Alexandria.

Strabo wrote a treatise on geography in 17 volumes. The first 2 volumes served as an introduction; 8 volumes dealt with Europe; 6 with Asia; and 1 with Africa. Strabo correctly pictured the earth as round, although he wrongly placed it at the center of the universe. Following the practice of earlier geographers, Strabo divided the world into five climate zones. Although he gave lip service to mathematics and astronomy, he was more interested in the history and physical description of the places he wrote about than in their precise locations. He popularized the notion of the *oikumene,* or inhabited part of the world, a known or knowable region that was surrounded by wilderness and ocean.

Strabo's geography was printed in many editions after about the 6th century. Its chief importance is the information it gives us about earlier scientists, such as Eratosthenes and Hipparchus, whose own works have vanished.

SEE ALSO
Ancient and medieval mapmakers

Strait of Anian

SEE Northwest Passage

Sturt, Charles

BRITISH EXPLORER OF AUSTRALIA

- *Born: Apr. 28, 1795, Bengal, India*
- *Died: June 16, 1869, Cheltenham, England*

BORN IN India to English parents, Charles Sturt went to school in England and then joined the British army, which sent him in 1827 to the British colony in Australia. At that time, the British colonists lived only in coastal settlements that formed a fringe around the outer edge of the island continent. Australia's interior was entirely unknown to the British. In 1828 Sturt began a series of exploring missions into that mysterious interior. Like many explorers, he used rivers as his guides into new country. He began by tracing the course of the Macquarie River in southern Australia and then turned his attention to the Murray-Darling river system, opening up new areas to farming and sheep ranching.

Sturt spent a few years in England and then returned to Australia in 1835. He shared a widespread theory about the still-unexplored interior of the continent. Large lakes, inland seas, and navigable rivers had been found in the interiors of Asia, Africa, and North and South America. Sturt was convinced that the center of Australia, too, must have an inland sea or waterway. In 1844 he led an expedition inland to answer what he called "the most important and the most interesting geographical problem in the world": the question of whether Australia's vast, unexplored interior was fertile and well-watered, as the settlers on the rim of the continent hoped. But instead of the rivers and meadows he hoped to find, Sturt discovered only "gloomy and burning deserts," desolate expanses now known as the Sturt and Simpson deserts. Ravaged by scurvy, the expedition returned to Adelaide, a city on Australia's south coast, in 1846. Sturt went back to England in 1853 and spent the rest of his life there. He had failed to find the inland sea he sought, but he had added important details to the maps of Australia made by later cartographers.

SEE ALSO
Australia, mapping of

Surveying

SURVEYING IS the science of determining the location and relative position of points on the earth's surface. It is the basis of all mapmaking. Maps are made either from surveys or from other maps, which in turn were based on surveys.

Surveyors use a combination of measuring instruments and mathematics to achieve results. Topographic surveys, from which many maps are made, are done through triangulation, the measurement of an interconnected set of triangles across the landscape. The mathematical relationships of angles in triangles allow the surveyor to start from a carefully measured baseline, whose latitude and longitude are known, and then move out from there in triangular steps. (See diagram on page 270.) The point of surveying is to cover the landscape with points whose position is known. The topography between those points can then be filled in with further surveying, and the positions plotted on a grid to produce a map.

A surveyor needs two basic instruments: one to measure distance and one to measure angles between lines of sight. The first surveyors measured distance by pacing. Then they discovered that by pushing a wheel along the ground and recording the number of times the wheel turned, they could calculate distance. Some 18th-century surveyors had special carriages in which they could ride while their wheels measured the journey.

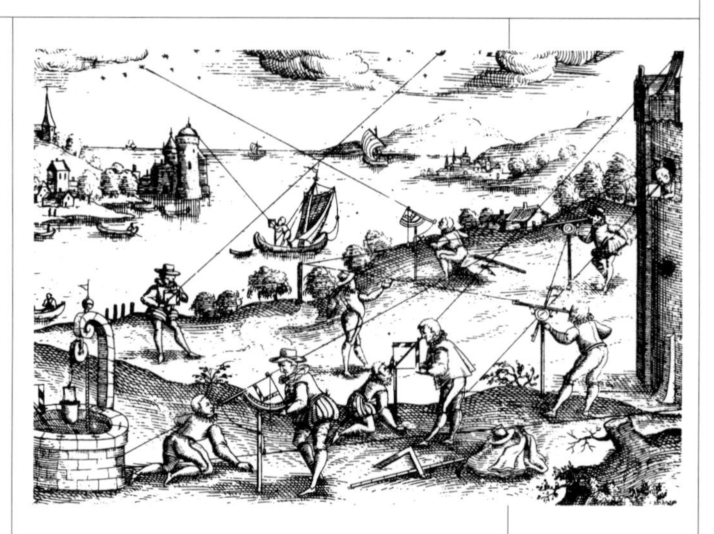

A group of surveyors demonstrates every aspect of their art in a drawing from a 1594 German textbook on surveying.

Chains were also used, laid out across the landscape; the distance would be given in so many "chains." Steel tapes replaced the chains, and in the 20th century electronic distance-measuring devices have come into use. These operate by sending radio waves, light waves, or laser beams from point to point. The speed at which these beams travel is known, so the distance they cover can be calculated with great precision.

The surveyor's second standby is an instrument for measuring the angles between measured baselines and lines of sight to new points. Sextants and theodolites are two of the instruments that can be used to measure these angles.

Most wide-area topographic maps made after the mid-20th century are based on photogrammetric surveying, or measurements made from photographs, usually aerial photographs. Ground surveying is still needed, however, to establish control points to anchor the survey.

SEE ALSO
Great Trigonometrical Survey of India; Photogrammetry; Sextant; Theodolite; Triangulation

T-and-O map

THE TERM *T-and-O map* (or T-O map) was coined by modern historians of cartography to describe a certain style of medieval map. These maps depict the earth as a round disk surrounded by a circular ocean; within that circle, the earth is divided into continents by straight lines that represent the Mediterranean Sea, the Nile River, and the Don River in Russia.

The first T-and-O maps were simple sketches or outlines—symbols rather than practical maps. The Romans, and possibly the Greeks, used them to illustrate broad geographic concepts; the Arabs, too, made diagrammatic maps on the T-and-O model.

Isidore of Seville introduced the T-and-O map into European cartography in the 7th century. His map was a circle crossed by two straight lines in the shape of a T; it was copied hundreds of times during the Middle Ages. East was at the top of the map, Jerusalem at its center. The vertical stroke of the T was the Mediterranean Sea, while the longer horizontal stroke represented the Don River, the Black Sea, and the Nile River.

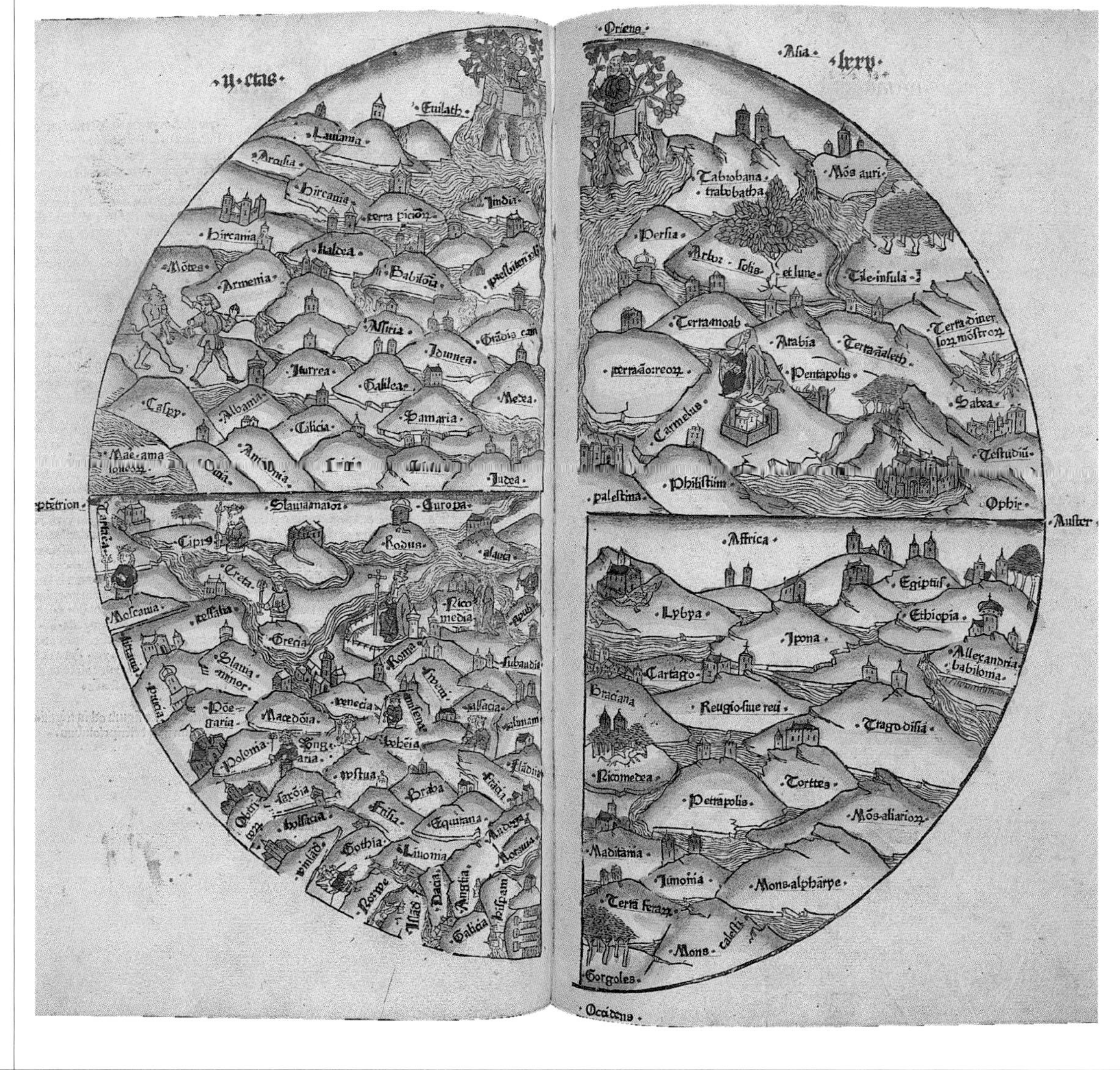

A T-and-O map in four parts: Asia (upper left), Arabia (upper right), Africa (lower right), and Europe (lower left). The walled garden at the top—in the region farthest from the known countries around the Mediterranean—represents the Garden of Eden.

The land surface of the world was thus divided into three parts, with Asia, the biggest, at the top, Europe on the lower right, and Africa on the lower left. Some variations used a cross rather than a flat-topped T to show the seas and rivers. This divided the world into four parts: Asia, Arabia, Africa, and Europe.

The first T-and-O maps were plain, with no geographic details and no labels other than a few words naming the bodies of water and the continents. Gradually, however, they became larger and more ornate, as mapmakers filled them with information and illustrations. They became part of the medieval mapmaking tradition that produced great *mappae mundi* such as the Hereford map and the Ebstorf map. A close look at these crowded, elaborate, decorative world maps shows that they are built around the same T-and-O structure used by Isidore of Seville.

SEE ALSO
Ancient and medieval mapmakers; Ebstorf map; Hereford map; Isidore of Seville; Mappa mundi

Tasman, Abel Janszoon

DUTCH NAVIGATOR AND EXPLORER OF AUSTRALIA

- *Born: about 1603, Groningen, Holland*
- *Died: about 1660, Indonesia*

IN THE early 1630s, Abel Janszoon Tasman went to work as a sea captain for the Dutch East India Company. He made several voyages in the East Indies, the islands that now make up the nation of Indonesia. He also sailed in the waters of Japan. In 1642 he was sent by Anthony Van Diemen, the Dutch governor of the East Indies, to explore a land called New Holland that had been sighted by earlier mariners southeast of the East Indies. No one knew whether New Holland was a continent or a group of islands. Tasman's major achievement would be to show that it was a single large island continent, the continent that today is called Australia.

Tasman's second-in-command was Frans Jakobzoon Visscher, a pilot and gifted chart maker who made maps of Tasman's discoveries. Sailing below the southern coast of Australia, Tasman came upon land, which he called Van Diemen's Land (in 1835 British settlers renamed this island Tasmania, the name it bears today). He sailed on and made the European discovery of New Zealand. He had seen no people in Van Diemen's Land, although he had observed signs of human habitation, but in New Zealand he encountered the well-organized and martial Maori civilization. Maori warriors killed some of Tasman's crewmen, and he named the site of their attack Murderers' Bay (today it is called Golden Bay). Tasman returned to the East Indies by way of several Pacific landfalls, including Tonga, the Fiji Islands, and New Guinea.

In 1644 Van Diemen sent Tasman out again, this time to see whether New Guinea was part of New Holland. Tasman explored the waters between the two islands. By the end of this trip, the northwestern coast of Australia had been traced on the map, and it was clear that New Holland was a substantial continent.

Tasman was rewarded with the rank of commander. He made a trading voyage to Thailand and led a war fleet against Spain before retiring into private life as a merchant in Java. His journal and papers vanished into the archives of the Dutch East India Company, which considered them of little value because they did not point the

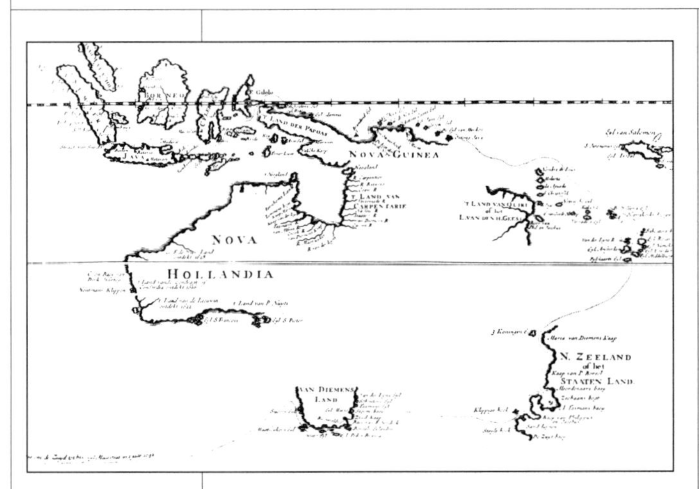

Some geographic errors die hard. Although this chart claims to show Tasman's discoveries, it does not show the sea-lane that Tasman found between Australia (Nova Hollandia) and New Guinea (Nova Guinea).

way to valuable resources. Tasman's account of his voyages was not published until the end of the 17th century. But in 1648, only four years after Tasman's second voyage, the Dutch cartographer Jan Blaeu drew a splendid world map that included the discoveries Tasman had made.

Surprisingly, however, other mapmakers continued to copy or reprint old maps of the Pacific that did not show Van Diemen's Land, New Zealand, or the north coast of New Holland. Only in the late 1650s and the 1660s did the majority of maps begin to reflect Tasman's voyages. Visscher's map of Tasman's voyages was published in 1670, and from that time on no real changes were made to the map of Australia until Captain James Cook charted part of the continent's east coast in the late 18th century. One record of Tasman's voyages was unusually beautiful, although it had no practical use. It was a mosaic floor map in the Dutch Royal Palace in Amsterdam.

SEE ALSO

Australia, mapping of; Cook, James; Pacific Ocean, mapping of

FURTHER READING

Beaglehole, J. C. *The Exploration of the Pacific.* 3 vols. Stanford, Calif.: Stanford University Press, 1966.

Dunmore, John. *Who's Who in Pacific Navigation.* Honolulu: University of Hawaii Press, 1991.

Terra Australis

THE TERRA AUSTRALIS (Southern Land) is a geographic phantom that appeared on maps for more than 1,600 years. In the 2nd century, the Greek geographer Ptolemy claimed that there had to be a large landmass in the Southern Hemisphere to balance the weight of Europe and Asia in the Northern Hemisphere—without such a landmass the world would be top-heavy and would tip over. Because Ptolemy was revered as the "father of geography," his ideas were accepted without question by many later mapmakers. This is why many maps from the 16th and 17th centuries show a huge continent just south of Africa and South America, often swelling far north into the Pacific. It is usually labeled Terra Australis Incognita (Unknown Southern Land) or Terra Australis Nondum Cognita (Southern Land Not Yet Known)—a true act of faith on the part of cartographers.

When Magellan rounded the tip of South America in the early 16th century, he thought the island now known as Tierra del Fuego was the northern tip of the Terra Australis. For several centuries thereafter, whenever land was sighted in the Pacific, it was believed to be part of the elusive unknown continent. But gradually explorers found that New Guinea, New Zealand, Australia, and the smaller islands were not connected to the landmass they sought. In the late 18th century, the British sent Captain James Cook on several Pacific voyages to look for the southern land. He crisscrossed the southern ocean and determined that the Terra Australis does not exist. He did, however, suggest that a smaller landmass might lie concealed in the iceberg-choked waters of the far south, and within a few decades explor-

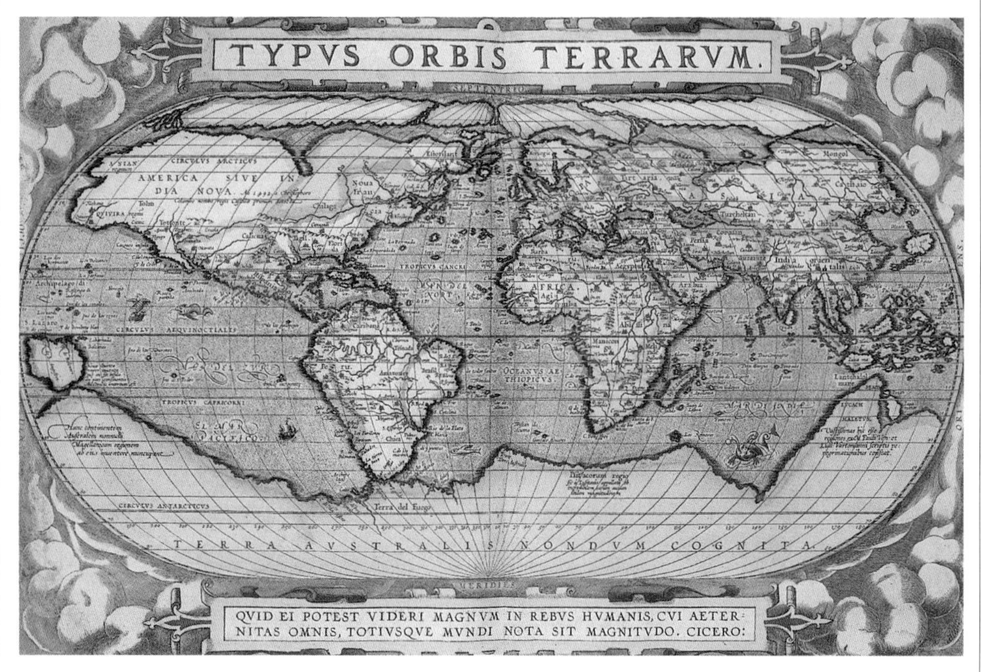

Like most 16th-century mapmakers, Abraham Ortelius added a huge continent to the Southern Hemisphere; he also filled the Arctic with non-existent landmasses.

ers proved him right by discovering Antarctica.

SEE ALSO

Antarctica, mapping of; Australia, mapping of; Cook, James; Dalrymple, Alexander; Pacific Ocean, mapping of; Ptolemy

Thales of Miletus

SEE Ancient and medieval mapmakers

Theodolite

THE THEODOLITE is an instrument used in surveying to measure the angle between two lines. It consists of a telescope mounted on a tripod, with movable measuring devices. The theodolite can measure either horizontal angles, such as the angle between two lines along the ground, or vertical angles, such as the angle between a line that runs straight from the theodolite to the horizon and a second line that runs from the theodolite to the top of a mountain.

Sightings from two previously measured points are taken on an unknown point. Once the angles between the known points and the unknown point have been correctly measured, the exact position of the third point can be determined. This type of surveying is called triangulation because it relies upon the geometric properties of triangles. The theodolite is one of the main tools of triangulation.

SEE ALSO

Surveying; Triangulation

Frontier photographer William Henry Jackson captured surveyors at work on the summit of Salton Mountain, Colorado, in the 1870s. The standing man is using a theodolite to take a sighting on a distant point.

Thompson, David

ENGLISH MAPMAKER AND EXPLORER OF CANADA

**Born: 1770, London, England*
**Died: 1857*

DAVID THOMPSON attended a London charity school for poor children and orphans. In 1784 the school apprenticed him to the Hudson's Bay Company, and he was sent to Canada. Soon he began making journeys into the wilderness of the Canadian interior, traveling from post to post of the trading company and occasionally living for months at a time with friendly Native Americans.

Thompson mastered the arts of navigation, surveying, and chart making largely on his own, with little formal training. He made very careful astronomical observations on all of his journeys, and thus he was able to produce highly accurate maps. He also kept notes on weather, climate, and natural history.

In 1797 Thompson joined the North West Company, a rival to the Hudson's Bay firm. Two years later he married a 14-year-old girl, the daughter of a trader; she and their large family of children accompanied him on many of his later journeys. Thompson explored and surveyed many parts of western North America: the Missouri and northern Mississippi rivers, the Canadian Rockies, and Lake Athabasca in Canada's far northwest. But he is most often remembered for his journey along the Columbia River to the Pacific Ocean in 1811. In 1812 he retired from the fur-trading business to Montreal, where he summed up his years of exploration in an excellent map of western Canada. He then served for 10 years on the U.S.–Canada boundary commission, which determined where to draw the border between the two countries. Sadly, Thompson's enormous contribution to geography and cartography did not guarantee prosperity. In his final years, he was so poor that he had to sell his scientific instruments and even his coat to buy food.

SEE ALSO

Americas, mapping of

FURTHER READING

Thomson, Don W. *Men and Meridians: The History of Surveying and Mapping in Canada.* 3 vols. Ottawa: Duhamel, 1966–69.

Topographic maps

TOPOGRAPHIC MAPS (also called topo sheets) show the topography, or configuration, of the earth's surface. The universal feature of topographic maps is that they show relief, presenting a picture of the elevation and depression of the area that has been mapped. Topographical maps, in fact, attempt to show the three-dimensional shape of the land on the two-dimensional surface of the map. Contour lines—sometimes accompanied by hypsometric shading—are used to

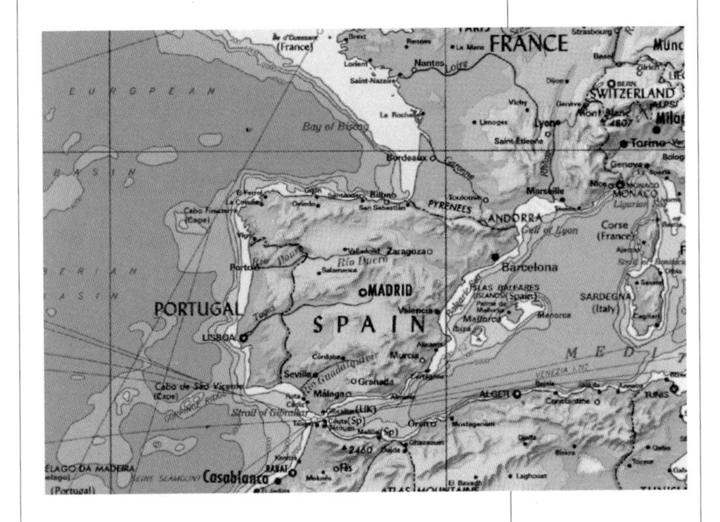

Color and shading are used to show the shape of the land in this map of Spain, Portugal, and southern France. Spain's mountainous terrain leaps into highlighted relief next to the flat French coast around Bordeaux. Shallow water offshore is indicated by lighter tints.

A contour map of the summit of Mount Rainier in the northwestern United States.

show both the elevation and the shape of the terrain. A properly prepared topographic map will show not just how high a mountain is but also its shape and how steeply or sharply it rises. From such a map, a climber can determine the easiest way to reach the mountaintop. Topographic globes show relief in three dimensions, with raised ridges for mountain ranges and grooves for canyons and ocean trenches.

In general, topographic maps show the location and shape of mountains, valleys, and plains; streams and rivers; and principal man-made features such as towns, roads, and dams. They make extensive use of lines, symbols, and often color to convey a great deal of information in a compact, easy-to-read way.

Such maps are produced on a variety of scales. For example, the U.S. Geological Survey (USGS) publishes topographic maps of the United States in five different series, each drawn to a different scale. From largest to smallest, these scales are 1:24,000; 1:62,500; 1:100,000; 1:250,000; and 1:1,000,000. The larger the scale, the more detail is shown but the smaller the area that is covered. Each map in each series covers a quadrangle, or quad, which is a rectangle bounded by lines of latitude and longitude. The Ordnance Survey of Great Britain produces an even wider range of topographic maps. Beloved by hikers in the British Isles, the larger-scale Ordnance Survey maps include details as minute as windmills, public telephones, and large rocks.

Topographic maps have many uses. Topos are the maps of choice for recreational hikers and campers. They also play a vital role in military operations—a general cannot give the order "Take that hill!" without knowing where the hill is located and how high it is. Land-use planners rely on topographic maps to help them decide where (or where not) to build dams, highways, satellite relay stations, and pipelines.

In many countries, topographic maps are produced by government mapping agencies: the USGS in the United States, the Ordnance Survey in Britain, and the Surveys and Mapping Branch in Canada. Topographic maps for nearly everywhere in the world are available from a variety of government and commercial sources.

SEE ALSO
Contour lines; Hypsometric shading; Orienteering; Relief

Topography

A PLACE'S TOPOGRAPHY is the sum of its shape and features. Topography always includes elevation—the heights and depths, hills and valleys, of the terrain. It also includes major natural features such as rivers and lakes. In addition, smaller natural features and some man-made features such as dams, gravel pits, and stone quarries may be considered part of the topography.

SEE ALSO
Topographic maps

Town plans

A TOWN PLAN is a map of a town or city. Town plans appeared as decorative insets or panels on many 16th-century maps. These images were pictures or panoramas as well as maps. They were so popular that starting in 1572 German cartographer Georg Braun published an atlas of more than 500 of them, covering cities all over the world. Many city and town atlases have been produced in the years since then. Some cities have been mapped at almost every stage of their history. The history of large cities, especially old cities such as Paris and London, can be traced through old town plans from different eras.

Town plans are still produced for a variety of purposes. Some are architectural plans, produced by surveyors. Some are pictorial, with drawings of buildings and landscape features; others are simple diagrams of intersections and rivers. Extremely precise and detailed plans, often generated from databases using computer modeling, are used by fire departments, city planning offices, and the like. Some fire departments have computers that produce street maps highlighting the best routes to locations where fire alarms have been activated.

Many modern town plans are general and informal, designed to keep people from getting lost. Instead of showing every single building in its proper scale and position, they show major streets and buildings, as well as other points of interest. These city maps are distributed by civic groups and tourism associations and are often printed in guidebooks.

Maps of existing towns and cities are only one type of town plan. Maps have also been used as tools in the discipline called urban planning, which is concerned with planning new buildings and neighborhoods, or even whole new towns and cities. Urban planning is not new; archaeological evidence suggests that some very ancient communities, such as the Maya settlements in Mexico and Guatemala, may have been planned before they were built. Among the signs that a city might have been planned are rectangular, orderly street systems and wide boulevards in symmetrical patterns.

Most towns and cities grew gradually and somewhat haphazardly over

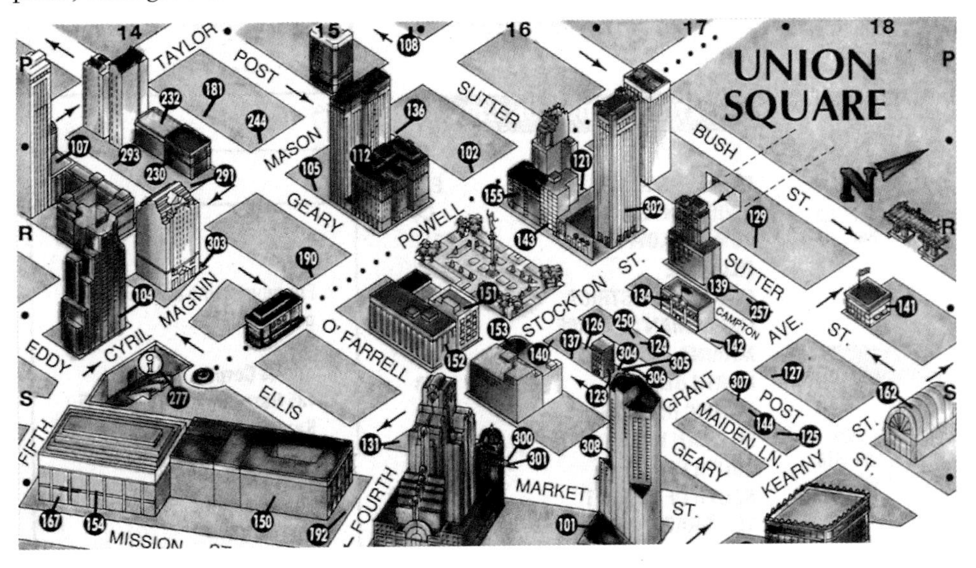

A plan of the Union Square district in San Francisco, showing streets and major buildings; the numbers are keyed to information about restaurants and other features.

time, but a number of cities that started as colonies and were built fairly quickly were planned in considerable detail before they were built. Philadelphia, Pennsylvania, for example, was built to a plan developed by the city's founder, William Penn. He envisioned a layout of streets at right angles to one another, enclosing rectangular plots of land. Pioneers who moved west from Pennsylvania carried this easy-to-use grid pattern with them, and it became the basis for many American towns and cities. The best-known example of urban planning in the United States is the nation's capital, Washington, D.C., which was planned in 1791 by a French engineer named Pierre L'Enfant. Another national capital, Brasília, was built from a blueprint. Designed by a team of American and Brazilian architects, Brasília was begun in 1957 and became Brazil's capital in 1960.

In the early 20th century, local governments began taking responsibility for urban planning—not just building new communities but creating long-term plans for cities that already existed. Maps have played an important role in urban planning. Planners who hope to solve urban problems such as congestion or poor traffic flow, as well as those who envision new office buildings, parks, highway access ramps, or whole communities, use maps to study not only the city as it exists but also the city as it might be.

SEE ALSO

Braun, Georg; Decoration

FURTHER READING

Reps, John William. *Cities of the American West: A History of Frontier Urban Planning.* Princeton, N.J.: Princeton University Press, 1979.

———. *The Making of Urban America: A History of City Planning in the United States.* Princeton, N.J.: Princeton University Press, 1965.

Treasure maps

"THERE'S NO getting away from a treasure that once fastens upon your mind," the 19th-century author Joseph Conrad said about the lure of treasure hunting. Many stories have been told about treasure maps—dramatic stories in which grizzled prospectors or retired pirates on their deathbeds hand over a map that is supposed to lead to a hoard of buried treasure.

Many such tales cluster around the figure of Captain William Kidd. Kidd was hanged for piracy in London in 1701, but just before his death he promised to lead British officials to a great treasure if they would spare his life. The government ignored his offer, but many people over the centuries have searched for Kidd's hoard. Several maps of dubious origin turned up in the 20th century, and treasure hunters say they have found clues to Kidd's hoard in Canada, New York, and the Caribbean. One searcher may have stumbled on the treasure in a cave in Japan in the 1950s. Or perhaps Kidd's hoard never existed at all.

Another legendary treasure map was said to point the way to a lost gold mine in Arizona. In the 1860s, two German men saved a Mexican man named Peralta from being beaten up in a Wild West saloon brawl. In gratitude, Peralta showed them the location of an old mine—with plenty of gold still in it—that his family had found years before. One of the Germans, Jacob Waltz, plundered the mine for years, killing anyone who came too close to his secret, even his own nephew. In 1891, while he was dying, Waltz confessed his crimes and allegedly drew a map showing the location of the mine. Since then, many people have ventured into the harsh Superstition Mountains near Phoenix, Arizona, following the clues he left, but no one has yet found his trove, which has come to be called the Lost Dutchman Mine. Tales of violence, mystery, and even the supernatural cluster around the legend of the mine.

One of the most famous treasure maps of the 19th century probably inspired a classic adventure story. The map was of Cocos Island, a tiny speck of land in the Pacific Ocean off Costa Rica. Starting in the 16th century, this uninhabited island was a hideout for pirates who terrorized the Spanish settlements in South America. The island's biggest treasure hoard, however, dates from the 19th century. In 1823, Spanish officials in Lima, Peru, feared that rebels would seize the city's treasures. They asked a Scottish captain named William Thompson to take the treasure to safety on his ship, the *Mary Dear*. Thompson loaded the ship with tons of treasure, including a life-size statue of the Virgin Mary made of solid gold. But once out of sight of land, Thompson killed the Spanish guards and made for Cocos Island. He hid the treasure, only to be captured by a Spanish captain, who left him on the island to die. Thompson was finally rescued from the island by a whaling ship, but he never managed to return to Cocos for the treasure.

In 1844 Thompson was dying in Canada. In the grand tradition of treasure lore, from his deathbed he passed a map to a friend named John Keating, who is said to have gone to Cocos Island and brought home part of the hoard. Keating's partner died mysteriously—perhaps murdered to keep the secret. Keating later passed the secret to another friend, but this friend could not find the treasure. Hundreds of other searchers have also failed, maybe because a flood or a rockslide changed Keating's landmarks. A few pieces of treasure have been found in the 20th century. One of them was a two-foot-high golden statue. But in 1978 the government of Costa Rica, which owns the island, outlawed treasure hunting.

The story of Cocos Island was widely known in the 1880s, when the Scottish writer Robert Louis Stevenson wrote his tale of piracy, marooned men, murder, and buried treasure on a deserted isle, and literary scholars speculate that Cocos Island was one of the inspirations for Stevenson's map and the story that he wrote to accompany it—the classic adventure novel *Treasure Island.*

FURTHER READING

Coffman, F. L. V. *Atlas of Treasure Maps.* New York: Nelson, 1957.

Wilson, Derek A. *The World Atlas of Treasure.* London: Collins, 1981.

Treaty of Tordesillas

THE TREATY of Tordesillas was a 1494 agreement between Spain and Portugal to divide up the Americas and the Indies for purposes of exploration, exploitation, and colonization. Portugal had pioneered the sea route east around Africa to India and beyond, and Spain had opened the sea route west to the New World. Spain and Portugal were squabbling over how far their respective realms of influence extended, so Pope Alexander VI stepped in and negotiated

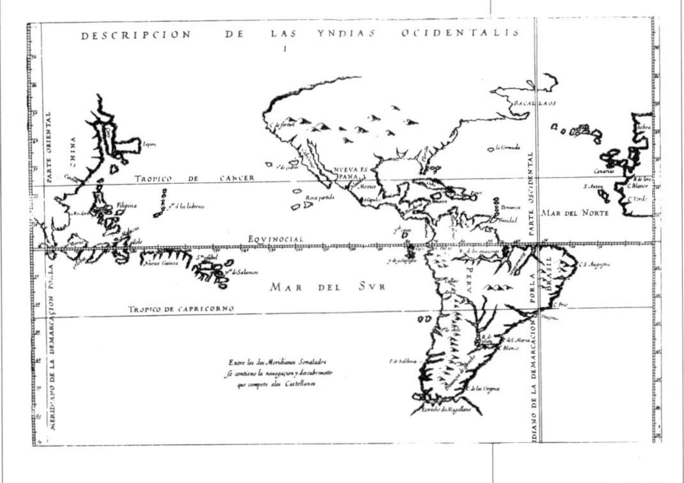

In 1494, Pope Alexander VI divided the world between Portugal and Spain along the vertical line near the right edge of this map.

the treaty between the two nations.

The treaty line, which is marked on many 16th- and 17th-century maps, ran longitudinally through the middle of the Atlantic Ocean and the eastward bulge of Brazil. Spain received all rights to the lands west of the line, which included nearly all of the Americas. Portugal got everything east of the line, including Brazil, Africa, and India. No other nation's ruler ever acknowledged the Treaty of Tordesillas; the French, British, and Dutch calmly proceeded to stake their own claims in Asia and the Americas, and whatever influence the treaty had possessed in its early years had faded away by the middle of the 17th century.

Triangulation

TRIANGULATION IS a method of surveying land that is based on a branch of mathematics called trigonometry, which deals with the properties of triangles. In triangulation, a baseline is measured, and then sightings are taken from both ends of the baseline on a predetermined point in the distance. By measuring the angles between the baseline and the sight lines to the new point, the surveyor can fix the new point's exact position. That point then becomes one of the anchor points for a new baseline, as another triangle is constructed. In this way accurate lines of measurement can be extended for a long distance in the form of chains of interlocked triangles. Triangulation is the principal method of making topographic maps.

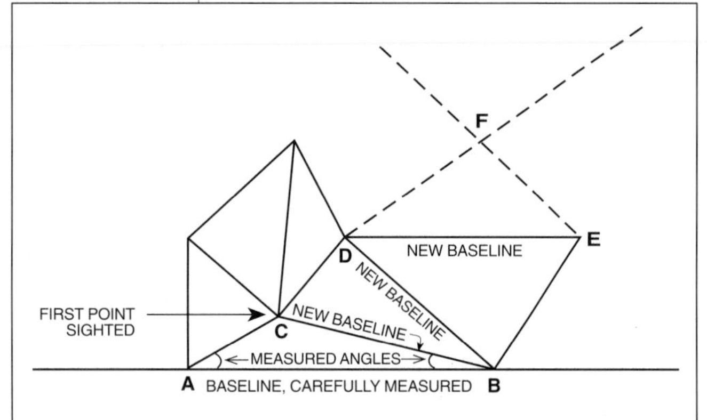

Triangulation is the only way to measure a straight line over the earth's curved surface. It begins with a baseline of known length; here, the line from A to B. When point C is sighted from points A and B, the measured angles allow the surveyor to calculate point C's exact position. Then the surveyor measures the distances along lines A-C and B-C. These lines become the new baselines for sightings on farther points.

SEE ALSO

Great Trigonometrical Survey of India; Topographic maps

Tropic of Cancer

THE TROPIC of Cancer is an imaginary line on the earth's surface, usually represented by a dotted line on maps. It is parallel to the equator at about latitude 23.5 N. It marks the northernmost point at which the sun appears directly overhead at noon. On June 21, the summer solstice in the Northern Hemisphere, the sun's rays shine vertically on the Tropic of Cancer. The name refers to the constellation Cancer (the Crab), which first becomes visible in the Northern Hemisphere on that date. The part of the earth's surface that lies between the Tropic of Cancer and the Tropic of Capricorn is called the "tropics."

Tropic of Capricorn

THE TROPIC of Capricorn is an imaginary line on the earth's surface, usually represented by a dotted line on maps. It

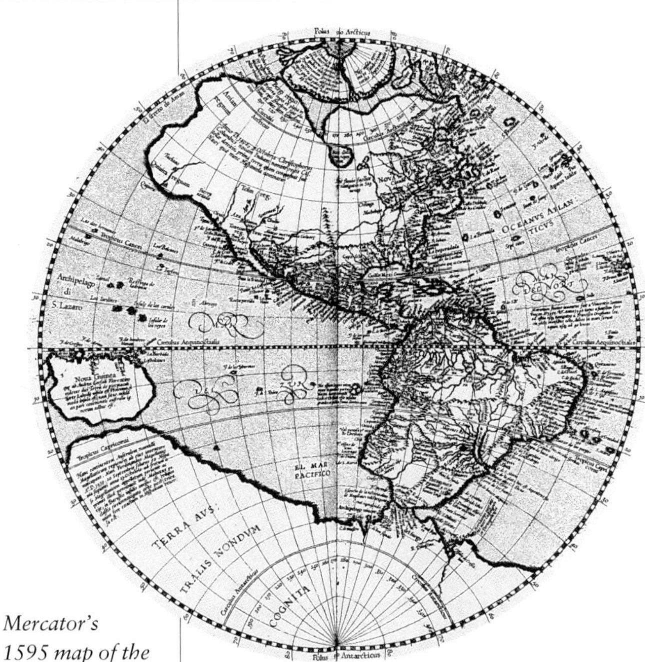

Mercator's 1595 map of the Western Hemisphere. Curved lines on either side of the equator represent the tropics of Cancer and Capricorn.

is parallel to the equator at about latitude 23.5 S. It marks the southernmost point at which the sun appears directly overhead at noon. On December 21, which is the summer solstice in the Southern Hemisphere and the winter solstice in the Northern Hemisphere, the sun's rays shine vertically on the Tropic of Capricorn. The name refers to the constellation Capricorn (the Goat), which first becomes visible in the Northern Hemisphere on December 21. That part of the earth's surface that lies between the Tropic of Cancer and the Tropic of Capricorn is called the "tropics."

Turin papyrus

THE ANCIENT Egyptians are known to have made maps and town plans, although few of these relics survive. The oldest known Egyptian map is the Turin papyrus, so called because it is preserved in a museum in Turin, Italy. It is drawn on papyrus, the writing surface that the Egyptians made from reeds, and was probably made during the reign of the pharaoh Ramses II, around 1300 B.C. It shows a valley and several roads in a gold-mining region, most likely in the mountains of Nubia (part of present-day Sudan).

Undersea mapping

THE MAPPING of the seafloor is a modern endeavor. Before the 19th century, no one knew how deep the oceans were; they might have been hundreds of miles or kilometers deep. People cared about the waters close to shore, the shallows and sandbars and reefs that could destroy ships. These were charted as carefully as possible. Occasionally, a mariner made an attempt to sound—that is, to measure—depth far out at sea. The Portuguese navigator Ferdinand Magellan is said to have tied all his ship's ropes together, dropped one end over the side into the Pacific Ocean, and then, when it failed to reach bottom, announced that the oceans were too deep to measure.

In the 19th century, scientific curiosity and practical considerations (the

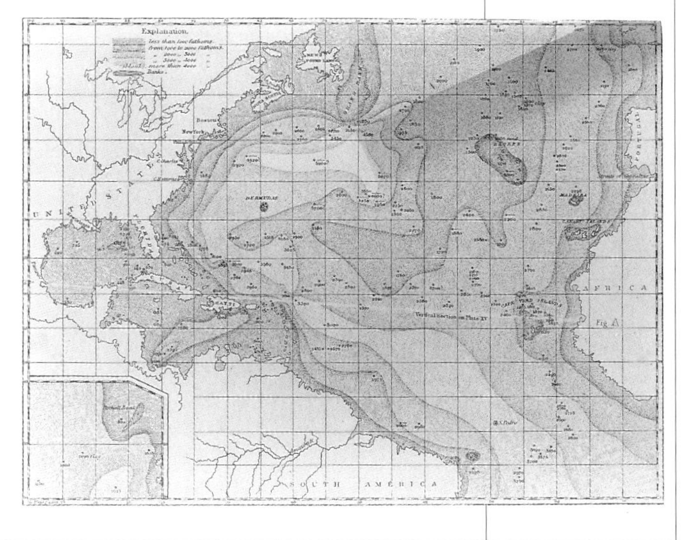

Matthew Fontaine Maury's 1855 chart revealed for the first time that the seafloor is a landscape.

The North Atlantic seafloor was first mapped by Marie Tharp and Bruce Heezen in 1957; this revised map was made in 1972. Spain and Africa are to the right. The Azores and Canary islands are mountains jutting up from the abyssal plain.

desire to connect North America and Europe with a telegraph cable on the Atlantic Ocean floor) stimulated a flurry of deep-ocean studies. Matthew Fontaine Maury of the U.S. Navy launched the science of oceanography and supervised a series of ocean depth soundings, or measurements of the distance from the surface to the seafloor. From these, in a connect-the-dots manner, he made the first contour map of the Atlantic Ocean floor; this bathymetric chart was published in 1855. For the first time, people realized that the seafloor was not a flat, featureless plain, as had long been supposed, but a landscape with heights and depths.

Later in the century, the British Challenger Expedition took many more soundings. Scientists aboard the *Challenger* probed the deepest point in the world's oceans, the 35,810-foot-deep (10,915 metersp) Marianas Trench in the western Pacific. They also discovered a mountain ridge running through the length of the Atlantic Ocean—a discovery that suggested to some people that the "lost continent" of Atlantis had been found. Further exploration by the German Atlantic Expedition and other research teams showed that this mountain chain winds for 46,000 miles (74,030 kilometers) through all the world's oceans. It is called the Mid-Ocean Ridge and is both the longest and the tallest mountain range on earth.

The next big advance in undersea mapping occurred in the years after World War II (1939–45). Using echo-sounding instruments—sonar and other devices that bounce sound waves off the seafloor and measure the time it takes them to return—scientists were able for the first time to obtain depth readings for more than just isolated points. They obtained profiles of long strips of the seafloor and then joined these together to make the first undersea maps that did not consist almost entirely of guesswork. The most striking maps were made by Marie Tharp and Bruce Heezen of Columbia University's Lamont-Doherty Laboratory. In the 1950s they produced a series of physiographic diagrams, pictorial maps that show the earth's oceans as they would look if all the water had drained away. Published by the National Geographic Society and the Geological Society of America, these maps startled viewers with their images of stark, dramatic landscapes hidden below the waves.

The Tharp-Heezen maps provided a key piece of evidence to support a geologic theory known as continental drift, or plate tectonics. Since the early 20th century, scientists had speculated that the continents have changed position over the earth's history, but no one knew how the continents could move. Studies of the seafloor revealed that the continents rest on top of large, slow-moving plates of the earth's crust. The ridges and rifts in the seafloor, like those on land, mark the places where those plates have met or parted.

In 1978, the American *Seasat* satellite measured the ocean's floor in still another way: it used microwaves to record variations in the level of the sea's surface. Dips or rises in sea level are caused by the gravitational pull of the earth. Dips in the sea's surface indicate valleys or rifts beneath the surface, where the mass of the seafloor has receded, drawing the water down slightly; bulges in the sea's surface indicate massive concentrations of matter, such as undersea mountains, that attract the water toward them. William Haxby of Lamont-Doherty used the gravity data from *Seasat* to make the most detailed map yet of the ocean floor, revealing hundreds of previously unknown seamounts. *Geosat,* launched in 1985, has yielded even more information about the structure of the seabed, enabling Haxby to make three-dimensional computer-modeled maps of regions such as Antarctica's Weddell Sea. It is now possible for cartographers to make topographic maps and relief globes of the earth's entire surface—maps in which the border between land and sea vanishes, and the true borders are those between the plates of the earth's crust.

SEE ALSO

Challenger Expedition; German Atlantic Expedition; Maury, Matthew Fontaine

FURTHER READING

Gaines, Richard. *Explorers of the Undersea World.* New York: Chelsea House, 1994.

Hall, Stephen. *Mapping the Next Millennium: The Discovery of New Geographies.* New York: Random House, 1992.

Marx, Robert F. *A History of Underwater Exploration.* New York: Dover, 1990.

U.S. Geological Survey

THE U.S. GEOLOGICAL SURVEY (USGS) was formed by an act of Congress in 1879. Before that time, four different government-sponsored surveys had been mapping the country. These were combined into the USGS, whose job is to collect and publish topographic and geologic information. It is part of the U.S. Department of the Interior.

The USGS is in charge of the country's National Mapping Program. It produces local, regional, and country maps, and some world maps, based on its own surveys and on data from other agencies, including the National Oceanic and Atmospheric Administration (NOAA), the Department of Defense, and the Central Intelligence Agency (CIA). The USGS also collaborates in map-publishing activities with the Smithsonian Institution and other organizations.

The single most important part of the National Mapping Program is the many series of topographic maps that are produced by the USGS.

The USGS operates the Earth Science Information Center (ESIC), which provides information about a wide range of maps available from the USGS and other government agencies.

SEE ALSO

Appendix 2: Organizations and Publications Related to Maps

FURTHER READING

Bartlett, Richard. *Great Surveys of the American West.* Norman: University of Oklahoma Press, 1962.

Goetzmann, William H. *Exploration and Empire: The Explorer and the Scientist in the Winning of the American West.* New York: Norton, 1978.

Manning, Thomas. *Government in Science: The U.S. Geological Survey, 1867–1894.* Lexington: University Press of Kentucky, 1967.

Vancouver, George

BRITISH NAVIGATOR AND PACIFIC EXPLORER

- *Born: June 22, 1857, King's Lynn, England*
- *Died: May 10, 1798, Richmond, England*

GEORGE VANCOUVER joined the British Royal Navy and served on two of Captain James Cook's voyages to the Pacific. On one occasion, when Cook sailed farther south into the waters near Antarctica than any ship had gone before, Vancouver climbed out on the ship's bowsprit and clung there, as close to the front of the ship as he could get, so that he could claim to have gone farther south than any other person.

Vancouver's zest for exploration stayed with him throughout his career as a naval surveyor. He spent the years 1792–95 making a painstaking survey of the northwestern American coast, especially British Columbia, Washington, and part of Oregon. This survey, published in 1798, formed the basis of Great Britain's claim to the region. Vancouver Island and the city of Vancouver, both in British Columbia, as well as the city of Vancouver, Washington, are named for the surveyor.

SEE ALSO
Americas, mapping of

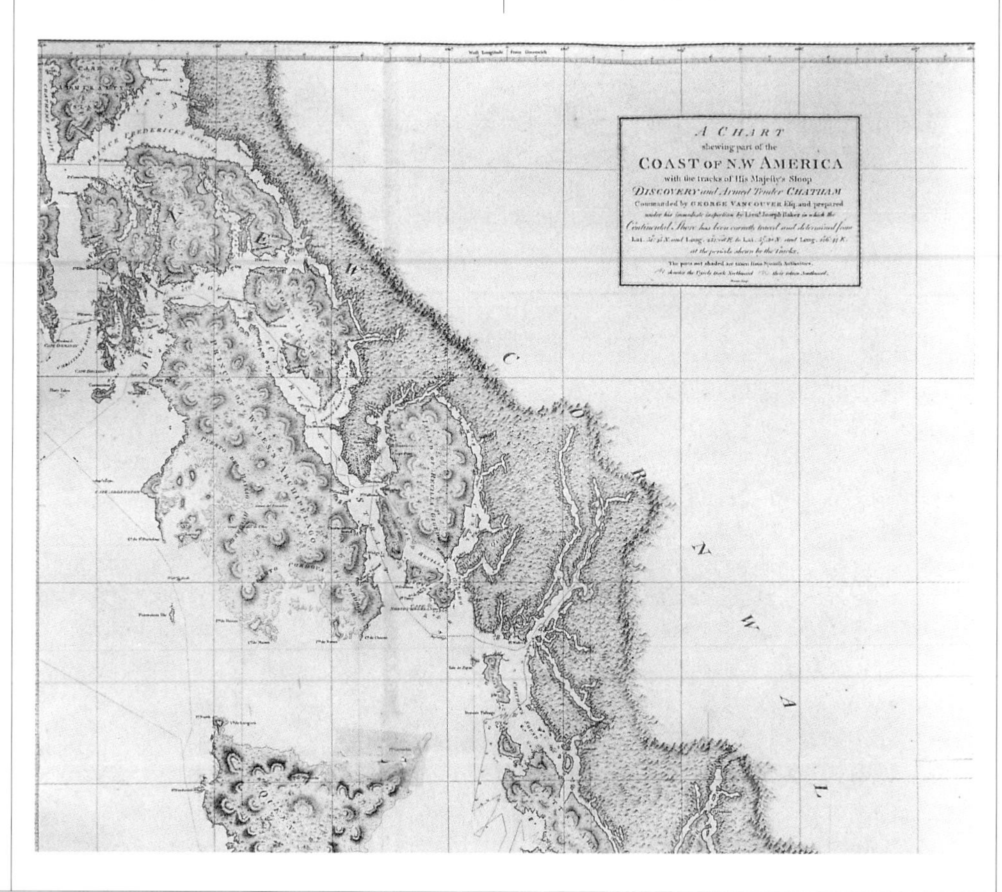

A map of the southeastern Alaskan coastline made under Vancouver's command. The Queen Charlotte Islands are at the bottom left; the zigzag lines record the track of Vancouver's vessels.

Vespucci, Amerigo

ITALIAN NAVIGATOR

- *Born: 1454, Florence, Italy*
- *Died: 1512, Seville, Spain*

AMERIGO VESPUCCI is the only person in history for whom two entire continents—North and South America—have been named. He did not name them himself; instead, he owes the honor to the German geographer and mapmaker Martin Waldseemüller.

Born into a prominent family in Florence, Vespucci became an agent for an Italian bank. In 1491 he was sent to Seville, Spain, where he became acquainted with Christopher Columbus, who was about to set out on the first of his voyages. In 1497 Vespucci was part of a commission sent to Hispaniola by the Spanish crown to investigate charges that Columbus had mismanaged the colony there. Two years later, he took an expedition to the region that Columbus had begun to explore. At this time, Columbus was still insisting that he had reached the coast of Asia, and geographers did not yet realize that, in fact, he had encountered a different continent altogether.

Vespucci claimed that in several later voyages undertaken for Portugal he explored the coast of Brazil. Historians have been unable to verify his claims because no other reports of these voyages have survived. Vespucci's boastfulness, too, has caused his version of events to be questioned; he blustered, "I was more skillful than all the shipmasters of the whole world." But Vespucci circulated a letter about his travels under the title *Mundus Novus* (*The New World*), and it was widely read. He declared that the lands he and Columbus had explored should be regarded as a "new world, because none of these countries were known to our ancestors." He went on to say, "I have found a continent in that southern part more populous and more full of animals than our Europe or Asia or Africa." This is the first known public recognition of the fact that the lands encountered by Columbus were not part of Asia at all but were a previously unknown continent.

In an illustration on Martin Waldseemüller's 1507 world map, Amerigo Vespucci prepares to measure the continents that now bear his name.

A few years later, in 1507, Martin Waldseemüller published an important book called *Introduction to Cosmography*. Waldseemüller proposed that because Vespucci had added a "fourth part" to the world, which had previously been considered to consist of only three continents, the new land should be named after Vespucci. On a map published that same year, Waldseemüller wrote "America" (a Latin form of Amerigo) across a crude outline of South America. This was the first time the new continent had been named. Waldseemüller also started the custom of showing the Eastern Hemisphere (Europe, Asia, and Africa) on one part of the map and the Americas on another; this became standard cartographic practice within a few years and reinforced the notion that the Americas were indeed a "new world."

In 1508, Vespucci was appointed pilot major, or head official of all ships' pilots, in Spain. This position placed him in control of all new geographic information about the Western Hemisphere; he held the post until his death. The year after Vespucci died, Waldseemüller issued a new map without the name America. Apparently, he had reconsidered his earlier suggestion. But it was too late. In just half a dozen years, cartographers, geographers, pilots, and historians had adopted the name that Waldseemüller had given in 1507, and "America" could not be removed from the map—a testimony to the power of maps and mapmakers to shape our view of the world. In 1515, the name America was given to the northern part of the hemisphere as well. Had he lived a few years longer, the vain Vespucci would no doubt have been delighted to see his name written across half of the world.

SEE ALSO

Americas, mapping of; Columbus, Christopher; Waldseemüller, Martin

Viking explorers

THE VIKINGS, seafarers from Scandinavia, were one of the dominant forces in Europe from the 8th to the 11th century. They raided coastal towns in England, Ireland, and France, and they carried commerce and war through present-day Poland and Russia as far as the gates of Constantinople (now Istanbul) in Turkey.

The Vikings were also explorers who pushed westward across the North Atlantic Ocean in their very seaworthy ships. Much of what we know about these western journeys is seasoned with guesswork. The main source of information about them is the Viking sagas, traditional narratives in which history is laced with liberal amounts of legend and myth. One thing, however, we know for certain: the Vikings reached the North American continent five centuries before Christopher Columbus did.

The Vikings crossed the North Atlantic Ocean in a series of migrations. Venturing west from Scandinavia, they found the Shetlands, Orkneys, and Faeroes, three island groups north of Great Britain between Norway and Iceland. They established outposts in these islands, and during the 9th century several Scandinavian mariners bound for ports in the islands were blown off course to westward. These navigators sighted a larger island, which came to be called Iceland, and by the late 9th century the Norse people, Vikings from Norway, had begun to settle Iceland. In the early 10th century, a Norse ship bound for Iceland was blown off course by a storm, and its captain glimpsed a rocky shore west of Iceland. This is the first recorded European sighting of Greenland, the largest island in the Atlantic Ocean and the Vikings' bridge to North America.

The first Viking expedition to Greenland took place in 982–85 under the leadership of Erik Thorvaldsson, a quarrelsome Viking who was called Erik the Red. Exiled from Iceland for committing murder, Thorvaldsson explored the southern coast of Greenland and founded a colony there. In 986 a ship bound for that colony sailed too far to the southwest and sighted an unknown coastline. Four or five years later, that coastline was explored by Thorvaldsson's son Leif Eriksson.

Eriksson sailed west from Greenland until he found the unknown coast, then followed it south. He landed at three places, which he called Helluland (Land of Flat Stones), Markland (Wooded Land), and Vinland (Wine Land or Fruitful Land). Scholars now believe that

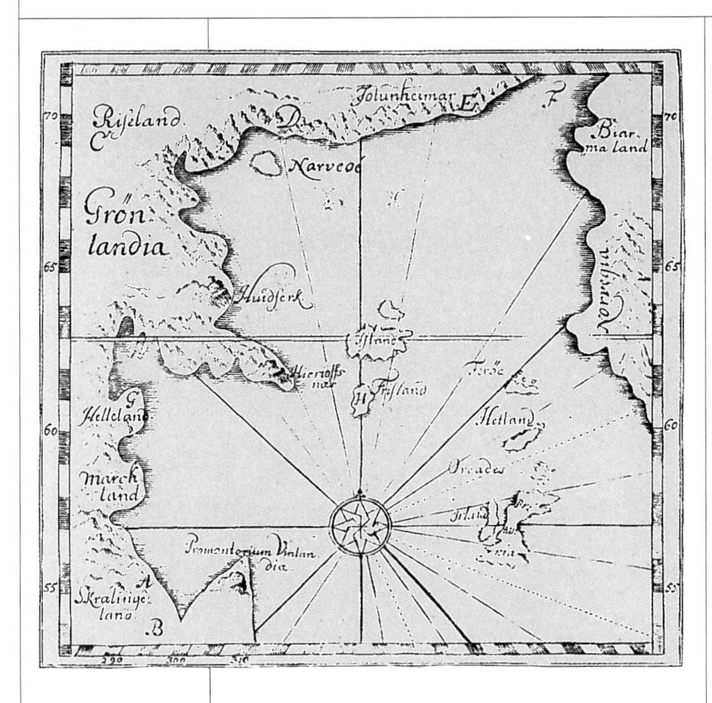

The Skalholt map, the oldest known genuine map of the Vikings' landings in North America (lower left).

Helluland was the southeastern tip of Baffin Island and that Markland was the coast of the Labrador Peninsula. The location of Vinland, however, has been the subject of much debate among historians. Arguments have been made in favor of sites all along the east coast of North America, from North Carolina to Canada's Gulf of St. Lawrence. Most experts now agree that Vinland was probably Newfoundland Island, where the ruins of a Norse settlement were uncovered in the 1960s. The settlement dates from about A.D. 1000 and was probably a seasonal shelter for Norse hunters or woodcutters from Greenland. It seems to have been in use for only a few years. There is no evidence that the Vikings ever settled anywhere else in North America, and although they may have explored parts of New England or Canada, there is no firm evidence of that either.

The Viking colony in Greenland prospered for a time and then fell into decline as trade and communication between Greenland and the rest of the world came to an end. By the late 15th or early 16th century, the Norse Greenlanders had died out. Few people in Europe knew of Greenland and Vinland as anything other than dim, half-forgotten legends.

It is not known whether the Vikings themselves made maps of their western explorations. Such maps, if they ever existed, have not been preserved. The oldest known map that shows Greenland and Vinland is called the Skalholt map. It was made about 1570 by Sigurdur Stefansson, a schoolmaster in the Icelandic town of Skalholt. Stefansson's own map has been lost, but a copy dating from 1590 survives. It shows all the North American sites mentioned in the Viking sagas, but its geography is not very accurate; for example, Greenland is not an island but part of a giant continent, which also contains Jotunheim, the mythical home of the Norse gods. (Greenland and Vinland are more accurately depicted on another map that appears to have been made in the mid-15th century, but this Vinland map, as it is called, is widely thought to be a forgery.) Not until the 17th and 18th centuries did geographers and mapmakers turn their serious attention to the places mentioned in the Viking sagas, and not until the 19th century did they determine beyond doubt that Greenland was an island.

SEE ALSO

Americas, mapping of; Vinland map

FURTHER READING

Jones, Gwyn. *The Norse Atlantic Saga.* New York: Oxford University Press, 1986.

Roesdahl, Else. *The Vikings.* New York: Penguin, 1987.

Stefoff, Rebecca. *The Viking Explorers.* New York: Chelsea House, 1993.

Wernick, Robert, and Time-Life editors. *The Vikings.* Alexandria, Va.: Time-Life Books, 1979.

Wilson, David M. *The Vikings and Their Origins.* London: Thames & Hudson, 1989.

Vinland map

THE VINLAND MAP is the source of a 20th-century cartographic controversy. It appears to be a mid-15th-century map of the Norse discoveries in North America. If it is genuine, it is the only known map made before Columbus's voyages to show the places visited by the Norse in the Americas.

The map's tangled history has not yet been fully unraveled. It turned up in 1957 in the possession of a book dealer in New Haven, Connecticut, who claimed to have obtained it in Barcelona, Spain. The map, drawn on a patched and ragged sheet of parchment, was bound into a manuscript written by a Franciscan monk about a journey into Asia. But an inscription on the back of the map says that it is supposed to illustrate another book: *The Mirror of History*, by a 13th-century writer named Vincent of Beauvais.

The map passed into the hands of scholars at Yale University, who studied it for about eight years before making it public. The Vinland map, as it came to be called, presented a challenge. For one thing, it showed much more advanced geographic knowledge than other maps of its time.

Parts of the map seemed very much like a world map made in 1436 by Andrea Bianco, which suggests that the Vinland map was made around 1440. But the Vinland map contained some surprises. Unlike all other maps of the period, it showed Greenland as an island, not as part of a continental landmass attached to Scandinavia. This piece of geographic knowledge is not confirmed by any other 15th-century source; not until the 19th century did mapmakers and geographers know for sure that Greenland is an island. Yet on the Vinland map its outline is remarkably accurate. The Vinland map also shows an island called Vinlanda, which has roughly the same position and shape as Baffin Island. A note on the map says

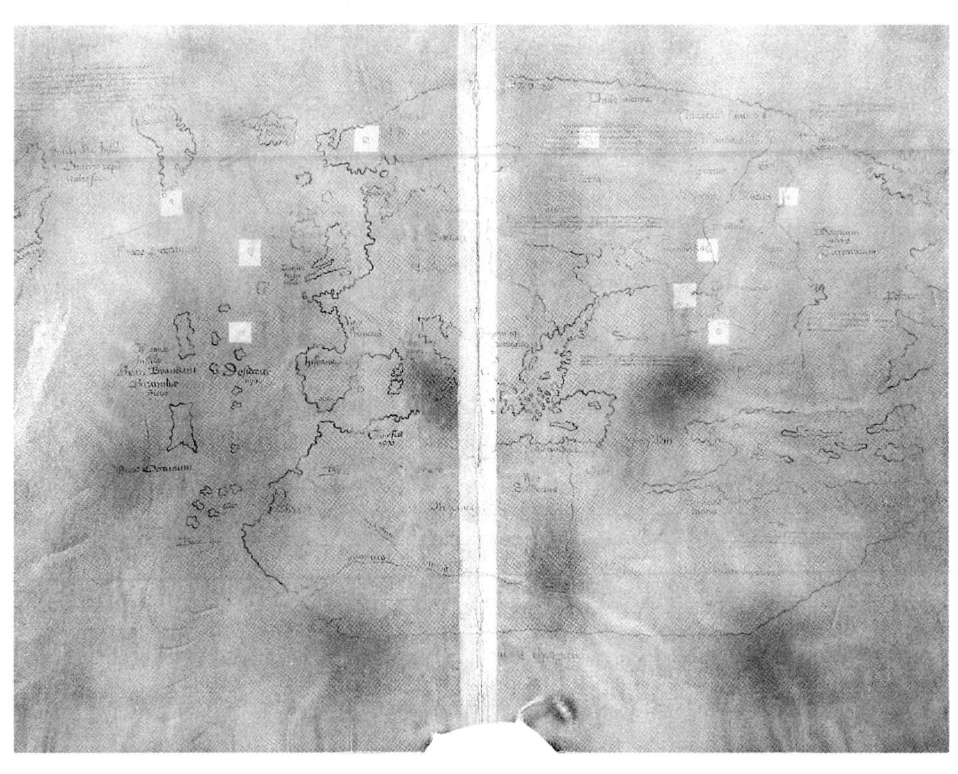

The Vinland map is almost certainly a hoax. If it were genuine, it would be the oldest known world map to show North America (here called Vinland, in the upper left).

that Vinlanda is the land discovered in the west by Leif Eriksson and describes it as "a new land, extremely fertile, and even having vines." In short, the Vinland map seemed to offer confirmation of the Viking sagas and also to present a new, surprisingly high level of geographic awareness in the 15th century.

The map passed various tests of authenticity. The paper dated from the 15th century, as did the style of writing. In 1965, Yale University presented the map to the world and announced that it was authentic, although some scholars disagreed. Then, in 1974, researchers at Yale applied a new test and found that the yellowish-brown ink used on the map contained anatase, a pigment that was not invented until the 1920s. The map, declared Yale, was a clever forgery. Most scholars now share this view, and the Vinland map has largely been discredited. Yet a few researchers now argue that a form of anatase might have occurred naturally in 15th-century ink under certain conditions. The Vinland map will probably continue to be the subject of debate for some time to come, as new tests for its authenticity are devised.

SEE ALSO
Bianco, Andrea; Viking explorers

Waggoner

SEE Portolan; Waghenaer, Lucas Janszoon

Waghenaer, Lucas Janszoon

DUTCH HYDROGRAPHER AND CHART MAKER

- *Born: 1533, Holland*
- *Died: 1606*

Navigational instruments adorn the title page of the first printed sea atlas. The two figures are taking soundings, or measuring the water's depth.

LUCAS JANSZOON WAGHENAER was a collector of maritime taxes in Holland, now known as the Netherlands. Apparently, he was not very good at his job, for he was asked to resign. He went into publishing and issued the first printed marine atlas, a collection of nautical charts in book form called *Spieghel der Zeevaerdt* (*The Mariner's Mirror*). The atlas contained 23 engraved charts; it also included a summary of all known astronomical and navigational information that was of use to seafarers.

The atlas was a tremendous success. It was translated into English in 1588 as *The Mariners Mirrour* and immediately became the most popular sailing guide among British navigators. They called it the "Waggoner," an English version of its maker's name, and soon British seamen were calling all sailing guides "waggoners."

In 1592, Waghenaer published a second, much larger sea atlas, which contained directions for reaching such remote ports as the Faeroe Islands, located in the North Atlantic Ocean, and Novaya Zemlya, north of Russia.

SEE ALSO
Dutch mapmakers; Portolan

Waldseemüller, Martin

GERMAN GEOGRAPHER AND MAPMAKER

- *Born: about 1470, Radolfzell, Germany*
- *Died: 1521*

MARTIN WALDSEEMÜLLER was a professor of geography in Lorraine, now part of France. In 1507 he published a book on geography and a world map printed from 12 wood blocks. In his book he suggested that the new lands encountered across the Atlantic Ocean by Christopher Columbus and other voyagers should be named after Amerigo Vespucci. The first use of the name America was on Waldseemüller's map, on South America. Later, however, Waldseemüller seems to have changed his mind, for when he issued a new edition of Ptolemy's *Geography* in 1513 he gave the credit for discovering South America to Columbus; that volume contains a map of the New World that, according to Waldseemüller, was based on information from an "admiral," possibly Columbus himself. But the name he had coined in 1507 stuck, and the newly discovered continents became the Americas.

Waldseemüller's final contribution to cartography was a second world map called the *Carta Marina,* or Marine Chart. It was published in 1516 and covered the entire known world, land and sea, in 12 sheets. Europe, Africa, and Asia are filled with place names, mountains, and pictures of throned monarchs that reveal Waldseemüller's geographic imagination at work—the monarchs far from Europe, in parts of Asia and Africa that were almost unknown to Waldseemüller and his contemporaries, have the biggest thrones and the most splendid pictures. The Americas, by contrast, seem rather barren, although the coast of South America is decorated with a picture of cannibals and another of a wild animal that resembles a cross between a wolf and a ram. The *Carta Marina* influenced some later mapmakers, who copied both its geography and its decorations. Peter Apian and other German scholars and cartographers copied it in various world maps from the 1520s through the 1540s.

Waldseemüller printed as many as a thousand copies of both the 1507 and the 1516 maps. Only one copy of each, however, is known to survive. The two maps were discovered in 1901 in a German castle.

SEE ALSO

Americas, mapping of; Columbus, Christopher; Globe; Vespucci, Amerigo

Way finders

WAY FINDERS are location maps designed to help people find their way to particular locations within a building, neighborhood, town, or transportation system. All unnecessary information is eliminated, so that the way finder is a streamlined, simplified map for a very specific purpose. The diagrams posted in many public places, with a dot labeled "You Are Here" and clear markings for features such as elevators, stairways, and fire exits, are examples of how way finders are used in everyday life.

One way finder, the London Underground Map, ushered in a new era of way finding in 1933. Before that time, London's subway system was charted on a map that retained the correct relationships of distance and direction between subway stations. The map was geographically accurate, but it was also

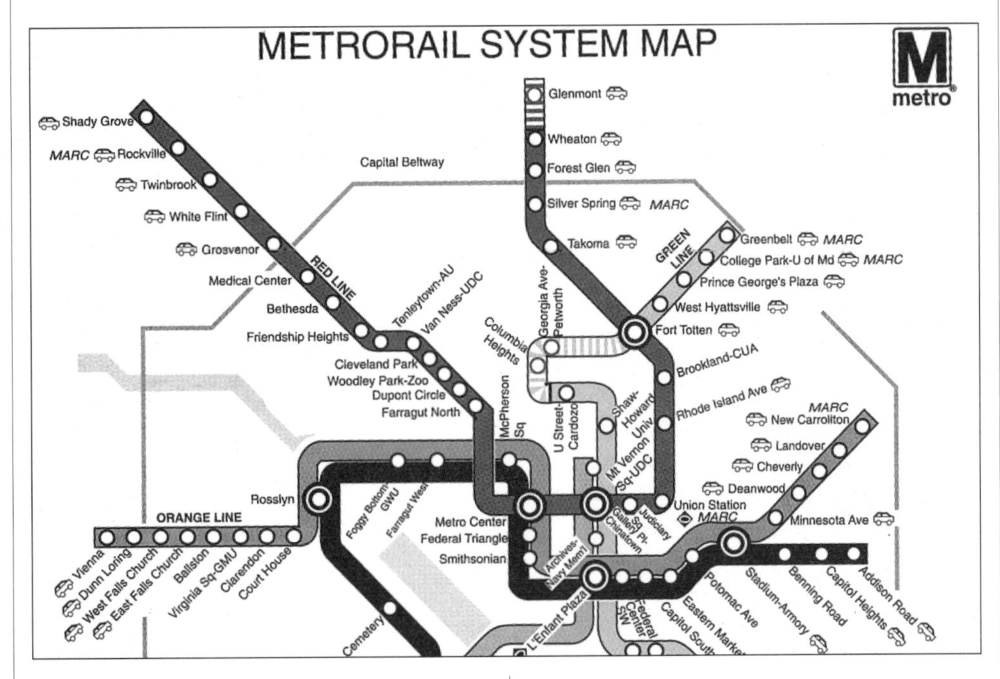

A map of the Washington, D.C., subway system uses Henry Beck's revolutionary, simplified way finder design.

crowded and hard to read. Subway lines meandered all over it, and in some places the stations were so close together that their names got mixed up. A draftsman named Henry Beck designed a new subway map with an entirely different approach. Instead of worrying about distance and exact direction, he focused on the one thing most commuters wanted to know: the sequence of stations on each line. Using only straight lines and angles of either 90 or 45 degrees, he created an easy-to-read map that gave the necessary information and eliminated geographic clutter. His way finder was so successful that its style was soon copied all over the world. Map experts at the British Library have called it "perhaps the single most famous map ever made."

Weather maps

WEATHER MAPS, also called meteorological maps, are special-purpose maps that record the weather over a given part of the world. They are of two types, usually called climate maps and meteorological maps to distinguish them from one another. Climate maps are the domain of climatologists, scientists who study the long-term patterns that make up a place's climate—that is, its average

Continent outlines have been added to satellite imagery to map changes in the ozone layer. The greatest degree of change is shown in bright red, at the bottom of the map—over Antarctica, where ozone is thinning at an alarming rate.

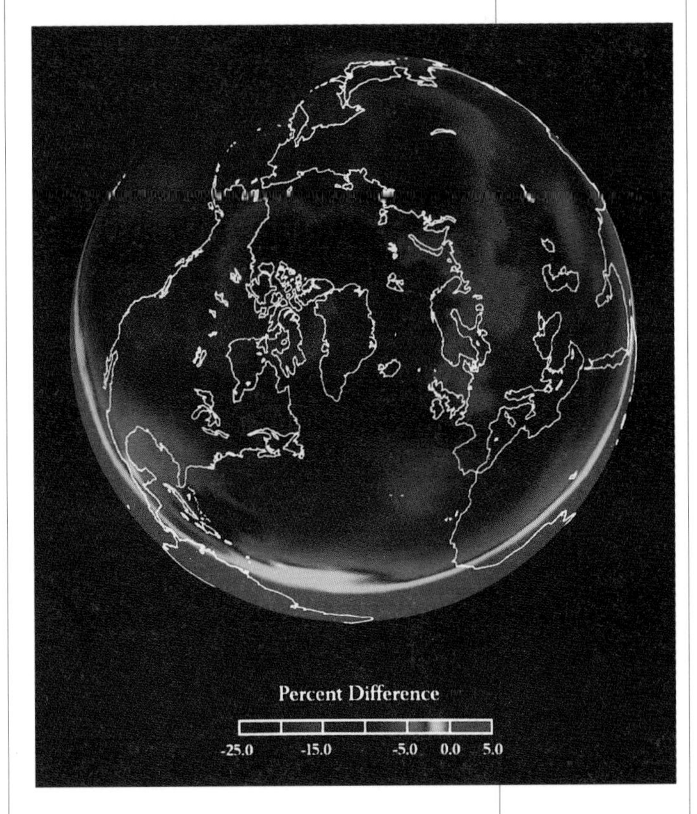

or standard weather over years or centuries. Meteorological maps are used and made by meteorologists, scientists who study short-term weather patterns that last a day, a week, or a few months at most.

Climate maps

Climate maps, unlike meteorological maps, are not designed to report or show predictions of short-term changes in the weather. Instead, they deal with the climate: the characteristic patterns of temperature, wind, cloud cover, rainfall, and other phenomena that, taken together, make various parts of the world tropical, temperate, arid, subarctic, and so on.

Some of the ancient Greek and Roman geographers divided the world into climate zones, freezingly cold in the north and unbearably hot in the south, toward the equator. A few visionaries even pictured a Southern Hemisphere on the other side of the hot zone. These climatologic notions were quite similar to the maps used today in geography books to show the broad zones of the earth's climates. Many modern climate maps, however, are far more detailed than these simple outlines. Some climate maps show standard weather patterns; others relate climate to vegetation, wildlife, and human population distribution.

In the United States, the National Climate Data Center (part of the National Oceanic and Atmospheric Administration, or NOAA) has published a set of historical climate atlases that show climate changes over long periods of time. The Canada Map Office publishes various climate maps as part of the National Atlas of Canada. Climate maps of the world and of various regions have been published by government agencies in many countries, as well as by commercial map publishers and scholarly journals.

Climate maps now play an increasingly important role in education and environmental policy. Discussions of controversial issues such as global warming draw upon maps to show current conditions—and future ones. Maps are part of the new science of climate modeling, in which climatologists program supercomputers to process millions of bits of weather information and come up with maps of possible future climates. Climate modeling allows a scientist to ask, "What would happen if the earth's average temperature rose 3 degrees Celsius in 15 years?" and see the answer in the form of a map that shows the effects of that particular change in the global climate, with new deserts, altered rainfall patterns, and new sea levels and shorelines—perhaps with Florida and Bangladesh completely submerged. Climate modeling is still in its infancy and is not yet foolproof, but it offers a powerful new tool for studying and protecting the environment. Maps are the central element of that tool.

Meteorological maps

To most people, the words *weather map* suggest a meteorological map, the sort of map they expect to find in their daily newspaper or see on their television screen during the weather segment of the nightly news program. These maps sum up the day's weather with notes on high and low temperatures, rainfall, wind direction and speed, and barometric pressure. Such maps are also used in short-term forecasts to show the weather that is likely to be caused in the next few days by areas of high and low air pressure, fronts of moving cold or warm air, precipitation, storms and hurricanes, and other meteorological phenomena.

Most countries have their own meteorological services. In the United States, weather maps are produced by the National Weather Service (NWS), part of NOAA. In addition, commer-

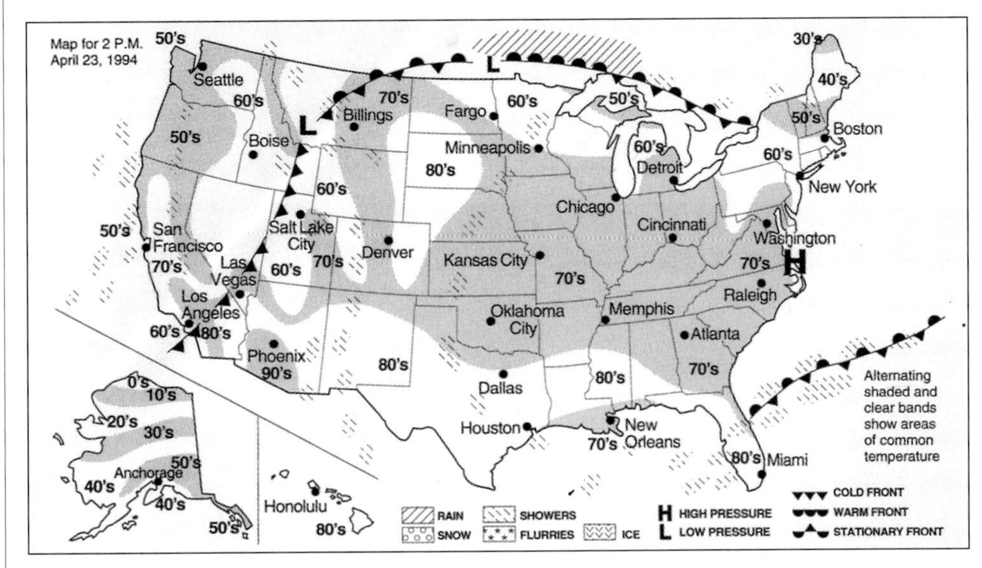

An enormous amount of data goes into the weather maps that are a feature of almost every newspaper or news broadcast.

cial firms such as Accu-Weather package raw meteorological data into weather reports and maps tailored for television, radio, and classroom use.

FURTHER READING

Hall, Stephen S. *Mapping the Next Millennium: The Discovery of New Geographies.* New York: Random House, 1992.

Wildlife maps

WILDLIFE MAPS are special-purpose maps that show the distribution of wildlife across a local region, a country, a continent, or the whole world. They have taken on new importance in the later years of the 20th century as part of the urgent environmental struggle to protect species threatened with extinction.

A wildlife map may focus on a single species, showing where it lives and, by using symbols to indicate population size, how many members of the species live in each part of the range. The World Wildlife Fund's map of the Sumatran rhinoceros, for example, shows the former range of the species in gray shading across southeastern Asia and Borneo, sprinkled with a pitifully few dots to indicate places where the Sumatran rhino still lives. Or a wildlife map may show the distribution of many different kinds of wildlife across a particular area such as a national park, a country, or a continent.

Wildlife biologists who work in the field, banding birds and animals with radio transmitters and then following their movements from afar, use maps both to select the best sites for productive research and to document their findings.

In the growing debate over environmental protection and habitat loss, wildlife maps play an important role in both education and decision making. Issues such as the controversy over protecting spotted owls versus logging in the old-growth forests of the Pacific Northwest of the United States are made clearer—although no easier to resolve—with the help of maps.

Wildlife maps are produced by many organizations, including international agencies such as the World Wildlife Fund and the World Conservation Union, government agencies such as the U.S. Fish and Wildlife Service and the Canada Map Office, and commercial firms such as American Nature Parks and the National Geographic Society.

FURTHER READING

Attenborough, David. *Atlas of the Living World.* Boston: Houghton Mifflin, 1989.

Burton, Robert, ed. *The Illustrated Encyclopedia of World Geography.* Vol. 6, *Animal Life.* New York: Oxford University Press, 1991.

Collins, Mark, ed. *The Last Rainforests: A World Conservation Atlas.* New York: Oxford University Press, 1990.

Nayman, Jacqueline. *Atlas of Wildlife.* New York: John Day, 1972.

Rand McNally editors. *Discovery Atlas of Animals.* Chicago: Rand McNally, 1993.

Scott, Sir Peter, ed. *World Atlas of Birds.* New York: Crescent Books, 1974.

Wind rose

SEE Compass rose

Woodcut

A WOODCUT is a print made from a carved block of wood. Woodblock printing is a type of relief printing, which means that the image is printed from a raised surface, not from lines carved into the surface. The printer prepares the wood block by carving the wood away from the desired image; then he coats the raised surface with ink and presses it to a sheet of paper. The earliest printed maps, dating from the late 15th century, were woodcuts. Some woodblock maps tended to remain in use even after their geography was outdated because, being carved in wood, they were almost impossible to revise. In the 16th century, the woodcut map was gradually replaced by the engraved map printed from a copper plate.

SEE ALSO

Engraving; Printing

Wytfliet, Cornelis

DUTCH MAPMAKER

• *Active: 16th century*

DURING THE mid-16th century, Cornelis Wytfliet served as the secretary of the council of Brabant, a duchy or dukedom in the Netherlands. His official position kept him in touch with the newest developments in exploration, and in 1597 he published the first printed atlas devoted solely to the Americas. Its 19 engraved maps sum up what Europeans knew—or thought they knew—about the American continents after a century of exploration.

Wytfliet's atlas contains much that is speculative, based only on rumors and hopes. For example, it includes a detailed map of the kingdom of Anian, complete with place names, rivers, and mountains, said to be located in the northwestern part of North America near the Arctic Circle—in spite of the fact that no one had ever seen such a place. Other parts of the atlas, however, were more accurate. The atlas contained the only map of

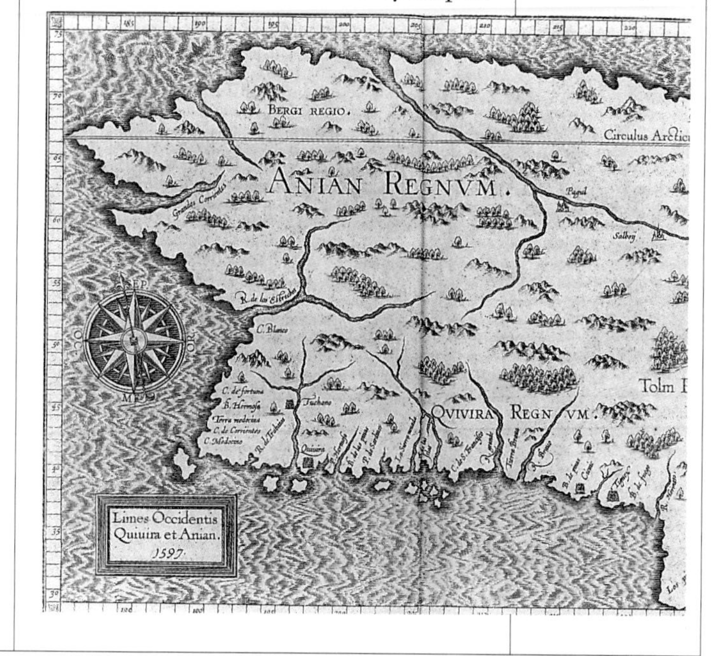

Wytfliet rightly guessed that there is a strait (left) between northwestern North America and Asia—but he placed the mythical kingdoms of Quivira and Anian nearby.

southern California and New Mexico to be printed in the 16th century; it also contained the most accurate map of Virginia and New England that had yet been published.

Wytfliet's atlas was reprinted in French and Latin in various editions until 1615.

Zacuto, Abraham

CATALAN ASTRONOMER

• *Active: late 15th century*

ABRAHAM ZACUTO was a Jewish scholar living in Salamanca, Spain, in 1492. That year the Spanish Inquisition, an arm of the Catholic church, ordered all Jews to convert to Christianity or leave the country. Zacuto left and settled in Portugal, where his pupil Joseph Vizinzo was established as an adviser to the court. Zacuto provided navigational instruction to the Portuguese mariners of the day. When Vasco da Gama set out in 1497 to sail around Africa, he carried mathematical tables and instruments made by Zacuto to help him measure his latitude.

Zenith

THE ZENITH is the point in the sky or the celestial sphere that is directly overhead from any given point on earth. The word comes from the Arabic language and was first used by Arab astronomers. The term is used in cartography to describe a kind of map projection, the zenithal projection (more commonly known as the azimuthal projection). Zenithal or azimuthal projections are projections in which the map is centered on a particular point. Some zenithal projections are centered on the north or south pole, but the center does not have to be a pole. It can be any point on earth: the mapmaker's hometown, a military base or airport, or a spot in the Pacific Ocean. The zenithal projection shows the world as it would appear to an observer in space directly above the center point—in other words, to an observer at the zenith.

SEE ALSO
Projections

Zorsi, Alessandro

SEE Americas, mapping of

APPENDIX 1

IMPORTANT DATES IN THE HISTORY OF MAPMAKING

2300 B.C.	The oldest known map is made in Mesopotamia (Iraq).
1300 B.C.	The Turin papyrus, the oldest surviving Egyptian map, is made.
6th century B.C.	The oldest surviving world map is made in Babylonia (Iraq); the Greek colony of Miletus is a center of geography and mapmaking.
200 B.C.	Eratosthenes measures the distance around the world.
2nd century B.C.	The oldest surviving Chinese maps are made.
20 B.C.	Strabo writes his *Geography*.
1st century A.D.	Roman emperor Augustus commissions a map of the entire Roman Empire.
2nd century	Claudius Ptolemaeus, called Ptolemy, publishes important works on astronomy and geography.
3rd century	Pei Hsiu publishes a handbook for Chinese cartographers.
750	The Albi map, the oldest known map of western Europe, is made.
around 1000	Vikings reach the coast of North America.
1154	Islamic geographer Idrisi makes a world map on 70 sheets for King Roger of Sicily.
1271–95	Marco Polo travels in Asia.
1280	The Hereford *mappa mundi* is made in England.
1290	The Pisan chart, the oldest surviving portolan, is made in Italy.
1311–12	Chu Ssu-pen makes an atlas of China.
1375	The Catalan atlas of manuscript maps is completed.
1425	Claudius Clausson makes the first known map to show Scandinavia and Greenland.
mid-15th century	Prince Henry of Portugal sponsors voyages of exploration along the African coast.
1459	Fra Mauro completes his world map.
around 1477	Printed maps begin replacing hand-drawn maps.
1482	Dominus Nicolaus Germanus publishes the first edition of Ptolemy's *Geography* to be issued outside Italy.
1487–88	Bartolomeu Dias sails around Africa.
1492	Martin Behaim makes what is now the oldest surviving globe.

1492–93	Christopher Columbus's first voyage to the Caribbean launches the European conquest of the Americas.
1494	Spain and Portugal sign the Treaty of Tordesillas, agreeing to divide newly discovered lands in America and Asia.
1498–1502	Vasco da Gama leads the first recorded voyage from Europe to India by sailing around Africa.
1500	Juan de la Cosa makes the oldest known map that shows the Americas.
1501–2	Amerigo Vespucci calls South America a "new world."
1502	The Cantino map is made, showing Portuguese discoveries such as Pedro Álvars Cabral's 1500 landing in Brazil.
1506	Giovanni Matteo Contarini makes the first known printed map to show the Americas.
1507	Martin Waldseemüller coins the name *America.*
1508	Johan Ruysch makes a map showing Vasco da Gama's discoveries in southern Africa and India.
1513	Vasco Nuñez de Balboa becomes the first European to see the Pacific Ocean from the west coast of America; Turkish cartographer Piri Reis makes a map that may have been based on Columbus's own maps.
1520	Ferdinand Magellan rounds the southern tip of South America.
1528	Benedetto Bordone publishes his *Isolario,* a collection of maps of the world's islands.
1533	Regnier Gemma Frisius publishes a handbook on triangulation, which becomes a guide for surveyors.
1544	Sebastian Münster publishes his *Universal Cosmography,* with influential maps of the continents.
1569	Gerardus Mercator introduces the Mercator projection.
1570	Abraham Ortelius publishes the first modern atlas, the *Theatrum Orbis Terrarum* (*Theater of the World*).
1572–1617	Georg Braun issues the *Civitates Orbis Terrarum* (*Cities of the World*), the first atlas of town plans from all parts of the world.
1585	Gerardus Mercator publishes the first edition of his *Atlas.*
1595	John Davis invents the backstaff for measuring latitude.
1597	Cornelis Wytfliet publishes the first atlas devoted to the Americas.

1646–47	Robert Dudley publishes a sea atlas that is the first compiled by an Englishman, the first to use the Mercator projection, and the first to cover the whole known world.
1675	John Ogilby publishes road maps of England in a strip format that is widely copied and remains in use today.
1688	Edmund Halley issues the first weather map.
1728	Vitus Bering discovers the strait that separates northeastern Asia and northwestern North America.
1752	Philippe Buache makes the first bathymetric map.
1765	John Harrison invents a marine chronometer for measuring longitude at sea.
late 18th century	James Cook explores the Pacific Ocean.
1784	Abel Buell issues the first map of the United States made entirely in America.
1791	The Ordnance Survey is founded to map Great Britain; the survey of France, carried out by four generations of the Cassini family, is completed.
1795	Matthew Carey publishes the first American atlas.
1814	William Clark's map of the Lewis and Clark Expedition is published.
1815	William Smith publishes the first geologic map.
1820	Fabian von Bellingshausen is the first explorer to see the Antarctic continent.
1830	The Royal Geographical Society is founded in London.
1876	The Great Trigonometrical Survey of India is completed.
1879	The U.S. Geological Survey is established.
1884	An international committee agrees that the prime meridian will run through the Royal Observatory in Greenwich, England.
1891	The International Map of the World is proposed.
1913	Standards are set for the International Map of the World.
1950	The first computer-generated map is published.
1950s	Marie Tharp and Bruce Heezen issue pictorial maps of the ocean floors.

1957–59	The polar regions are studied and mapped during the International Geophysical Year.
1970s	Radar is used to map the Amazon River basin in Brazil.
1972	The United States launches the first *Landsat* satellite to study the earth from space.
1978	The United States launches the *Seasat* satellite to study the sea, resulting in new maps of the seafloor.
1985	The United States launches the *Geosat* satellite for continued study of the oceans.
1990s	France's *SPOT* satellite is the most powerful earth orbiter, producing detailed photographs for mapping purposes.

APPENDIX 2

ORGANIZATIONS AND PUBLICATIONS RELATED TO MAPS

The organizations listed below include government departments whose publications are available to the public, organizations that offer membership to the general public, and professional associations whose publications are available to nonmembers. In addition, many national and international organizations around the world serve the needs of professional mapmakers, surveyors, geographers, and map dealers.

Canada

Canada Map Office
615 Booth Street
Ottawa, Ontario K1A 0E9
613-952-7000

Government agency that distributes a variety of maps of Canada.

Canadian Hydrographic Service
Department of Fisheries and Oceans
Ottawa, Ontario K1A 0E6
613-998-4391

Government agency that prepares and distributes maps of Canadian coastlines and waters.

Canadian Institute of Surveying and Mapping
P.O. Box 5378, Station F
Ottawa, Ontario K2J 3CI
613-224-9851

Professional organization for surveyors and mapmakers; promotes public education about maps.

Royal Canadian Geographical Society
488 Wilbrod Street
Ottawa, Ontario K1N 6M8

Promotes geographic research and education; membership open to the public; members receive *Canadian Geographic* magazine six times a year.

England

Hakluyt Society
c/o The Map Library
British Library
Great Russell Street
London WC1B 3DG
091-868-6359

Founded in 1846, the Hakluyt Society is dedicated to reprinting classic narratives of exploration, travel, and discovery. Its publications are available to members and, at somewhat higher prices, to nonmembers.

International Map Collectors' Society
29 Mt. Ephraim Road
Streatham
London SW16 1NQ
081-769-5041

Promotes map collecting and the study of maps and their history; publishes the *International Map Collectors' Society Journal* four times a year.

Ordnance Survey
Romsey Road
Maybush, Southampton S09 4DH
071-820-0127

The surveying and mapmaking branch of the British government; noted for its many maps, especially topographic maps, that are available to the public and distributed through many regional and local outlets.

Royal Geographical Society
1 Kensington Gore
London SW7 2AT
071-589-5466

Sponsors geographic research and maintains a library for the use of members and visiting scholars; together with the British Broadcasting Company, publishes a monthly magazine called *Geographical;* also publishes the *Geographical Journal* three times a year.

Sweden

International Orienteering Federation
Box 76
S-191 21, Sollentuna
8-353455

Federation of orienteering organizations from many nations; promotes founding of new organizations and growth of orienteering as an international sport.

United States

American Geographical Society
156 Fifth Avenue, Suite 600
New York, NY 10010
212-242-0214

Promotes geographic research; works to exchange information among geographers in various nations; publishes reports, maps, a monthly newsletter, and a quarterly journal called *Geographic Review*. The AGS archives are stored at the University of Wisconsin library in Milwaukee and are available to scholars and other researchers.

Association of American Geographers
1710 16th Street, N.W.
Washington, DC 10009
202-234-1450

Founded in 1904 to promote geography education and support the profession of geography; membership open to students; publishes a monthly newsletter and two quarterly journals.

Association of Map Memorabilia Collectors
8 Amherst Road
Pelham, MA 01002
413-253-3115

Publishes quarterly newsletter *Cartomania* to exchange news among collectors of maps and map-related objects.

Carto-Philatelists
c/o Clifford Mugnier
University of New Orleans
Department of Civil Engineering
New Orleans, LA 70122
504-286-7095

Organization for collectors of stamps with maps on them; issues quarterly newsletter.

Chicago Map Society
c/o Secretary-Treasurer
60 W. Walton Street
Chicago, IL 60610
312-943-9090

Holds monthly meetings for anyone interested in maps and mapping; membership open to the public; members receive *Mapline*, a magazine published four times a year by the Hermon Dunlap Smith Center for the History of Cartography at Chicago's Newberry Library.

Earth Science Information Center (ESIC)
507 National Center 1C402
Reston, VA 22092
703-648-6045

Government service that operates regional and state offices across the country to provide maps and cartographic information produced by the U.S. Geological Survey and other federal agencies; call for location of nearest regional center.

Geography Education National Implementation Project
1710 16th Street, N.W.
Washington, DC 20009
202-234-1450

Joint project of the American Geographical Society, the Association of American Geographers, the National Geographic Society, and the National Council for Geographic Education; founded in 1984 to improve geographic education in the United States; produces materials for teachers; newsletter, published three times a year, is available to the public.

Map Online Users Group
Map Library
University of South Carolina
Columbia, SC 29209
803-777-2802

An organization for institutions and individuals who use cartographic information in computer databases; publishes a newsletter four times a year.

National Council for Geographic Education
Leonard 16A
Indiana University of Pennsylvania
Indiana, PA 15705
412-357-6290

Founded in 1915 to promote geographic education; trains teachers and develops geography education programs for schools; publishes *Journal of Geography* six times a year and *Perspectives* newsletter monthly.

National Geographic Society
17th and M Streets, N.W.
Washington, DC 20036
202-921-1200

Promotes geographic research and education; sponsors events such as National Geography Bee for students and geographic education seminars for teachers; membership open to the public worldwide; members receive monthly *National Geographic* magazine and maps produced by NGS; also publishes books, television documentaries, and a children's magazine called *World.*

New York Map Society
c/o Map Division, New York Public Library
Fifth Avenue and 42 St.
New York, NY 10018
212-930-0587

Holds monthly meetings September through May at the American Museum of Natural History for those interested in studying and preserving maps; special focus on antique maps; publishes monthly newsletter called *Rhumb Line;* membership open to the public.

North American Cartographic Information Society
American Geographic Society Collection
University of Wisconsin–Milwaukee
P.O. Box 399
Milwaukee, WI 53201
800-558-8993

Professional organization for map collectors and dealers, cartographers, and teachers and students who use maps. Publishes journal called *Cartographic Perspectives* four times a year.

Special Libraries Association, Geography and Map Division
1700 18th Street, N.W.
Washington, DC 20009
202-234-4700

Professional organization for librarians specializing in maps and geographic materials; publishes *Bulletin* four times a year; also produces reference guides for cartographic studies; membership open to students.

U.S. Defense Mapping Agency
DMA Combat Support Center
Washington, DC 20315
202-227-2495

Produces maps for the Department of Defense; provides catalogs of aerospace, hydrographic, and topographic maps that are available to the public. Includes Board on Geographic Names, which establishes place names and answers inquiries.

U.S. Geological Survey
USGS Map Sales
P.O. Box 25286
Denver, CO 80225
303-236-7477

Principal mapmaking agency of the federal government; produces a wide range of maps and map-related information; maps and other materials are available to the public from either the national USGS center or regional ESIC offices; call toll-free 800/USA-MAPS for list of maps and order form.

U.S. Orienteering Federation
P.O. Box 1444
Forest Park, GA 30051
404-363-2110

Promotes orienteering, helps establish local clubs; sponsors competitive orienteering events in the United States.

APPENDIX 3

COLLECTIONS AND EXHIBITS OF MAPS

Listed below are some of the major collections and exhibits of maps and map-related materials around the world. Maps are also on display at thousands of other locations, either as part of permanent collections or featured in temporary displays. To find out about interesting maps on view in your area, check with public and university libraries, state and local historical societies, and museums (especially science or history museums).

Australia

Bligh Museum of Pacific Exploration
Adventure Bay
Bruny Island, Tasmania
Collection focuses on the history of Pacific Ocean exploration, including Antarctica, and features many maps.

Nautical Museum
135 St. Vincent Street
Port Adelaide, South Australia 5000
Collection includes marine maps and navigational charts.

Austria

Peter-Anich-Museum
Schulhaus, A-6173
Oberperfuss
A small museum with a collection of astronomical instruments, maps, and globes.

Brazil

Museu de Colono
Rua Presidente Vargas s/no.
Santa Leopoldina
Collection contains maps and other artifacts from Brazil's colonial era.

Canada

Marine Museum of Upper Canada
Exhibition Place, Lakeshore Boulevard West
Toronto 1, Ontario M6K 3B9
Collection focuses on the exploration of central Canada; includes many maps.

Public Archives of Nova Scotia
6016 University Avenue
Halifax, Nova Scotia B3H 1W4
Museum of maritime history; contains maps, charts, and navigational instruments.

England

British Library, Map Library
Great Russell Street
London WC1B 3DG
071-323-7700
One of the world's largest and most important collections of maps and globes, including many one-of-a-kind items, some on permanent display to public; special exhibits change frequently; facilities are available for researchers and visiting scholars.

Museum of the History of Science
University of Oxford
Old Ashmolean Building, Broad Street
Oxford OX1 3AZ
065-277280
Large collection of scientific instruments, pocket globes; includes world's largest collection of astrolabes.

National Maritime Museum
Romney Road
Greenwich, London SE10 9NF
201-763-6000
Contains many maps, charts, portraits, and instruments relating to British maritime history and exploration; includes the Royal Observatory, through which the prime meridian passes.

France

Bibliothèque Nationale
58 rue de Richelieu
75002 Paris
France's national library; contains many antique and medieval maps as well as samples of work (some on permanent display) by great French cartographers.

Musée Naval
Arsenal de Lorient, F-56100
Lorient
A museum of naval history that contains sea maps.

Musée International du Long Cours
F-35400
Saint-Malo
Located in a harbor town from which many voyages of exploration set forth; focuses on the history of sea exploration.

India

Geological Museum
Gandhi Bhawan
Lucknow, Uttar Pradesh
Contains materials from the Survey of India in the 18th and 19th centuries.

Italy

Museo della Navi
Via Zamboni 33, I-40126
Bologna
Naval museum; collection includes antique maps and ship models.

New Zealand

Hocken Library, University of Otago
Great King Street
Dunedin, Otago Province
Contains early books and maps relating to New Zealand and the Pacific Ocean.

Spain

Museo Maritimo
Puerto de la Paz
Barcelona
Museum of maritime history; features maps from the age of Spanish exploration and colonization.

Museo Naval
Calle de Montalban 2
Madrid 14
Museum of Spanish maritime history; collection includes historical maps and navigational instruments.

United States

Christian Science Publishing Society
250 Huntington Avenue
Boston, MA 02115
617-450-3790
Houses the Mapparium, a unique walk-through globe of stained glass.

Daily News Building
220 East 42nd Street
New York, NY 10017
Lobby contains a rotating globe 12.21 feet (3.7 meters) in diameter.

Explorers' Hall
National Geographic Society
17th and M Streets, N.W.
Washington, DC 20036
202-857-7456
Museum focuses on history of exploration from early times to the Space Age; many exhibits feature maps.

Library of Congress, Geography and Map Division
101 Independence Avenue, S.E.
Washington, DC 20540
202-707-6277
World's largest cartographic collection: 4 million maps, 53,000 atlases, 350 globes, and 8,000 reference works; publishes lists of holdings; collection open only to researchers, but telephone reference service answers inquiries from the public and provides reproductions of maps for a fee.

New York Public Library
Research Libraries, Map Division
Fifth Avenue and 42nd Street
New York, NY 10018
212-930-0587
Items from the collection frequently appear in library exhibitions; collection open to students and researchers.

Smithsonian Institution
1000 Jefferson Drive, S.W.
Washington, DC 20560
202-357-2700
Large collection includes many books and artifacts related to travel and exploration; maps frequently appear in exhibitions; collection open to researchers.

United Nations Map Collection
Dag Hammarskjöld Library
New York, NY 10017
212-754-7425
Items from collection frequently used in library exhibitions; collection open to students, researchers.

U.S. Department of Interior Museum
1849 C Street, N.W.
Washington, DC 20240
202-208-4743
Collection includes maps, charts, and mapping and surveying equipment.

FURTHER READING

The books and articles listed below contain a wealth of information about many aspects of mapmaking, exploration, and geography. You will find other references, usually more specific, listed at the end of many of the entries in the *Companion.* Sources for maps, map-related computer software, and other products can be found in *The Map Catalog,* edited by Joel Makower, listed below.

General Reference

Adams, Percy G. *Travelers and Travel Liars, 1660–1800.* Berkeley: University of California Press, 1962.

American Library Association, Map and Geography Round Table. *Guide to U.S. Map Resources.* 2nd ed. Chicago: American Library Association, 1990.

British Library. *What Use Is a Map?* London: British Library, 1989.

Carrington, David K., and Richard W. Stephenson. *Map Collections in the United States and Canada: A Directory.* New York: Geography and Map Division of the Special Library Association, 1985.

Greenhood, David. *Mapping.* 3rd ed. Chicago: University of Chicago Press, 1964.

Hill, Gillian. *Cartographical Curiosities.* 1978. Reprint. London: British Library, 1984.

Makower, Joel, ed. *The Map Catalog.* 3rd ed. New York: Vintage, 1992.

Mango, Karin N. *Map-Making.* New York: Messner, 1984.

Monmonier, Mark, and George Schnell. *Map Appreciation.* Englewood Cliffs, N.J.: Prentice Hall, 1988.

Munk, Nina. "Map Mania." *Forbes,* August 17, 1992, p. 14.

Peters, Arno. *The New Cartography.* Translated by Ward L. Kaiser and D. G. Smith. New York: Friendship Press, 1984.

Raisz, Erwin. *General Cartography.* New York: McGraw-Hill, 1948.

Robinson, Arthur. *The Look of Maps: An Examination of Cartographic Design.* Madison: University of Wisconsin Press, 1986.

Robinson, Arthur, et al. *Elements of Cartography.* 6th ed. New York: Wiley, 1993.

Robinson, Arthur, and Barbara B. Petchenik. *The Nature of Maps: Essays Toward Understanding Maps and Mapping.* Chicago: University of Chicago Press, 1976.

Thompson, Morris. *Maps for America.* Washington: U.S. Geological Survey, 1979.

Thrower, Norman J. W. *Maps and Man.* Englewood Cliffs, N.J.: Prentice Hall, 1972.

Tooley, Ronald V. *Tooley's Handbook for Map Collectors.* Chicago: Speculum Orbis, 1985.

Tyner, Judith. *An Introduction to Thematic Cartography.* Englewood Cliffs, N.J.: Prentice Hall, 1992.

Weiss, Harvey. *Maps: Getting from Here to There.* Boston: Houghton Mifflin, 1991.

Wilford, John Noble. "Enormous Changes on a Small Planet." *New York Times Book Review,* December 20, 1992, p. 1.

Woodward, David, ed. *Art and Cartography: Six Historical Essays.* Chicago: University of Chicago Press, 1987.

History of Mapmaking

Allen, Phillip. *The Atlas of Atlases: The Map Maker's Vision of the World.* New York: Abrams, 1992.

Bagrow, Leo. *History of Cartography.* Revised and enlarged by R. A. Skelton. 2nd ed. Chicago: Precedent, 1985.

Brown, Lloyd A. *The Story of Maps.* 1949. Reprint. New York: Dover, 1979.

Campbell, Tony. *The Earliest Printed Maps, 1472–1500.* Berkeley and Los Angeles: University of California Press, 1987.

———. *Early Maps.* New York: Abbeville, 1981.

Crone, Gerald Roe. *Maps and Their Makers: An Introduction to the History of Cartography.* London: Hutchinson, 1966.

Dilke, O. A. *Greek and Roman Maps.* Ithaca, N.Y.: Cornell University Press, 1985.

Goss, John. *The Mapmaker's Art.* Chicago: Rand McNally, 1993.

———. *The Mapping of North America: Three Centuries of Map-making, 1500–1860.* Secaucus, N.J.: Wellfleet, 1990.

Greenblatt, Stephen. *Marvelous Possessions: The Wonder of the New World.* Chicago: University of Chicago Press, 1991.

Harley, J. B., and David Woodward, eds. *The History of Cartography.* Vol. 1, *Cartography in Prehistoric, Ancient, and Medieval Europe and the Mediterranean.* Chicago: University of Chicago Press, 1987.

Harley, J. B., and David Woodward, eds. *The History of Cartography.* Vol. 2, *Cartography in Traditional Islamic and South Asian Societies.* Chicago: University of Chicago Press, 1992.

Junger, Sebastian. "The Last Mapmakers." *American Heritage,* September 1991, p. 94.

Lister, Raymond. *Antique Maps and Their Cartographers.* Hamden, Conn.: Archon, 1970.

Monmonier, Mark. *Maps with the News: The Development of American Journalistic Cartography.* Chicago: University of Chicago Press, 1989.

Murray, Jeffrey S. "The County Map Hustlers: Early Canadian Map Publishers and Dealers." *Canadian Geographic,* December 1989–January 1990, p. 76.

———. "Fanciful Worlds: How Strange Beasts and Myths Found Their Way onto Early Maps." *Canadian Geographic,* September-October 1993, p. 64.

Nordenskiold, Adolf E. *Facsimile Atlas to the Early History of Cartography: Reproductions of the Most Important Maps Printed in the Fifteenth and Sixteenth Centuries.* 1889. Reprint. New York: Dover, 1973.

Potter, Jonathan. *Antique Maps: An Introduction to the History of Maps and How to Appreciate Them.* London: Hamlyn, 1988.

Putman, Robert. *Early Sea Charts.* New York: Abbeville, 1983.

Robinson, Arthur. *Early Thematic Mapping in the History of Cartography.* Chicago: University of Chicago Press, 1982.

Shirley, Rodney W. *The Mapping of the World: Early Printed World Maps, 1472–1700.* London: Holland Press Cartographica, 1983.

Sudo, Phil. "Charting the Course: Historical Views of the World." *Scholastic Update,* September 20, 1991, p. 21.

Thomson, Don W. *Men and Meridians: The History of Surveying and Mapping in Canada.* 3 vols. Ottawa: Duhamel, 1966–69.

Tooley, Ronald V. *Dictionary of Mapmakers.* Amsterdam: Liss, 1979. Supplement published 1985.

———. *Maps and Map-Makers.* 1949. Reprint. New York: Dorset, 1990.

Tooley, Ronald V., and Charles Bricker. *Landmarks of Mapmaking: An Illustrated History of Maps and Mapmakers.* 1976. Reprint. New York: Dorset, 1989.

Wilford, John Noble. *The Mapmakers.* New York: Knopf, 1981.

Exploration and Mapping

Skelton, R. A. *Explorers' Maps: Chapters in the Cartographic Record of Geographical Discovery.* London: Routledge & Kegan Paul, 1958.

Stefoff, Rebecca. *Accidental Explorers: Surprises and Side Trips in the History of Discovery.* New York: Oxford University Press, 1992.

———. *Scientific Explorers: Travels in Search of Knowledge.* New York: Oxford University Press, 1992.

Suárez, Thomas. *Shedding the Veil: Mapping the European Discovery of America and the World.* River Edge, N.J.: World Scientific, 1992.

World Explorers and Discoverers. New York: Macmillan, 1991.

Geography

Abler, Ronald F., et al., eds. *Geography's Inner Worlds: Pervasive Themes in Contemporary American Geography.* New Brunswick, N.J.: Rutgers University Press and the Association of American Geographers, 1992.

Demko, George J. *Why in the World: Adventures in Geography.* New York: Doubleday, 1992.

Gould, Peter, and Rodney White. *Mental Maps.* 2nd ed. Boston: Allen & Unwin, 1986.

Livingstone, David N. *The Geographical Tradition: Episodes in the History of a Contested Enterprise*. Cambridge, Mass.: Blackwell, 1992.

Marshall, Bruce, ed. *The Real World: Understanding the Modern World Through the New Geography*. Boston: Houghton Mifflin, 1991.

Technology and Mapmaking

Hall, Stephen A. *Mapping the Next Millennium: The Discovery of New Geographies*. New York: Random House, 1992.

Monmonier, Mark. *Technological Transition in Cartography*. Madison: University of Wisconsin Press, 1985.

Parry, R. B., and G. R. Perkins, eds. *World Mapping Today*. London: Butterworth, 1987.

Pope, Gregory. "The New Cartographers: Satellite and Computer Technology." *Omni*, December 1991, p. 74.

Using and Understanding Maps

Bolz, Diane M. "Follow Me . . . I Am the Earth in the Palm of Your Hand." *Smithsonian*, February 1993, p. 113.

Carey, Helen H. *How to Use Maps and Globes*. New York: Franklin Watts, 1983.

Gould, Peter. *The Geographer at Work*. London: Routledge & Kegan Paul, 1985.

Keates, J. S. *Understanding Maps*. New York: Wiley, 1982.

Kjellstrom, Bjorn. *Be Expert with Map and Compass*. New York: Scribners, 1975.

Lobeck, Armin. *What Maps Don't Tell Us: An Adventure into Map Interpretation*. Chicago: University of Chicago Press, 1993.

Monastersky, Richard. "The Warped World of Mental Maps: Students Worldwide Share a Skewed Vision of the Continents." *Science News*, October 3, 1992, p. 222.

Monmonier, Mark. *How to Lie with Maps*. Chicago: University of Chicago Press, 1991.

Murray, Jeffrey S. "Maps that Deceive." *Canadian Geographic*, May-June 1992, p. 82.

Wood, Denis. *The Power of Maps*. New York: Guilford, 1992.

Atlases

Atlas of the Oceans. Chicago: Rand McNally, 1994.

Atlas of the World. New York: Oxford University Press, 1992.

Baker, D. James. *Planet Earth: The View from Space*. Cambridge: Harvard University Press, 1990.

Discovering Maps: A Young Person's Atlas. Maplewood, N.J.: Hammond, 1989.

Facts on File editors. *The Geography on File Collection*. New York: Facts on File, 1990.

Fernandez-Armesto, Felipe, ed. *The Times Atlas of World Exploration: 3,000 Years of Exploring, Explorers, and Mapmaking*. New York: HarperCollins, 1991.

Garrett, Wilbur E., ed. *Atlas of North America: Space Age Portrait of a Continent*. Washington, D.C.: National Geographic Society, 1986.

Hammond Atlas of the World. Maplewood, N.J.: Hammond, 1993.

Hammond Atlas of World History. Rev. ed. Maplewood, N.J.: Hammond, 1990.

Harper Atlas of World History. New York: HarperCollins, 1992.

National Air and Space Museum and Smithsonian Institution. *Looking at Earth*. Washington, D.C.: Smithsonian Books, 1994.

National Geographic Atlas of the World. 6th ed. Washington, D.C.: National Geographic Society, 1990.

The New International World Atlas. Chicago: Rand McNally, 1994.

New York Times Atlas of the World. 3rd. rev. ed. New York: Times Books/Random House, 1992.

Times Atlas of the World. 9th ed. New York: Times Books/Random House, 1992.

World Atlas for Students. Maplewood, N.J.: Hammond, 1993.

INDEX

References to illustrations are indicated by page numbers in *italics*. References to main article entries are indicated by page numbers in **boldface.**

Picture Credits

American Philosophical Society: 19; Arni Magnusson Institute, Reykjavik: 277; Austrian National Library, Vienna; 135, 216; Bancroft Library, University of California, Berkeley: 171-T, 172, 253-T; The Beinecke Rare Book and Manuscript Library, Yale University: 278; The Bettmann Archive: 149; Biblioteca Nationale Marciana, Venice: 191; Bibliothèque Nationale de France: 66, 71, 81, 83, 164, 219, 223; The Bodleian Library, Oxford. MS. Gough Gen. Top. 16: 138-T; The British Library: 17 (Maps C.6.c.4), 20 (Maps 184.f.1), 29 (Add. MS. 28681), 37 (Maps C.45.f.2), 40 (Add. MS. 10049.f.64), 45, 57 (Maps 4.Tab.8), 61-B (MS. Cott Tib.Bv.f56v), 63 (Maps C.44.f.7), 80 (Maps C.6.d.8), 97 (Maps C.46.i.18), 100 (Maps C.2.cc.4), 110-B (Maps C.7.e.2), 120-R (Maps C.2.a.7), 130-T (Maps C.9.d.1), 131 (Maps C.7.e.1), 136 (Maps C.1.c.2), 139 (Maps Ref.K.5), 143-T (Maps*69810.18), 154 (Maps C.3.d.1), 160-T (Maps 856.6), 160-B (MS. 11654.cc.16), 189-T (Add. MS. 27376. 187v-8), 190, 207 (Maps C.6.d.8), 212 (Add. MS. 7085), 213 (Royal MS.14C. viii.ff.4v-5) 220 (569.g.10), 225 (Add. MS. 27376.180v-1), 233 (1C.9304), 249 (maps C.7.c.1), 256 (Maps C.7.c.20), 261 (Maps C.3.d.11), 264-T (Maps C.2.d.7), 289 (Maps C.46.i.18); The British Library, Oriental and India Office Collection: 88-T (15201.d.4), 241 (x595/1); copyright © British Museum: cover-top left, 211-T, 238-B; courtesy of the John Carter Brown Library at Brown University: 211-B; Christian Science Publishing Company: frontispiece; Fotomas Index: 14; Free Library of Philadelphia, Rare Book Department: 206; Germanisches National Museum, Nürnberg: 53-B; Government Printing Office: 196; The Dean and Chapter of Hereford and the Hereford Mappa Mundi Trustees: 151; by permission of the Houghton Library, Harvard University: 128-T; Hudson's Bay Company Archives, Provincial Archives of Manitoba: 148; Imperial War Museum, London: 9; courtesy, Independence National Historical Park, Philadelphia: 90-B; Kloster Ebstorf, Ebstorf, Germany: 118; Library of Congress: cover, 11, 13, 16, 22, 23, 25, 32, 43, 44, 49-T, 49-B, 51, 53-T, 65, 72, 75, 86, 88-B, 90-T, 93, 94, 102-L, 102-R, 103, 107, 108, 114, 116, 117, 119, 121, 126, 128-B, 132, 144, 153, 167, 176-177, 180, 181-T, 181-B, 186, 189-B, 194, 195, 202, 204, 210, 215-R, 218-B, 224, 226, 227, 245, 238-T, 239, 247, 250-B, 251, 257-L, 257-R, 271-T, 271-B, 274, 275, 279, 284, 304; Metropolitan Museum of Art, Harris Brisbane Dick Fund, 1951 (51.537.2): 254; Museo Capitolino, Rome: 28; Museo Naval, Madrid: 171-B; Nakajo Town Office, Niigata Prefecture, Japan: 169; NASA: cover-top right, 218-T, 248, 255, 281-B; National Archives: 125; National Archives of Canada: 155 (NMC 6172), 200 (NMC 117726); National Gallery of Canada, Ottawa: 184; National Maritime Museum, London: 47, 50, 85, 92, 95-T, 95-B, 109, 123, 124, 145, 185-B, 237, 244; photos courtesy of the Newberry Library: 61-T (The Edward E. Ayer Collection), 163, 199 (The Edward E. Ayer Collection), 205 (The Edward E. Ayer Collection), 215-L; The New York Public Library, Map Collection, Astor, Lenox and Tilden Foundations: 77, 174, 197; The New York Public Library, Rare Books and Special Collections, Astor, Lenox and Tilden Foundations: 42, 64, 68, 106, 137-B, 141, 253-B; The New York Times Company. Reprinted by permission: 258 (copyright © 1992), 283 (copyright © 1994); Ninna-ji, Nara, Japan: 143-B; Peabody Institute Library: 30, 269; courtesy Replogle Globes, Inc. Broadview, Illinois: 137-T; The Royal Library, Stockholm: 56; Smithsonian Institution, Neg. No. 82-11338: 69; copyright © 1994 Swanston Publishing Limited: 78; *Physiographic Diagram of the North Atlantic*, by Bruce C. Heezen and Marie Tharp, revised 1972. Copyright © Marie Tharp, 1972. Reproduced by permission of Marie Tharp: 272; R. V. Tooley: 36, 39, 41, 48, 59, 94, 99, 122, 158, 185-T, 242, 260, 263; copyright © 1992 by Travel Graphics International ®, Minneapolis: 267; U.S. Department of Agriculture: 15; U.S. Geological Survey: 10, 52, 55 (Instruments–Topography 122), 98, 99, 101, 105, 133, 147 (Jackson, W.H.–282), 166, 217 (Walker, W.C. 70), 229, 230, 231, 236, 240, 250-T, 264-B (Jackson, W.H. 1111) 265, 266; U.S. Naval Historical Center: 140, 192; University of Pennsylvania, University Museum: 27; University of Texas at Austin, Harry Ransom Humanities Research Center: 156; courtesy Washington Metropolitan Area Transit Authority: 281-T.

Rebecca Stefoff has written more than 50 books for young adults, specializing in geography and biography. Her lifelong interest in reading and collecting travel narratives is reflected in such titles as *Lewis and Clark, Magellan and the Discovery of the World Ocean, Marco Polo and the Medieval Travelers, Vasco da Gama and the Portuguese Explorers, The Viking Explorers,* and numerous books on China, Japan, Mongolia, the Middle East, and Latin America. Books on exploration include *Accidental Explorers, Women of the World,* and *Scientific Explorers.* Ms. Stefoff has served as editorial director of two Chelsea House series, *Places and Peoples of the World* and *Let's Discover Canada,* and as a geography consultant for the *Silver Burdett Countries* series. She earned her Ph.D. at the University of Pennsylvania and lives in Portland, Oregon.